Biomedical Visualization

Series Editor

Paul M. Rea, Anatomy Facility, School of Medicine, Dentistry and Nursing, College of Medical, Veterinary and Life Sciences, University of Glasgow, Glasgow, UK

The Biomedical Visualization book series invites contributions related to visualization and imaging in the biomedical sciences and related fields such as medicine, dentistry, veterinary surgery, informatics, and the allied health professions.

This can encompass work in image analysis, workflow methodologies, photogrammetry in science, animations, digital reconstructions and applications, big data and visualizations, educational methodologies, usability and evaluations, augmented and virtual reality, and 3D and 4D technologies, including 3D printing, as well as informatics, e-tutorials, MOOCs, HCI and public engagement. It can be from macroscopic to microscopic but show how we can view data and information related to the biomedical field in a much more accessible, innovative and engaging way using technology.

This series is intended for researchers, clinicians, students, and teaching staff in biomedical and medical schools who use and develop visualization and imaging techniques in medical education, patient care, and biomedical and medical research.

Mark Roughley
Editor

Approaches for Science Illustration and Communication

Springer

Editor
Mark Roughley
Liverpool School of Art & Design
Liverpool John Moores University
Liverpool, UK

ISSN 2731-6130 ISSN 2731-6149 (electronic)
Biomedical Visualization
ISBN 978-3-031-41651-4 ISBN 978-3-031-41652-1 (eBook)
https://doi.org/10.1007/978-3-031-41652-1

Cover Illustration: Catherine MacRobbie.

This Springer imprint is published by the registered company Springer Nature Switzerland AG
The registered company address is: Gewerbestrasse 11, 6330 Cham, Switzerland

Paper in this product is recyclable.

Preface

In today's rapidly advancing world, the importance of effective science illustration and communication cannot be overstated. Biomedical science, with its intricate concepts and complex ideas, requires visual aids and clear explanations to reach wider audiences and foster understanding. Whether it's visualising 'invisible' health conditions such as endometriosis using interactive 3D models, explaining the impact of the antimicrobial resistance 'hidden pandemic' through public engagement activities, or demystifying biotechnologies used in organ transplantation via collage, the art of science illustration plays a pivotal role in bridging the gap between the scientific community and the general public.

Advancements in scientific knowledge have driven the need for new visual representations. Scientists often rely on illustrators to visualise their research findings, allowing them to present their work in a visually appealing and comprehensible manner. Similarly, communicators work closely with scientists and illustrators to craft narratives that engage diverse audiences, from expert to non-expert. Excitingly, this work is increasingly interdisciplinary in nature and features novel collaborations between scientists, illustrators and communicators.

This edited volume aims to inspire both aspiring and established science illustrators, scientists and educators interested in effective science communication. Ten chapters provide an in-depth introduction to the approaches, techniques and challenges faced by science illustrators and communicators and uncover the multifaceted nature of this field and its crucial role in making science accessible, engaging and impactful. Together, we embark on a journey to explore the power of visual storytelling in conveying the wonders and complexities of the biomedical science, fostering curiosity and inspiring a new generation of thinkers and creatives.

Each chapter explores the techniques, tools and methodologies employed by artists, designers and scientists to convey scientific information, often in unexpected ways. We delve into a range of approaches for effective science illustration and communication, which include 2D drawings, diagrams and animations, through to interactive participatory tabletop games.

The volume begins by examining the historical roots of anatomical illustration and its symbiotic relationship with medical education in chapter 1. The incorporation of art-based workshops in medical curricula as a means to teach soft skills is described and the chapter highlights the potential of drawing in developing these crucial attributes.

With the prominence of digital technologies in our daily lives, contemporary science illustrators are expanding the possibilities of digital science communication across various platforms. Endometriosis, a debilitating condition affecting millions of women worldwide, serves as a compelling case study in chapter 2, which discusses the importance of accurate and visual representation online. The chapter reveals the lack of illustrations and educational tools that properly depict the disease and emphasises the need for accessible information to improve understanding and increase awareness.

Chapters 3, 4, 5 and 6 explore the fundamental principles that underpin effective science communication. Communicating complex scientific ideas to a non-expert audience requires careful consideration of visual elements, storytelling techniques and the appropriate level of detail. These chapters demonstrate the importance of the role of science illustrators in ensuring that scientific information is not only visually captivating but also scientifically accurate. Editorial illustrations, 2D animations, infographics and virtual reality experiences have the potential to visually present narratives that make complex information accessible and meaningful to a range of audiences.

Chapter 7 builds on these storytelling techniques to explore the intersection of comics and medicine, known as graphic medicine, in the creation and reception of auto-ethnographic pathographies. Here, personal narratives help us understand the ways in which being a patient shapes individual identity and patient understanding of their own body and health conditions.

Games, both digital and traditional, prove to be effective tools for science communication, and chapter 8 acknowledges their popularity and discusses their potential in engaging audiences and facilitating learning, with useful recommendations for those thinking about designing their own science communication game. This playful approach to science commutation is further described in chapter 9, which suggests the integration of art activities as science public engagement tools becomes crucial when communicating the urgent global health threat of antimicrobial resistance and promoting microbiology literacy.

Lastly, the final chapter dives into the realm of bioengineering and the ethical debates surrounding bioengineering of hybrid human and non-human animals for human organ donation. The chapter encourages critical reflection on emerging bioengineering practices by mimicking bio-tech practices such as CRISPR through cut-and-paste methods used in analogue collage.

Through case studies and research projects, this volume showcases how art-science collaborations harness the creativity and enhance the impact of science communication, while debating ethical considerations and emphasising the importance of accuracy, accessibility and responsible representation of scientific information. Whether we are working with researchers in the laboratory or collaborating with educators in the classroom, science illustrators play a vital role in bridging the gap between different domains of expertise and fostering a deeper appreciation for scientific discoveries.

If you are a scientist seeking to enhance your ability to communicate your research or an illustrator interested in biomedical visualisation, this chapter serves as an introduction to the contemporary approaches in science illustration and communication. By understanding the creative methods and techniques employed in this field, we can collectively work towards fostering a deeper appreciation of art in science, continuing to captivate and inspire audiences worldwide.

Liverpool, UK Mark Roughley

Acknowledgements

I would like to thank the wonderful authors who have dedicated their time in sharing their work in this volume. Their contributions showcase the range of approaches undertaken by contemporary science illustrators and communicators across the globe. The featured interdisciplinary art-science practice, projects and case studies are truly impactful and have never been discussed together in one book before. Their research and professional practice have the possibility to change how we work as inspiring or established science illustrators and communicators.

I would also like to thank my colleagues at Face Lab LJMU for their peer support while editing this volume, and the series editor, Prof Paul Rea, for his unwavering encouragement and guidance. Thanks to the Springer Nature team; without them, this volume would not have been realised.

Contents

Editors and Contributors

About the Series Editor

Paul M. Rea Paul is Professor of Digital and Anatomical Education at the University of Glasgow. He is Director of Innovation, Engagement and Enterprise within the School of Medicine, Dentistry and Nursing. He is also a Senate Assessor for Student Conduct, Council Member on Senate and coordinates the day-to-day running of the Body Donor Program and is a Licensed Teacher of Anatomy, licensed by the Scottish Parliament.

He is qualified with a medical degree (MBChB), an MSc (by research) in craniofacial anatomy/surgery, a PhD in neuroscience, the Diploma in Forensic Medical Science (DipFMS) and an MEd with Merit (Learning and Teaching in Higher Education). He is a Senior Fellow of the Higher Education Academy, Fellow of the Institute of Medical Illustrators (MIMI) and a registered medical illustrator with the Academy for Healthcare Science.

Paul has published widely and presented at many national and international meetings, including invited talks. He has been the lead Editor for Biomedical Visualiz(s)ation over 15 published volumes and is the founding editor for this book series. This has resulted in almost over 110,000 downloads across these volumes, with contributions from over 450 different authors, across approximately 100 institutions from 23 countries across the globe. It has over 500 citations from these volumes. He is Associate Editor for the European Journal of Anatomy and has reviewed for 25 different journals/publishers. He is the Public Engagement and Outreach lead for anatomy coordinating collaborative projects with the Glasgow Science Centre, NHS and Royal College of Physicians and Surgeons of Glasgow. Paul is also a STEM ambassador and has visited numerous schools to undertake outreach work.

His research involves a long-standing strategic partnership with the School of Simulation and Visualisation, The Glasgow School of Art. This has led to multi-million-pound investment in creating world leading 3D digital datasets to be used in undergraduate and postgraduate teaching to enhance learning and assessment. This successful collaboration resulted in the creation of the world's first taught MSc Medical Visualisation and Human Anatomy combining anatomy and digital technologies, for which Paul was the Founding Director having managed this for 12 years. The Institute of Medical Illustrators also accredits this postgraduate degree. Paul has led college-wide, industry, multi-institutional and NHS research linked projects for students.

About the Volume Editor

Mark Roughley is a Reader in Interdisciplinary Digital Visualisation at Liverpool School of Art & Design, Liverpool John Moores University, and is a researcher in the Face Lab research group that explores faces and art-science applications. Mark trained as an illustrator (BA(Hons)) and subsequently as a medical artist (MSc). He is a professional member of the Institute of Medical Illustrators (UK), a Section Editor for the Journal of Visual Communication in Medicine and a Fellow of the Higher Education Academy (UK).

Mark specialises in visualising anatomy through 3D data acquisition, modelling and fabrication. His research focuses on the affordances that 3D digital technologies allow for both digital and haptic interaction with anatomical and cultural artefacts. He has published widely in relation to medical illustration and 3D facial depiction of people from the past.

Mark is the Programme Leader for the MA Art in Science programme at Liverpool School of Art & Design, which brings artists and scientists together to explore collaborative approaches in art and science. Additionally, Mark is the host of the art-science Liverpool LASER Talks (Leonardo Art Science Evening Rendezvous); Leonardo: Art Science and Technology's international public engagement series.

List of Contributors

Stefan Bruckner Institute for Visual and Analytic Computing, University of Rostock, Rostock, Germany

Raechelle D'Sa University of Liverpool, Liverpool, UK

Ingri Dahi Faculty of Medicine and Health Sciences, University of Antwerp, Antwerp, Belgium

Joachim Deblock Royal Academy of Fine Arts Antwerp, Antwerp, Belgium

Yuki Dejonckheere Faculty of Medicine and Health Sciences, University of Antwerp, Antwerp, Belgium

Johan Devrome Royal Academy of Fine Arts Antwerp, Antwerp, Belgium

Tom Dietvorst Royal Academy of Fine Arts Antwerp, Antwerp, Belgium

Caroline Erolin Reader in Medical Art, School of Science and Engineering, University of Dundee, Dundee, UK

Jennifer E. Fairman Art as Applied to Medicine, Johns Hopkins University School of Medicine, Baltimore, MD, USA

Laura A. Garrison Mohn Medical Imaging and Visualization Centre, Department of Radiology, Haukeland University Hospital, Bergen, Norway
Bouvet ASA, Bergen, Norway

Lou Gils Royal Academy of Fine Arts Antwerp, Antwerp, Belgium

Francis Van Glabbeek Faculty of Medicine and Health Sciences, University of Antwerp, Antwerp, Belgium
Biological and Medical Art in Belgium (BIOMAB), Antwerp, Belgium

Aimée Hutchinson Independent Medical Artist and Instructional Design Illustrator, Cambridge, UK

Monique Meuschke Department of Simulation and Graphics, University of Magdeburg, Magdeburg, Germany

Natasha Niethamer Liverpool School of Tropical Medicine, Liverpool, UK

Tim Phelps Art as Applied to Medicine, Johns Hopkins University School of Medicine, Baltimore, MD, USA

Tony Pickering Graphic Illustrator, Greater Manchester, UK

Hana Pokojná Faculty of Informatics, Masaryk University, Brno, Czech Republic

Pascale Pollier Biological and Medical Art in Belgium (BIOMAB), Antwerp, Belgium
Medical Artists' Association of Great Britain (MAA), Edgware, Great Britain
Association Européenne des Illustrateurs Médicaux et Scientifiques (AEIMS), Paris, France

Bernhard Preim Department of Simulation and Graphics, University of Magdeburg, Magdeburg, Germany

Tom Quisenaerts Faculty of Medicine and Health Sciences, University of Antwerp, Antwerp, Belgium
Biological and Medical Art in Belgium (BIOMAB), Antwerp, Belgium

Babette Van Rafelghem Faculty of Medicine and Health Sciences, University of Antwerp, Antwerp, Belgium
Biological and Medical Art in Belgium (BIOMAB), Antwerp, Belgium
Association Européenne des Illustrateurs Médicaux et Scientifiques (AEIMS), Paris, France

Farhan Rasheed Department of Science and Technology, Linköping University, Norrköping, Sweden

Mark Roughley Liverpool John Moores University, Liverpool, UK

Konrad J. Schönborn Department of Science and Technology, Linköping University, Norrköping, Sweden

Pawel Stachyra Włodawa, Poland

Alice Thickett Manchester Metropolitan University, Manchester, UK

Mhairi Claire Towler Vivomotion, Dundee, UK

Ann Van de Velde Faculty of Medicine and Health Sciences, University of Antwerp, Antwerp, Belgium
Biological and Medical Art in Belgium (BIOMAB), Antwerp, Belgium
Association Européenne des Illustrateurs Médicaux et Scientifiques (AEIMS), Paris, France

The Importance of Integrating the Visual Arts into the Medical Curriculum

1

Tom Quisenaerts, Yuki Dejonckheere, Ingri Dahi, Francis Van Glabbeek, Pascale Pollier, Babette Van Rafelghem, Tom Dietvorst, Johan Devrome, Joachim Deblock, Lou Gils, and Ann Van de Velde

Abstract

Medical education has long been considered a mostly scientific education process, with a clear focus on knowledge of biomedical processes, pathologies, diagnostics and therapy. In more recent years, there has been an increased appreciation for learning soft skills in the medical curriculum, including communication and empathy. How these soft skills should be taught, however, is less clearly established. One of the possible approaches is to use art and art-based workshops to learn and acquire these soft skills, which are necessary for clinical practice.

To explore the integration of the visual arts in the medical curriculum, this chapter will begin by examining the interaction of art and science to achieve a better understanding of how these two seemingly distant worlds can come together. The properties of art, which make it a suitable candidate for helping in the development of soft skills, will then be discussed. Then, a reflection on the learning outcomes expected from art-based workshops will be presented and discussed by taking pedagogical concerns into consideration required to achieve these outcomes.

After presenting the 'why' and 'how', the challenges of implementing art-based workshops in the medical curriculum are discussed, including ongoing research by BIOMAB and the Faculty of Medicine and

T. Quisenaerts (✉) · F. Van Glabbeek
Faculty of Medicine and Health Sciences, University of Antwerp, Antwerp, Belgium

Biological and Medical Art in Belgium (BIOMAB), Antwerp, Belgium
e-mail: francis.vanglabbeek@uza.be

Y. Dejonckheere · I. Dahi
Faculty of Medicine and Health Sciences, University of Antwerp, Antwerp, Belgium
e-mail: Yuki.Dejonckheere@student.uantwerpen.be; Ingri.Dahi@student.uantwerpen.be

P. Pollier
Biological and Medical Art in Belgium (BIOMAB), Antwerp, Belgium

Medical Artists' Association of Great Britain (MAA), Edgware, Great Britain

Association Européenne des Illustrateurs Médicaux et Scientifiques (AEIMS), Paris, France
e-mail: pascale@artem-medicalis.com

B. Van Rafelghem · A. Van de Velde
Faculty of Medicine and Health Sciences, University of Antwerp, Antwerp, Belgium

Biological and Medical Art in Belgium (BIOMAB), Antwerp, Belgium

Association Européenne des Illustrateurs Médicaux et Scientifiques (AEIMS), Paris, France
e-mail: Babette.VanRafelghem@student.uantwerpen.be; Ann.vandevelde@uza.be

T. Dietvorst · J. Devrome · J. Deblock · L. Gils
Royal Academy of Fine Arts Antwerp, Antwerp, Belgium
e-mail: tom.dietvorst@ap.be; johan.devrome@ap.be; joachim.deblock@ap.be; lou.gils@ap.be

M. Roughley (ed.), *Approaches for Science Illustration and Communication*, Biomedical Visualization 4,
https://doi.org/10.1007/978-3-031-41652-1_1

Health Sciences of the University of Antwerp, focussing on how medical students perceive the use of art in their anatomy lessons. This research promises to be an opening for the further integration of the visual arts into the medical curriculum. A brief excursion is taken to discuss the *art* of anatomical drawings themselves. To close this chapter some concluding remarks are postulated.

Keywords

Human anatomy · Drawing · Medical illustration · Medical education · Science communication · Educational methodology

1.1 Introduction

There is a noticeable increase in the integration of the visual arts in scientific research and education. One example is the current trend to integrate art-based workshops into the medical curriculum, however, the visual arts are not the only non-medical field of study being looked at to complement the medical curriculum. The visual arts are only a part of a movement to integrate a wider spectrum of health humanities into the education of medical doctors.

The health humanities have an indirect influence and may offer new insights into the study of the health sciences. They encompass a wide variety of disciplines such as the fine arts, philosophy, law, history and more. The goal of the health humanities is not only to offer new insights but also to approach health and healthcare from a more holistic approach. They offer points of view that are often overlooked in the classical approach to medicine. Next to that they also offer solutions to problems that are hard to answer from a purely biomedical scientific point of view.

This chapter focuses on the role of the visual arts in the holistic education of medical doctors and by extension, health care professionals in general. Visual images and anatomical drawings assist students in improving their knowledge and perceptions of the human body, considering both the objective and subjective side of medicine (Reid et al. 2019; He et al. 2019; Dalia et al. 2020; Styn et al. 2023). In this way, the arts can grant a deeper scientific insight and a more fundamental understanding of the dynamics of the human body.

The medical curriculum is one based on objective knowledge, to make a diagnosis as accurately as possible. Subjective perceptions take the backseat here, especially in the initial phase of medical training. This makes bachelor's and parts of master's programs mostly academic and scientific minded, with little attention to subjective and often subtle perceptions which play a role in both art and clinical practice.

Biological and Medical Art in Belgium (BIOMAB) is an organisation which aims to improve the cooperation between scientists and artists in medicine and anatomy, but the scope is not limited to these fields. One of the methods of achieving this goal is to organise anatomical dissections of fresh frozen specimens of human and animal limbs. These dissections are organised and performed by medical doctors and aimed towards an audience of artists who come to these events in order to view the anatomy up close and in real life. They provide the perfect opportunity for the artists to observe and record anatomy from real-life specimens, instead of anatomical books or other forms of image-based media. An example of such a cooperation is shown in Fig. 1.1, an aquarelle painting of such a dissected arm. It is not only a unique experience for people outside of the medical field but also an enormous boost to the artist's understanding of 3D anatomy, for example, how the muscles and tendons interact with the bones in order to create movement, and how the superficial anatomy observed in living subjects comes to be from the structures which are situated under the skin.

These events are also a great opportunity for people from both the medical and artistic fields to interact with each other. During such interactions, medical doctors and medical students who understand anatomy in a medical and scientific sense get the opportunity to observe the artists and how they create their art from this input. They also get the opportunity to talk with these artists about how they perceive anatomy and what this means

Fig. 1.1 An anatomical study of hands [aquarelle painting] by Frederik Rietdijk

for them. This can serve as a great way of increasing cross-pollination between both worlds. It is from these experiences that we are both intrigued by and convinced of the added value of integrating art into the sciences and more specifically the medical curriculum.

In the following sections, the added value of integrating art into the medical curriculum is explored, first, taking a look at the synergy of the two seemingly different entities of art and science. This is followed by a detailed analysis of how art can stimulate learning, how it achieves this and how it can be implemented. Then follows a discussion of ongoing research in Belgium on the views of medical students on the integration of art into their curriculum, which will help in identifying the perceptions, preferred strategies and roadblocks in implementing art into the medical curriculum in Antwerp. Finally, the topic of anatomical sketching is touched upon. This text tries to capture the essence of this form of art and its method of observing the human form.

1.2 Art and Science Are Not that Distant as Thought

In medicine, there is a long-standing tradition of using visual representation to depict anatomy, physiology and procedures. 2D drawings, sketches and prints serve to translate a 3D reality, reinforced by insight and understanding, into a compact, timeless and portable medium. They assist in passing on this knowledge from the author to whoever might need it. These visual mediums excel in this goal over the written medium as visual representation leaves little room for misinterpretation as opposed to the written word on its own. This is a tradition that stems from the first use of visual depictions by our cave dwelling ancestors such as the depiction of a pregnant woman in a Spanish cave (Tsafrir and Ohry 2001), and the ancient Egyptian hieroglyphs in the Djedkare pyramid depicting the procedure of a circumcision (Mohamed and Vymazalov 2011). One of the first known anatomical depictions of the heart was created by the Olmecs (Bendersky 1997), and the Mawangdui medical texts from 168 BC in China is probably the oldest surviving anatomical atlas (Harper 2015). The use of numerical references in an image accompanied by a legend was introduced by Andreas Vesalius (Vesalius 1543), and served as a precursor to the modern-day plethora of anatomical textbooks building on the shoulders of giants in the field of anatomy.

This also works the other way around, as art has a tradition of depicting issues which plague human existence. As a result of this, medicine has often been a source of inspiration for artists, for example a Roman fresco found in Pompeii dating from the first century CE was shown to depict the Roman physician Iapyx removing an arrowhead from Aeneas's thigh with Venus looking upon the scene (Raddato 2015). It can also depict a range of pathologies, for example, there is a theory that the Mona Lisa by Leonardo Da Vinci has a xanthelasma next to her left eye and a lipoma on her right hand. Then there is the self-portrait by Vincent van Gogh in which he depicts himself with a bandaged ear after losing it, which is alleged to have happened after he cut it off by accident when he had a seizure while threatening his roommate with a razor. In more modern times there are continued examples depicting illness, such as the series by Shawn Coss in which he depicts different forms of mental illness (Coss 2017).

But the delineation between whether something is mainly artistic or scientific can sometimes fade, as in the case of 'the anatomical Venus'. This is an eighteenth century concept in which the human anatomy is displayed by using artistically crafted and aesthetically pleasing wax models of beautiful women. The idea behind these models was to showcase the human anatomy in the form of a partial dissection of the body while the rest of the model is crafted in order to be pleasant to look upon. This served the purpose of bringing medical knowledge to the wider public without putting them off too much (Ebenstein and Morbid 2016).

The same concept is displayed in Valverde's anatomical tome, which shows images by Becerra, and where anatomy is showcased using a Venus image to display the anatomy of pregnancy or an artistically embellished Roman cuirass that showcases the abdominal anatomy (Van Hee et al. 2020). Both are good examples of how the artist and author worked together to create something which represents both art and science. A more contemporary example is the work by Babette Van Rafelghem, which is shown here as Fig. 1.2 and sets out to show medical procedures to patients and the general public through artistic depictions of these procedures, with the goal of educating them while avoiding real-life images as these may be too graphic.

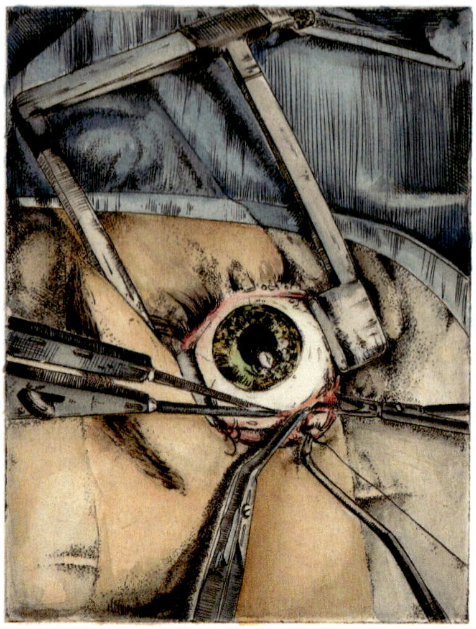

Fig. 1.2 *Eye surgery* (2018) [hand-coloured etching, 13 × 9.5 cm] by Babette Van Rafelghem

Art is capable of inspiring us, reminding us of tales, histories, other perspectives and the beautiful, but also the ugly sides of nature and life itself. It begs us to pay attention to detail in order to discern the similarities and differences between artistic depictions and real life, often with a message hidden in these similarities but also in the differences. Art makes us pay attention to hidden meanings and details, as is the case with the bustling sceneries, hidden parables and old adages in the works by Pieter Bruegel the elder. It is this effect of art, which forces us to take a step back and think outside of the box or from another perspective which is able to give the scientist a fresh look at their work, strive for greater detail and maybe try a new angle (Hanshumaker 1980).

The two entities of art and science may seem like opposites when seen in the light of popular opinion and pop-culture entertainment such as in movies and television series. Stereotypes of the strict, up-tight, book-devouring scientists posed against the liberal, 'carpe diem', unstructured artist reflect this perceived difference. These portrayals and thinking inside of boxes should not be allowed to mislead the public. There is plenty of room and opportunities for the arts and sciences to come together and create synergy, as proven by the ancient Greeks and the scientists during the age of enlightenment, each reflecting cultures in which a well-rounded education and approach to life was highly valued, a concept beautifully described by the term 'homo universalis'. Some notable examples from ancient Greece are Plato, Socrates and Aristotle, all of them can be considered true polymaths, dabbling in philosophy, science but also art such as poetry, drama and music.

One of the most prominent examples of how one person can excel in both science and art and how these can be integrated is Leonardo Da Vinci. He is famous as an inventor of various machines, based on observations from nature such as a flying machine called the 'Ornithopter' which was based on birds, for describing and drawing human anatomy, and for creating well-renowned artworks including the aforementioned painting of Mona Lisa. Taking these examples into consideration, it is safe to conclude that art and science are not mutually exclusive and often even synergistically.

1.3 Biomedical Illustration Schools

Using images for descriptive purposes has been a constant must in the history of humanity. This trend, as Marinella Pigoza describes in the catalogue of the Biomedical Art 2008 exhibition in Bologna, became stronger in the second half of the fifteenth century, as the innovative printing techniques and new ways to distribute books overcame former barriers. Images as a source of information acquired a primary and paramount role in scientific publishing. Sight aids memory more than any other sense and images are the most effective means of amplification. The illustrated book presents itself as a comfortable theatre of knowledge. An example of how this process of medical drawing comes to life is seen in Fig. 1.3, where a medical artist is drawing a human maxilla bone during a workshop.

Biomedical illustration schools and organisations such as the Association Européenne des Illustrateurs Médicaux et Scientifques (AEIMS), the Medical Artists' Association of Great Britain (MAA) and the American Association of Medical Illustrators (AMI), are exemplary long-standing instances of learning and of teaching the science of observing, representing and describing the three worlds—animal, plant and mineral. They focus on a scientific field endowed with a great past, spanning from early history to today's skilled biomedical artists. The later are now able to dialogue with the traditional techniques but are also ready to renew their art-embracing new computer-aided strategies.

Today's medical illustrators have the ability to weave a systematic and balanced dialogue with the world's centres of excellence in the biomedical field. Illustration always wants, and is able, to mix with art and science. Watercolour drawings often are of an extraordinary art and science concentrate. With their strong emotional impact, they immediately creep into the observer's memory, settling into a suspended dimension between history and eternity. They match scientific analysis and artistic awareness; the act of remembering brings knowledge, emotion and pleasure. They prove how fruitful it is to weave empirical verification. Alongside the physical horror instilled by illness and the coldness perceived whilst looking at the images, the high-quality artistic stroke and the full colours make a lasting impression in our minds.

However, images and words are an unbreakable bond in the medical field, pursuing a single analytical inquiry as written by Elisabetta Cunsolo (Biomedical Art 2008 exhibition catalogue). A poem by Bryan Green (Fig. 1.4) for the 27th AEIMS Congress (2018), called 'The Cutting Edge of Medical Art', beautifully illustrates this strong bond between image and word within the fields of Medicine and the Medical Sciences.

Throughout its history, from antique collections to the most recent digital illustrations, scientific images have acquired the ability to remind through demonstration and to confirm the message with which words are entrusted. Far from being a mere decorative distraction disjoined from the written work, illustration, as a synthesis of the knowledge to be transmitted, establishes, with a text, a relationship formed by

Fig. 1.3 Carbon dust drawing of a maxilla (after dry specimen) during a medical illustration workshop at the RSU anatomy museum of Riga, Latvia

factual references. This clearly shows that the scientific reasoning to which the illustration refers is complete and concrete, and schools of Biomedical Illustration aim to reinforce the role that anatomical drawing has, to this day, in the medical field.

1.4 Effects of Art on Learning in the Medical Curriculum

The positive effects of using art in the academic learning process have been thoroughly studied in different settings, populations and fields. Some examples as mentioned by Haidet et al. (2016) are: 'health care professionals, medical or nursing students, industrial design, students, pre-service education students and teachers, school leaders and administrators, corporate employees and community members'. This line of inquiry has led to the understanding that the integration of art in different forms has various positive effects on learning in an academic setting (Haidet et al. 2016). The following sections of this chapter are largely based upon the theoretical framework of Haidet et al. as they succeeded in capturing a very thorough conceptual model of how art education is able to interact with and improve the medical curriculum (Haidet et al. 2016). The framework is supplemented with our own thoughts and experiences of the authors as well as input from the existing literature.

In their systematic review, Haidet et al. (2016) identified four common themes and their respective subthemes pertaining to the ability of art to develop humanistic characteristics. These themes are, as mentioned by the following direct quote: '(i) unique qualities of the arts that can promote learning, (ii) particular ways in which learners engage with art, (iii) documented short- and long-term learning outcomes that arise from arts-based teaching, and (iv) specific pedagogical considerations for using the arts to teach in professional education contexts' (Haidet et al. 2016). These topics and the framework constructed from them were used in order to guide the reader through how art can be used in medical education.

There are various methods of using art in an educational setting, this should not come as a surprise since art is enormously heterogenic in nature, it can take the form of drawing, painting, dancing and poetry, to name just a few. Therefore, it would be impoverishing to set out a certain list of forms and thereby restrict other equally valid ways of using art in the medical curriculum. In this section, certain approaches are discussed but by no means are they intended to form an exhaustive summation. For example, a basic art interpretation class, in which students are taught to interpret different forms of art, promotes

the cutting edge of medical art
is the raw wound of thought
the studies the preparation
the research
come to this moment

and in a little corner of the mind is said
this is what I do
this is what I want to do
I can do this
and I can do it well
even though everything
in its own way
is unknown

a pencil or brush on a white page
a scalpel on unblemished skin
words that break the silence in a museum
and as the lights go out the photos are queueing to exist

the film hovers in your intent
the medium is mixed
the armature made
allowing the work to commence
whilst limiting the outcome
it is the cutting edge of medical art

you have trained yourself thoroughly
you have mastered hands and eyes and mind
as best you could

and what is more

you dare

to begin

Fig. 1.4 '*The cutting edge of medical art*' (2018) by Bryan Green [poem]

perception and understanding of the subjects at hand (Haanstra 1996; Haidet et al. 2016).

Art-based workshops can be tailored to specific topics. For example, the drawing of anatomical structures as the student in Fig. 1.5 is seen doing. Another example of how this can be used in the medical curriculum is by observing a ballet performance and interpreting the movements made by the dancers and to identify the muscles and neurological processes used in their execution. The workshops can also be off-topic, for example interpreting the details of a painting or script writing for a script for a theatrical play.

1.4.1 Qualities of Art

The first qualities of art which shall be discussed are its metaphorical and representational nature (Haidet et al. 2016; Peppler et al. 2021). In this context, it is the power of art to present themes such as romance, pivotal decisions, and life and death, which can be considered universal to the human experience. It does this implicitly in either the story being told or in the mood being presented. This makes the beholder perceive these themes in an often-indirect way, making them reflect on these topics and letting them make their own connections and interpretations, tailored to their own experiences. This metaphorical and representational nature of art is a thankful tool to broaden the scope of a topic and stimulate the learner to take other perspectives into consideration. It pulls them out of the standard way of learning and thinking where information is processed almost mechanically. Instead, it provokes reflection and broadens their scope.

Another quality of art is its subjectivity (Haidet et al. 2016). Art is projected upon the metaphorical eye of the beholder. As opposed to a plain PowerPoint slide, art is not projected on an empty white screen. Its intentions and exact meaning are rarely explicitly proclaimed and therefore it is up to the beholder to take it in and interpret its meaning. This interpretation is de facto coloured by personality traits, past experiences and learned cultural interpretations. From this shared stimulus stem as many different interpretations as there are people perceiving it. The subjective nature of art can be used by teachers to show students different ways of looking at the same thing, this promotes broad thinking, mutual understanding, the acceptance of multiple experiences and possibilities (Haidet et al. 2016). Embracing subjectivity helps to challenge the linear way of thinking which is often professed as the only possibility in scientific approaches.

A third quality is the ambiguity and complexity of art. It is observed that learners frequently turn to their teachers to express that they are unsure about how to understand and interpret a

Fig. 1.5 A photograph of a student drawing during a medical illustration workshop at the RSU anatomy museum of Riga, Latvia

certain work of art. The fact that art presents the beholder with an element of uncertainty leads to a thought-provoking process which forces the student to go on their own search for meaning and interpretation of what the artist intended to portray and convey to them. It is this self-initiated and intrinsically motivated process of searching and discovery, which is fuelled by the human reflex to give meaning to and understand our environment, that leads to a higher level of engagement in the learning process and promotes creative thinking.

Finally, the ability of art to create a levelled playing field is named as a contributing quality. When engaging with art in a classroom, students notice that the group is not divided by hierarchical features such as socio-economic status, culture and academic results to name a few. This is likened to the fact that art is open to interpretation, ambiguous and often a personal experience and such barriers inhibit the creation of art (Haidet et al. 2016). While we can agree with the gist of this statement, there are still attributes that can present as a confounding factor in this process, such as students being present with a wider background in the arts and cultural preferences, alongside an ability to adapt, life stresses preventing self-development (Maslow 1943) or just a disinterest in art. It is important to promote the equalising properties of art while being aware of the possible confounding factors when introducing art as a tool to boost the learning process.

1.4.2 Art Engages People

Both the observation of art and the participation in its creation can be very engaging experiences for those who are open to it. Studies have pointed out that there are different levels of engagement. On the level of reflection and interpretation, which refers to the capacity of art to elicit subjective experiences within a person and their interpretation of said experiences. This reflection and interpretation refers to a conscious and cognitive way of thinking about what is being experienced. Another form of engagement is affective or emotional engagement. This refers to the experiencing of emotions, feelings and more subjective experiences when enjoying art. These experiences are often times less explicit and more implicit and subjective.

Some terms to describe these emotional/affective experiences are: ‘empathetic imagination’, ‘embodiment’, ‘aesthetic knowing’ and ‘affective experience’ (Haanstra 1996; Haidet et al. 2016; Peppler et al. 2021). Moreover, art is a multisensory experience where the viewer is engaged through a combination of visual, auditory, tactile and sometimes even olfactory and gustatory stimuli. This multi-sensory experience helps to improve engagement since more neural pathways are activated in the acquisition and processing of knowledge, more specifically both the occipito-temporal and occipito-parietal pathways are being engaged more actively (Shams and Seitz 2008; Veelaert and Santens 2011; Hardiman et al. 2014). These experiences are believed to at least partially reflect the cognitive, somatic, affective and spiritual dimensions of human experiences.

In medical education, it is well known that the vast majority of courses are dedicated to theoretical biomedical knowledge and clinical skills. Therefore, we could state that in this curriculum the cognitive aspect of learning is considered to be the most important dimension to become a good medical doctor since it is expected that a physician can objectively and correctly diagnose and treat the condition with which a patient presents themselves. This is also reflected in medical art, such as in Fig. 1.6 where a scientifically correct drawing of the human heart is shown. However, there are good arguments to promote the development of other dimensions as well. Healthcare professionals who are 'in touch' with their somatic sensations will be better at using all sensory faculties and 'gut-feeling' to guide them in their clinical decision-making as well as improve their technical skills (Dhaliwal 2011; Wong et al. 2012).

Developing the empathic dimension promotes engagement with the patients and co-workers leading to more fulfilling relations and job experiences (Raab 2014). However, it is dangerous to let emotions cloud good clinical judgement, as it is dangerous to allow emotions from the daily reality of clinical practice to overwhelm you. It is a delicate balancing act that can be strengthened by developing the empathetic dimension.

Finally, the spiritual dimension is an exceedingly charged subject and often stems from personal experience. The general trend in the Western world is one of moving away from religion while there is little to fill the spiritual void, but the need for spiritual fulfilment remains present, especially in difficult times. Developing one's own spiritual dimension may help to understand this longing for spiritual development and promote connectedness with other people. A prime example where spiritual development can be of help is the doctor–patient relationship where topics of life and death are frequent (Perry et al. 2011; Haidet et al. 2016).

When looking at how art promotes reflection, interpretation and learning, it becomes clear that it is the ambiguity and complexity of art that triggers the human mind to give meaning to what is being perceived. This triggers new experiences and the recall of memories in the observer or creator, which in turn leads to them giving meaning to the stimulus. These processes of deriving information from an ambiguous stimulus are also known as 'generation of information from a cue' and 'semantic elaboration'. They translate to other contexts and lead to people thinking about novel information in new and different ways, which in turn improves retention and thus learning (Hardiman et al. 2014).

Working in healthcare is inadvertently linked to a higher level of exposure to emotionally charged and difficult situations (Delany et al. 2021). In practice there may not always be the appropriate time or place for the medical personnel to express and process these feelings. Art can serve to project clinical experiences and feelings and help reflect on and work with these emotions. This process of allowing emotions from past experiences to come to the surface can help in promoting empathy and self-awareness while also promoting the process of learning. On the other hand, art can be used to engage students and make them experience how patients feel about and live with their diagnosis, but also how the healthcare system and healthcare professionals are perceived by laymen (Perry et al. 2011; Haidet et al. 2016).

Art helps to engage people and improve learning through the concept of the 'enactment effect'. This is reflected in the finding that when people are acting out the information that they are thinking about or are studying, it improves both engagement and retention (Cohen 1981; Hardiman et al. 2014). This means that physical forms of art and physical engagement, if they are

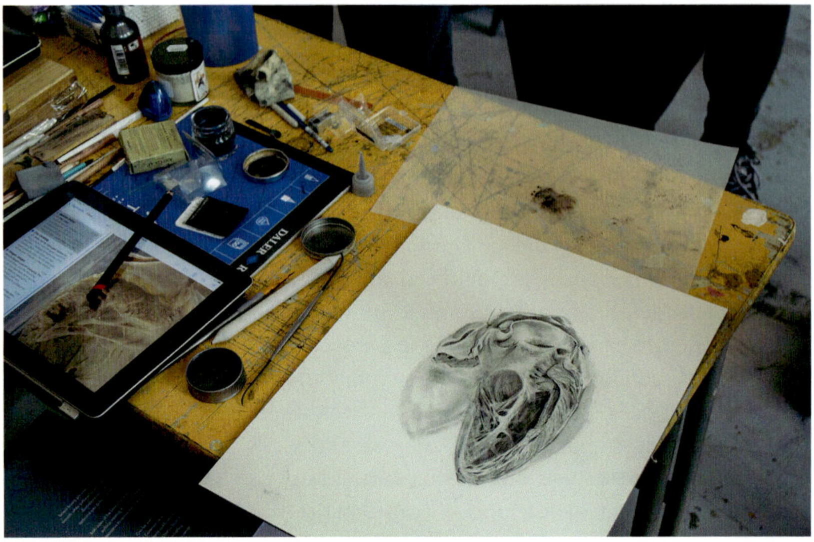

Fig. 1.6 Carbon dust drawing of a heart (wet specimen) during a medical illustration workshop at the RSU anatomy museum of Riga, Latvia

related to the topic, help to improve learning outcomes.

1.4.3 Learning Outcomes and Goals

There are many forms of art and each of these can be expressed in many ways and take on a wide variety of topics and interpretations. This heterogenicity can be considered both a blessing and a curse in the implementation of art courses in a medical curriculum. On the one hand, it is hard to make a uniform approach to achieve a certain goal, while on the other hand, this also means that there are different equivocal approaches to achieve the same goal. Therefore, it is more important to focus on the form of guidance and the process, rather than on the specific artform.

When planning the process of integrating art into the medical curriculum certain goals should be set out, so that they can serve as something to work towards and judge whether the approach is working or in need of adjustment.

1.4.3.1 Short-Term Goals

Goals that can be achieved in the short term, as named by Haidet et al. are: developing self-awareness, increased openness to other perspectives, an enhanced ability to cope with ambiguity and a more nuanced or deeper understanding of selected concepts/topics (Haidet et al. 2016). To achieve these goals, art-based workshops can be organised in a way to invite the participants to openly discuss, in a non-judgemental way, the thought process behind either the interpretation or creation of a work of art. Emphasis should be on the process of listening to the interpretations, ideas and experiences of fellow participants. This next step is crucial to disconnect from a self-centred view and to step into the perception of others. While this can be second nature for some people, it can take some more effort for others in order to achieve this. This exercise first involves exploring your own process of how you experience and handle feelings, experiences and emotions. In the next part you let go of your own perspective in order to intentively listen to others going through the same steps and trying to imagine their experiences. This process leads to more self-awareness and openness (Perry et al. 2011; Haidet et al. 2016; Peppler et al. 2021).

Art is capable of improving mental flexibility, being the 'the ability to shift a course of thought or action according to the changing demands of a situation' as defined by Loftis et al. in their Encyclopaedia of Clinical Neuropsychology (Loftis 2017). This was demonstrated when a group of academically high-performing students, that is, those that obtain high-grade points averages, were exposed to an experimental intervention in which they had to perform a problem-solving task using art in adjunct to a scientific approach. They outperformed their matched peers, who were not exposed to art, in both creativity and mental flexibility (Hanshumaker 1980). Perhaps it is the exposure to ambiguity in art which forces us to come up with solutions in order to give meaning to it, which in turn leads to improved creativity and mental flexibility. On the other hand, the exposure to art which uses different techniques, concepts and moods automatically also exposes us to a collection of new experiences, which might add to this effect.

1.4.3.2 Long-Term Goals

Short-term outcomes and the process of achieving them pave the way to some of the long-term goals. Some of these possible long-term goals which are named by Haidet et al. are: empathy, observation, communication, critical thinking, ethical reasoning and creative thinking (Haidet et al. 2016). There is one long-term goal for art-based workshops in the medical curriculum, which was named most often in literature and thereby stood out among the rest: *empathy*. This perhaps does not present itself as a surprise if you consider both the aspects of art and the power of empathy in medicine. Empathy in itself is a creative act because you must use your imagination to construct what you think the other person is experiencing based on the information that they provide. It is therefore proposed that creative acts promote the development of empathy and its use in our daily relationships (Hanshumaker 1980; Perry et al. 2011; Haidet et al. 2016; Peppler et al. 2021).

Following this reasoning, it can be posed that promoting the participation of medical students and trainees in creative workshops can function as training sessions to strengthen their empathy. The power of art education within the medical curriculum to improve interpersonal skills, humanistic approach to patient care and empathy in general has been widely discussed and studied. In the scientific literature, most authors agree that it is indeed capable of improving these outcomes. One of the critical reflections which are taken into consideration however is that the success or failure of these interventions will probably be largely dependent on the level of motivation of the participants. This implies that the approach to fostering engagement will be of great importance in achieving these outcomes.

Medical doctors and other healthcare professionals are often exposed to traumatic experiences, high patient loads and high levels of paperwork, which can possibly lead to a reduction of engagement and empathy. To counteract these effects it is important to pursue methods of improving and stimulating these empathetic skills (Passalacqua and Segrin 2012; Raab 2014). Therefore, if art-based workshops are capable of reigniting this flame of empathy within medical staff it is a path well worth exploring. The use of artwork by patients displaying their struggle with their medical conditions can act as a grounding experience and one can develop heightened sensitivity and empathy in future patient–doctor interactions (Perry et al. 2011; Haidet et al. 2016).

The development of observational skills does not only concern the recognition of physical properties but also emotions and contextual cues. Improved observational skills will help medical students become better physicians. The recognition of physical properties will lead to better recognition of medical situations, while the observation and recognition of both emotions and contextual cues improve the interaction with patients which can lead to patients feeling more at ease, comforted and better understood (Haanstra 1996; Perry et al. 2011).

A large part of art-based workshops needs to be dedicated to expressing one's own thoughts on the observed or created artworks as well as listening to the thoughts of others. This interaction can function as a simulation for communication in our daily lives. During this experience, the group may reflect on communication styles and so serve to improve each other's communication skills. This process forces the participant to engage in oral production and thus engages them to organise their thoughts and express them verbally, which leads to better retention of information (Hanshumaker 1980; Daykin et al. 2008; Perry et al. 2011; Hardiman et al. 2014; Haidet et al. 2016; Peppler et al. 2021).

When students are interpreting a piece of art, they are forced to look at the different meanings which may be woven into it. This in turn forces them to think creatively, but also critically about possible interpretations. When discussing ideas, they are forced to reflect back upon their own interpretations and analyse the interpretations of others. They may then come forward with supporting or disproving arguments. This is a process that forces them to think critically about both their own thoughts as well as those of other people.

Art often reflects the contemporary struggles and sensitive subjects of society, such as gender issues and racism or might even reflect more timeless and universal topics such as love and death. When discussing artworks, the discussion might go on a tangent and become more about these ethical issues than the artwork at hand. In doing so the artwork has accomplished what the artist may have intended it to do, to elicit certain thoughts, feelings and discussions on a certain topic. These discussions together with the development of other skills such as empathy, communication, creative and critical thinking help in the development of ethical thinking.

To conclude, it is important to realise that the effects which the qualities of art can potentially achieve are very much dependent upon the level of engagement and that this will, of course, dictate the outcome. Art is naturally capable of engaging people, yet due to the huge array of differences in personality traits, beliefs, past experiences and other promoting or confounding factors, it is hard to predict the outcomes of these interventions. The result will largely depend upon the individual participant's intrinsic motivation. Yet, it is possible to stimulate this process through pedagogical techniques such as those which will be discussed in the next section.

1.4.4 Pedagogical Concerns

As mentioned before, art is very heterogenous. However, there are certain pedagogic tools, techniques and considerations that can be used and implemented irrespective of the exact artform when teaching these art-based workshops to medical students. The systematic review by Haidet et al. (2016) offers a few main topics to start from, such as: *Engagement* strategies, *Meaning-making* strategies and *Translation* strategies.

Engaging students can sometimes be the hardest part, while success often depends on this crucial step. Investing a dedicated amount of time into engaging them, goes a long way. The first tool for engaging learners is transparency. It is crucial to build rapport with students through open communication and congruity. It can help if the teacher shares their experiences and struggles as it will make them more relatable and shows that engagement with the arts is linked with self-disclosure and that the art-based workshop is a safe place to open up (Haidet et al. 2016). Setting ground rules based on a feeling of mutual respect, respectful communication and the feeling of being part of a group, further supports the goals which were discussed in the previous paragraph. Another tool is to recognise factors that may hold someone back from engaging fully, discuss them openly with respect to privacy and finally address them by either reassurance or certain acceptable accommodations. It also helps when the teacher positions themself as a co-learner instead of a teacher. This can be achieved by participating in the activities and sharing their own experiences, struggles and findings. The common theme of these strategies is to lower the threshold and power difference between students and teachers as this will lead to higher levels of engagement and disclosure (Haidet et al. 2016).

After laying the groundwork with engagement strategies, meaning-making strategies can be introduced to stimulate participants to reflect on their own perspectives as well as understand and accept the perspectives of other participants. These strategies also help with handling ambiguity and deepen the understanding of the artwork and accompanying lessons. As for the techniques used to achieve this, the main focus should be to encourage students to reflect on how they experience observing or creating art and to foster communication between students concerning these topics. This can be achieved by assigning moments during the workshops to focus on reflection and discussion but also through engaging assignments, which do not need to be very taxing on the students but should be focussed on reflection, cooperation and communication (Haidet et al. 2016; Peppler et al. 2021).

The final category of strategies is the translational strategies. These strategies promote the application of new skills and insights into practice. A student is taught during the workshops to reflect upon art and discuss this with fellow students, without judgement, without power struggles and with unconditional acceptance. These experiences and skills should be promoted to be applied to clinical practice as well. These strategies can be used to reflect on experiences and what they meant, on how they handled certain situations and procedures, but also how to better cooperate with colleagues. An example of how this can be achieved is to incorporate reflection on clinical experiences during later sessions of the art-based workshops so that students are taught to reflect on these topics while they are still in the context of the art-based workshop, which they have learned to accept as a safe environment. The next step could be to implement these reflection sessions in the setting of clinical practice, this, of course, requires the participation of clinical personnel who should be capable and willing to emulate a similar environment in order to achieve translation (Haidet et al. 2016; Peppler et al. 2021).

Reflection and feedback should be a part of routine medical practice. While this is the case in some centres, it is not always organised in a constructive and organised fashion, and in most cases, there still is room for improvement and a lot can be learned from the arts and the techniques which have been outlined here. They can help structure feedback and reflection sessions while shaping the situation to be more stimulating than demotivating, which is a potential pitfall of giving feedback on things that could or should be improved.

1.4.5 Challenges in Implementing Art-Based Workshops in a Medical Curriculum

One could ask themselves, if art-based workshops seem to offer all these advantages, why then are they not yet implemented en-masse? As with any change to the norm, there is a certain level of resistance to this change and inertia offering resistance against it. However, with the medical curriculum there are some specific challenges as well.

When looking at the current state of medicine and comparing it to even just 150 years ago, it can be concluded that enormous strides have been made in our scientific knowledge of both the normal human body as well as the plethora of diseases, which can ail it. This is exemplified by looking at the history of surgery. The advent of general anaesthesia and antisepsis in 1846 and 1865, respectively, have made it possible for surgeons to start developing more advanced types of surgery. This led to a boom in new surgical techniques around the turn of the century and continued well into the twentieth century and even the modern-day twenty-first century. The first documented diagnostic laparoscopy was performed in 1901 by George Kelling, with the first cases of interventional laparoscopy, consisting of adhaesiolysis and biopsies, being performed in the 1930s. When in 1986, the video chip was introduced to laparoscopic surgery the field underwent a new wave of innovations (Spaner and Warnock 1997). Stemming from this line of surgery is the development of robotic surgery, which has its foundations in the 1980s, with the first robotic surgeries being performed in 1992 (Lane 2018). These advances came to a technological milestone with the first trans-Atlantic surgery, with a surgeon in New York and a patient in Strasbourg, being performed in 2001 (Gottlieb 2001).

This brief history of surgical developments spanning from 1846 to 2001 not only shows a clear crescendo trend, going from crude and fast surgery in the pre-anaesthetic era to trans-Atlantic robotic surgery, but also serves as an example of how the world of medicine has developed in just 150 years. Therefore, it is not hard to imagine that medical education programs are often very demanding of their students regarding the amount of knowledge that each student has to process in a relatively limited amount of time. This results in an educational program in which students receive many hours of lectures, are required to read through mountains of books and complete countless assignments. The workload which has already been assigned within the medical curriculum is not likely to be reduced in the coming years with new medical innovations and discoveries on the horizon. This workload, of course, forms a direct challenge for the implementation of any other form of education, which does not directly pertain to the advancement of medical knowledge within the medical curriculum.

While it makes sense to prioritise medical knowledge, it should also be taken into consideration that a physician should be more than just a medical expert, they should also develop interpersonal skills in order to be a proficient communicator and collaborator as described by the CanMEDS framework (Frank et al. 2015). This CanMEDS framework attempts to encompass all facets and roles which should be exemplified by a physician. This framework describes the abilities which a modern physician should poses in order to meet the healthcare needs of the modern patient. These abilities are divided into a set of roles, the framework poses that a physician should be able to act as a 'Medical expert', 'Communicator', 'Collaborator', 'Leader', 'Health advocate', 'Scholar' and a 'Professional'.Another advantage of integrating art-based workshops into the medical curriculum is to offer students a break from their standard type of lectures and have them engage in both a practical and expressive way. But also, to interact with each other on a more profound level.

Other than practical problems, there may be personal opposition from students or professors. There may for example be scepticism on the usefulness of implementing art-based workshops in a medical curriculum, some students may be frustrated to spend time allocated to their medical education on something else than medical courses or there may be disinterest from students regarding arts or participation in art-based workshops. These challenges of a personal nature may be the hardest to overcome since the success or failure of these interventions depends on intrinsic motivation. Therefore, it may be in the best interest of the students and the success of the workshops to make such art-based workshops an elective course within the curriculum and offer a scientific explanation of how these workshops can be in their best interest. If the community of medical students and the medical faculty notice a perceptible benefit from these art-based workshops a further implementation and integration could be considered.

Certain practical and personal challenges and opposition is to be expected in any attempt to change an existing status quo. Especially when trying to implement an art-based course within a heavily science-driven curriculum such as medicine. Nonetheless, there is reason for hopefulness and optimism concerning the real-world added value of this cooperation between art and science.

1.5 Anatomical Drawing Lessons: Physically Observing and Drawing the Human Body within the Medical Training

The human body as a living material has its own identity which is never the same and forever changing. Observing the human body is not a fully measurable cognitive process. It forces the student to look holistically. A process in which four components, visual perception, mental processing, emotional processing and thinking and putting into words what you see, all interact with each other. Visual perception is in fact nothing more than the act of light interacting with the rods and cones in our retinae, these physical processes are then translated into electrical impulses which are collected mainly in our occipital lobes. However, what the occipital lobe does next is to create emerging features from this raw input, which makes us perceive visual cues. This reconstructed image is processed by our brains, forming a mental representation of the objects we perceive. If our brain recognises it as something which we already know it is able to give meaning to it. However, when it doesn't know or recognise it, our brains will attempt to give meaning to these perceptions. But our brains also process this input emotionally, the fear response even foregoes the explicit perception by visual pathways leading straight to the amygdalae, which are triggered by any form of potential danger, while further and more complex emotional processing takes place after primary appraisal. When all this is done, we humans are finally ready to put into words our thoughts and feelings which are triggered by this light we perceived, projecting the piece of art and all its underlying intentions unto our eyes.

All these elements are poured into the end product, which represents more than a visual reality. This is exemplified by the drawing of the foetus in Fig. 1.7, with its calm demeanor and soothing impression of comfort within its amniotic sac, where the artists' interpretations and mental processes have led to the embodiment of emotions. To clarify this statement, should the added value of drawing in medicine be approached from a didactic point of view? What skills are developed through drawing and studying the human body? How do these skills position themselves in a cross-disciplinary context?

Drawing is not only a physical or visual effort but also an intellectual one. It learns how to look at something in a different way, which dictates the way of thinking about that subject and vice versa. No outside without an inside. By studying the human skeleton, the student literally learns to look through the form and to integrate the non-visible matter as a subcutaneous form. This skill is important and applicable to all disciplines. The student learns to gain insight into the process of observing and formal thinking. They discover the possibilities to develop a formal insight by reading the human skeleton as a construction that embodies both a static and a dynamic function. They do this by developing an understanding of the human body as a collection of surfaces, which are subject to the laws of the natural order. Through this insight, the student learns to understand natural harmony and translate it into a drawn image. They learn that the body is subject to gravity and gives formal meaning to the concept of stability of spatial forms and physical objects by bringing the human body as a form into balance in its totality.

Drawing is a performative act, learning to think and shape through embodied thinking or to give meaning through the way we physically relate to things. Drawing as a performative act can be illustrated with the proposition of the French philosopher Maurice Merleau-Ponty, who describes drawing as the inside of the outside and the outside of the inside (Maurice Merleau-Ponty 2012). This is not from purely cognitive thinking but shapes the experience of reality by means of his body, embodied thinking. The inner or emotional world of the medical student becomes externally perceptible through the material of the drawings. We must understand the eye as a window to the soul. The drawing sheet as a playing field, in which the medical student can physically position themself in a meta-position, and acts by looking at what is in front of them. Thinking from concrete action instead of rational thinking in the form of language.

Within anatomical drawing lessons, the emphasis is mainly on learning to draw the human anatomy through observation of the skeleton, the muscular man and the living model. Students are encouraged to draw from objective perception. In concrete terms: students are expected to draw what they see instead of what they think they see based on mental representation. An anatomical drawing course, such as those offered by BIOMAB, can be regarded as a supporting subject for other courses, such as anatomy, surgical courses, clinical skills and anatomical dissections. The course offered by BIOMAB, as demonstrated in Fig. 1.8, is a study moment in which the student derives the human figure from drawing. Through the drawing process, the student learns to understand human anatomy and translates it into a drawing. In the art of drawing there is the term 'study drawing', which can be used as a standardised approach to anatomical drawing. This term pertains to a drawing in which the human anatomy is represented

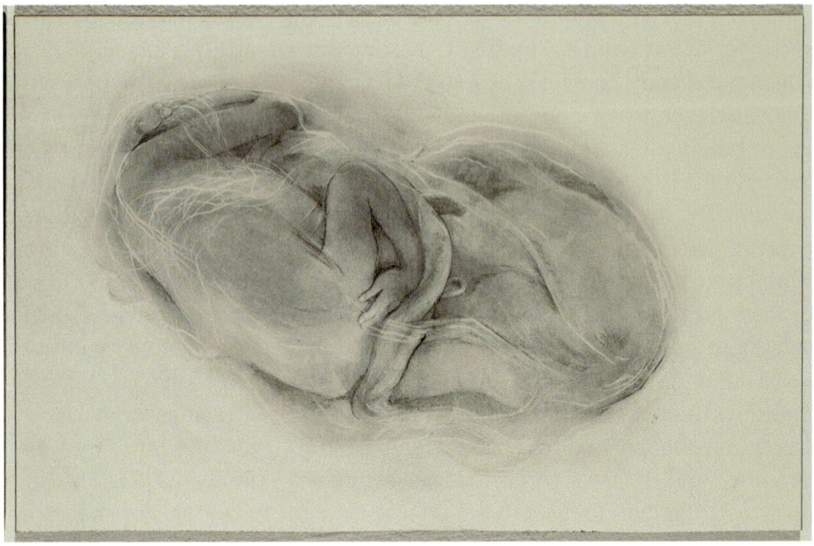

Fig. 1.7 Drawing of a foetus (wet specimen) during a medical illustration workshop at the RSU anatomy museum of Riga, Latvia by Ieva Vorkale

schematically by depicting the human figure as a musculoskeletal system. This is based on the methodology of Gottfried Bammes and George Bridgman (2017).

Anatomical drawing lessons are mainly divided into two parts. In the first part, students are introduced to the osteology of the human body. The emphasis is mainly on learning how to use an anatomical study drawing, an example of which is seen in Fig. 1.9, where the skeleton is built up graphically as a collection of shapes made up of different planes. This aims to offer a drawing tool that facilitates the activation of perspective and foreshortening perception, and thus improves the mental image and spatial understanding by adding a level of depth to a 2D image. Students learn to recognise bones on the body of the live model and use them as reference points for setting out a drawing frame. In the second part, students become acquainted with

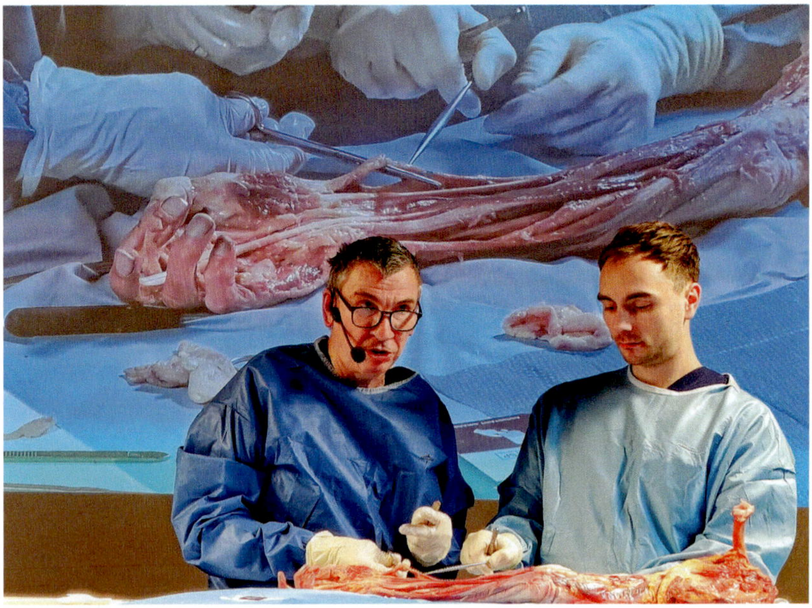

Fig. 1.8 Dissection class by BIOMAB and the Royal Academy of Fine Arts Antwerp. Photograph by the authors. Reproduced with permission

Fig. 1.9 Anatomical study drawing by Meri Toivanen

the superficial muscles and learn to understand the human body as a musculoskeletal system in order to give meaning to concepts such as form tension and stability. Such drawing workshops are certainly not a part of the classic medical curriculum, however, their integration should at least be considered.

1.6 Integrating Art in the Transfer of Anatomical Knowledge, an Ongoing Study

The integration of art in the transfer of anatomical knowledge has many benefits. It can aid in improving observational and analysis skills in students, as well as furthering the development of visual-perceptual ability and empathy (Bell and Evans 2014). Anatomical drawing can also lead to a deeper scientific insight and a thorough understanding of the dynamics of the human body (Anderson et al. 2019). Slowly building up the structures can be used in interactive lectures, for example by using screen capture software. Medical students often prefer using these reproduced anatomical sketches over plain text or a diagram (Pickering 2014). Wammes et al. (2018) note that drawing improved memory retention through a process of recognition and improved recollection (Wammes et al. 2018). The level of retention and recollection was better in the students who used drawing in their process of studying compared to those who did not.

A study performed by Alsaid and Bertrand (2016) compared the scores of three groups of medical students after they completed a course of anatomical dissections. Two groups were asked to draw an anatomical region either before or after the classes, and the third group of medical students participated in the dissection as normally without any drawing. When all students from these three groups were evaluated on their capacity to memorise the anatomical structures covered during the course, the two groups in which drawing before or after the classes was implemented showed significantly higher scores compared to the students from the group in which no drawing was linked to the classes (Alsaid and Bertrand 2016). These anatomical dissections form a core element in medical education, and techniques to improve the retention of anatomical knowledge during such dissections are of great value, as demonstrated in Fig. 1.10, which shows an arm that is about to be dissected.

Other than the positive effect on retention and recall of anatomical knowledge, drawing also allows students to utilise their imagination and express themselves (Grogan and Ferguson 2018). When analysing interviews with medical students regarding the creation of art and their professional development, Jones et al. identified three major themes in the responses. These were an enhanced reflection on the human dimensions of illness and medical care, a sense of personal growth and the development of interpersonal skills and collaboration. Further subtheme divisions were made, such as insights that students can use in clinical encounters, a heightened awareness of group dynamics and an increased self-discovery and pride (Jones et al.

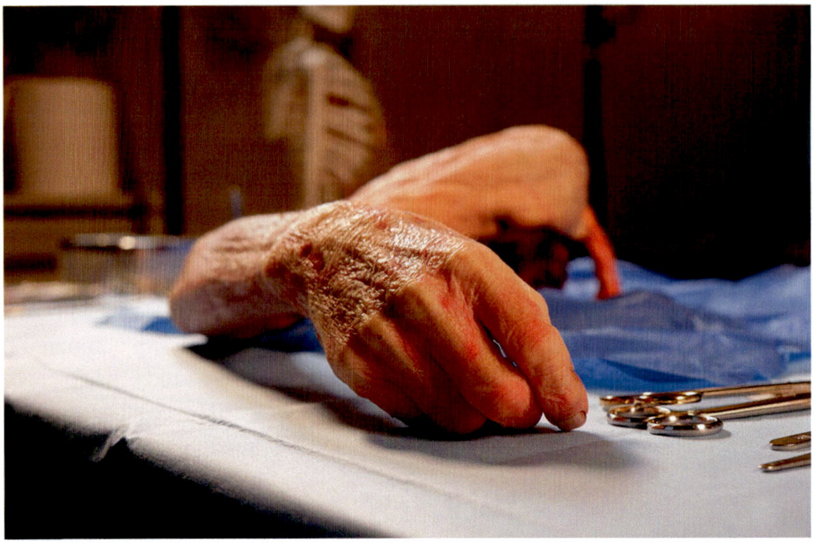

Fig. 1.10 Upper limb to be dissected at a dissection class by BIOMAB and the Royal Academy of Fine Arts Antwerp. Photograph by the authors

2017). This evidence shows the added benefit of integrating art and especially the act of drawing, into the process of transferring anatomical knowledge. These studies show us a glimpse of what can be achieved by combining art and science in this way, however, many questions still remain about how to implement this, for example: What are the perceptions of students? What roadblocks and challenges can be expected? While adding drawing as a method of studying in the anatomy courses of the medical curriculum can have other benefits, the main goal and focus should still be the improvement of anatomical knowledge in these students.

1.6.1 Setting up the Study

Understanding the anatomy of the human body is fundamental for future medical doctors. In order to improve the anatomical knowledge of future physicians it is important to investigate the way anatomical knowledge is transferred from teacher to student and how medical students process this information (Older 2004). To do this, BIOMAB in collaboration with the Faculty of Medicine and Health Sciences at the University of Antwerp has established an ongoing qualitative study that uses interviews to look at the perceptions of medical students towards the integration of art in the medical curriculum. More specifically the use of art in the transfer of anatomical knowledge is being investigated, and the opinions of medical students regarding the use of various forms of art in their anatomical education are explored. Their preferences and thoughts on the used mediums of anatomical knowledge transfer and studying are uncovered and the interviews evidence how to improve upon current methods.

The participant group consists of medical students from their first bachelor year to their third bachelor year. This is a population that has been scarcely studied regarding the topic of their perceptions of the integration of art into the medical curriculum. The opinions of this group are valuable because they are at a point in their career where they are forming the foundation of anatomical knowledge which is crucial for performing and excelling in their future career in medicine. Thus, if certain preferences dominate, these could potentially lead to new insights and teaching methods. If these results are similar to those of the previously consulted literature, and point to an added value of integrating art into the process of teaching anatomy, they could possibly lead to a further incorporation of the arts in the medical curriculum.

1.6.2 Developing a Theoretical Framework

At this point in the research, the prevailing hypothesis is that medical students find art valuable in both their own creative development as well as for studying and remembering human anatomy. A review of the scientific literature was performed using PubMed and Web of Science databases. These databases were searched using relevant MeSH terms such as 'art' and 'medical education'. The information from this review of the literature was used in the construction of a theoretical framework, which will be discussed in detail in the next paragraphs. In this framework, the most relevant and prominent factors are identified as well as how they interact in the process of knowledge transfer. The development of this framework is based on the theoretical frameworks of McIlrath et al. and Huitt et al., which were then adapted to serve our needs and supplemented with additional evidence and theoretical assumptions which need validation (McIlrath and Huitt 1995; Hoban 2000). The framework sets out to visualise and explain the relations between influencing factors and thereby helps to form interview questions, which gauge these factors and relationships in the study population. The results from the interviews will then allow a finetuning of this theoretical framework until data saturation is achieved and the framework is sufficiently capable of generating a testable hypothesis of how the manipulation of these factors can improve learning outcomes. However, this is a work in progress and future data collection and analysis is needed to tell how this framework will hold up against the results, and how insight will develop throughout the study as information is added from the analysis of the interviews.

The main concepts of the theoretical framework are best summarised as follows. According to the theoretical framework, it is predicted that the medical student will be influenced by several factors to attain and memorise anatomical knowledge. These factors are related to the student themselves, such as their personality, the environment in which they interact and their past experiences, but also how the student chooses to study and their learning process, which in turn is influenced by the previously named factors through their influence on the student's behaviour. Then there are the factors relating to the teacher, which also exert their influence on the student through their effects on the student's behaviour and learning process. Factors related to the teacher are the personality, environment and experiences of the teacher which also influence the behaviour of the teacher and this in turn dictates the teaching method which is used. Then there is the influence of the interaction between the behaviours of the student and the teacher, respectively, and the environment in which they preside. A more external factor is the collection of institutional policies implemented by the educational institution. All these factors and their relations are visually represented in Fig. 1.11.

In order to better understand these factors and how they come together to influence each other, an example is presented. Imagine a student with a high level of conscientiousness, who has had the experience of being punished by strict parents for performing poorly in school and who is being placed in a chaotic distracting classroom. This can result in the student being irritated and performing sub-par in their learning process, which eventually influences their retention of information. Here, elements of the personality, the experiences and the environment, are observed to influence the behaviour and learning process.

When looking at the role of the teacher, it could be that the teacher scores low on conscientiousness and is more aloof, has had the experience of having to teach disinterested students and therefore has become indifferent themselves, which influences teaching methods and causes them to be suboptimal. This in turn is detrimental to the learning process of the student.

When considering the factors in this part of the story the same factors seen in the student are observed here and a lot of processes ran in parallel. Then there is the interaction between these two sides and the environment in which the

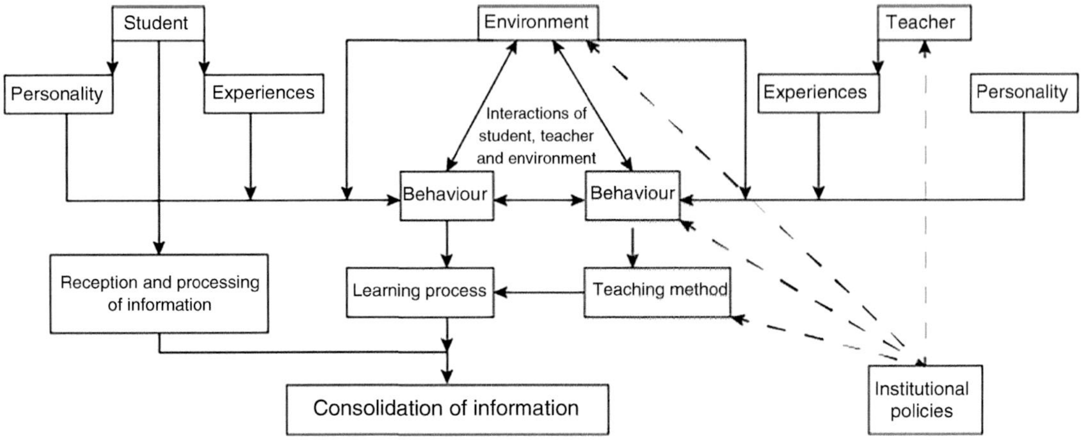

Fig. 1.11 Theoretical framework for the ongoing research into the effects of art on anatomical knowledge

interaction takes place, which can influence the process of learning. If for example the aloof behaviour of the teacher further irritates the highly conscientious student who then lashes out at the teacher, this surprises the classroom and could cause other students to retaliate against the student, potentially influencing the teacher's perception of the student further and their willingness to engage with them. This all has an effect on the learning process of said student.

All of these elements are fictional, serve as an example and are very generalistic, but they showcase the mechanisms of the theoretical framework. Factors such as art could influence this process by leading to higher levels of engagement, improved positive interactions, improved processing of information and offering additional learning processes such as adding drawing instead of just observing anatomy.

In this framework, there is the assumption that each student develops their own learning process, based on their characteristics. This takes into account positive characteristics such as prior knowledge and background, past experiences, interest, intrinsic motivation, confidence, personal reflection, preferences, studying techniques, positive interactions with the teacher or peers and environmental or external motivational factors. The negative characteristics which are taken into consideration are fatigue, feelings of frustration, boredom, confusion, negative interactions with the teacher or peers, lack of control and environmental factors such as noise.

Differences in these factors and characteristics, between people can lead to different preferences and needs. For example, research shows that the majority of students prefer a drawing like a comic, to a text or diagram (Pickering 2014). It is also assumed that the level of interest in art is a factor that may influence the effect of art on anatomical knowledge transfer. These factors are in turn influenced by family as well as community factors such as the level of education of the parents, family income and academic expectations at home, but also the effects that a community can have on a student such as the social support of the micro- and mesosystems, the local policy from the ecosystem or the cultural expectations dictated by the macrosystem according to the systems theory by (Bronfenbrenner and Cole 1981).

When looking at the critical components in this theoretical framework, the aforementioned characteristics of the student seem to converge as they influence the student's behaviour. The student's behaviour and the resulting studying method are two determining factors in whether or not the student improves his or her anatomical knowledge. Secondly, the teaching method and the behaviour of the teacher appear to be two other strong determinants of knowledge transfer. They place the teacher parallel to and on a quasi-equal level of importance with the student in this process. Interactions between teacher and student will also have an impact on the learning process. As shown by Xie et al. a student-teacher conflict can have a negative effect on the student's creativity (Xie and Jiang 2022). While Hernández et al. show that involvement and positive effort shown by the teacher leads to a positive effect on academic performances (Hernández et al. 2017).

A more externally located factor is the influence exerted by the policies of the institution, which controls the teacher, their behaviour and teaching methods but also dictates the environment in which classes are taught. The role institutional policies play is often more subtle and forms the cadre within which is operated. Though their influence can be less overt, they do lay out the groundwork and can therefore dictate whether there is a fertile soil to grow a positive learning culture or the lack of it.

Then there is the part of receiving and processing information, when information is presented to us, it has to be perceived through one of our senses and subsequently processed through neurological pathways in order for our brain to make sense of it. Presuming that the target population has normally functioning senses and processing pathways, this can also offer opportunities. For example, by viewing drawings, specific parts of the brain are activated more than when listening to auditory input or reading a text. In this case, the ventral or occipito-temporal pathway and dorsal or occipito-parietal pathway will both play a role in processing and later the storing of text and images. However, the ventral pathway will be more active when seeing an image as opposed to just text (Veelaert and Santens 2011). Also, the act of drawing, more specifically drawing the information which is being studied, is more effective for remembering the information than just listing or writing it down (Wammes et al. 2018). The quality of drawings does not matter in this process, so anyone studying anatomy could benefit from this, regardless of drawing skills (Wammes et al. 2016).

1.6.3 The Methodology of the Study

Since this study is of a qualitative nature, the methodology should be able to reflect this and therefore be able to to capture nuanced answers from the participants. The data collection method should have a reproducible set of questions with the opportunity to respond with open answers while also leaving room for further explorations based on the answers given by participants. Therefore, the method deemed most suited for these needs is the semi-structured, in-depth interview. Each interview is expected to last between 30–40 min.

Using the grounded theory approach, the results of these interviews will be used to refine the theoretical framework. This will be achieved by thematically coding and then analysing the data gathered from the individual interviews, until data saturation is achieved and a reliable hypothesis can be formed. The goal is to open the possibilities for further research, including quantitative research aimed to test parts of the resulting hypothesis.

Concerning the content of the interviews, a list of topics containing both questions and pictures is used. The latter includes anatomical drawings from Vesalius's 'De humani corporis fabrica libri septem', Gray's Anatomy as well as other atlases of human anatomy. Due to the iterative process of grounded theory, the questions can change during the course of the study, in accordance with the relevance of certain topics. A technique called 'member checking' will be used in order to ensure correct interpretation of the questions and credibility of the answers. The medical students will be interviewed individually, as the focus of the research lies not on the interaction between the participants, but rather on the personal thoughts regarding the subject.

1.6.4 Prospects of the Study

Though research has been conducted on the role of drawing in learning, additional exploration is necessary to clarify the way in which the incorporation of art can help medical students and solidify their knowledge. Art has been intertwined with anatomy for centuries, a cooperation which is unlikely to stop in the future, but still has opportunities for optimisation. Thus, the integration of art in the transfer of anatomical knowledge is a subject crucial for both medical students as well as for those in other fields such as medical illustrators, art students and artists. In conclusion, further investigation is indicated in order to optimise the potential of integrating art in the transfer of anatomical knowledge. The goal of this ongoing research project is to contribute to this optimisation process and offer insights that contribute to a better understanding of this interaction as well as promote further collaborations in the arts and sciences.

1.6.5 Preliminary Results

Though the qualitative characteristics of our research does not allow for generalising preferences, data analysis of the interviews to date showed recurring patterns. All students favoured colour over black-and-white drawings, both due to clearer visualisation as well as better retention of information. A downside of this, as mentioned by five students, was the misconception that the colours used in anatomical drawings (showcasing arteries as red, veins as blue et cetera) would be the same in the human body, leading to a sense of confusion. Most students spoke admiringly about the medical illustrations, such as shown in 'De humani corporis fabrica libri septem' from Vesalius and other anatomic atlases, with a partiality to modern art, due to the comprehensibility of the structures and the clear-cut lines. Pitfalls of the integration of art in the transfer of anatomical knowledge were also discussed, such as a lack of an accessible website containing all the necessary information, difficulty with localisation of structures in a sketch and the usage of a healthy human body in portraying different anatomical positions. Thus, transferring knowledge from theory to practice seemed to be a hurdle many students struggled with. In our research population, consisting of only medicine students, there also was a clear divide in opinions regarding whether anatomical drawings were considered as a form of art. Four students considered anatomical art not to be creative at all, instead insisting it was a form of science. However, five students believed the arts and anatomy to be integrated closely, with most referring to anatomical art as something to be appreciated as well as studied. When looking to the future, the focus lies on how to improve the anatomical transfer of art and for that, the voices of the students who are being introduced to the anatomy of the human body are essential to better both the memory and the comprehension.

1.7 Concluding Remarks

In recent years, there is an increasing appreciation of communication and empathy skills in the medical curriculum. This focus on the humanities is complementary to the importance of understanding complex biomedical processes, pathologies, diagnostics and therapeutic actions. Throughout this chapter, we have discussed examples of how art and science can come together to create synergy. Especially in the medical curriculum, where the incorporation of the arts and by extension, the medical humanities as a whole have proven themselves as a valuable partner and addition. This added value can be seen in aspects such as increased empathy, compassion, communication skills and connectedness.

The integration of the arts and health humanities into the medical curriculum is a trend that is gaining momentum on the world stage. It pays homage to the view that different fields of human knowledge are interconnected and that both individual patients and the health care system as a whole should be viewed from a holistic point of view. The medical humanities force medical personnel to take a step back from the dogmatic medical point of view and challenge them to look at the people they connect with in a different light, to take on their perspectives. As a result their perspectives on illness and health, enjoyment and suffering, life and death, may change. These different perspectives, offered by the arts and health humanities can serve as an inspiration to challenge the status quo and to eventually approach daily medical practice differently, perhaps even in a way that can be considered 'better'.

Taking global examples from history into consideration, from ancient rock paintings to Leonardo da Vinci's beautiful drawings based on detailed observations from nature, it becomes clear that art and science are not mutually exclusive and often highly synergistic. They are absolutely not that distant as thought. Biomedical illustration schools and organisations are exemplary longstanding instances of learning and teaching the science of observing, representing and describing the biomedical world. The poem 'The Cutting Edge of Medical Art' by Bryan Green beautifully illustrates the strong bond between image and word within the fields of medicine and the biomedical sciences.

Paul Haidet and his colleagues identified four common themes pertaining to the ability of art to develop humanistic characteristics. Art has unique qualities like its metaphorical nature, subjectivity, ambiguity and the ability to create a levelled playing field. Art helps to engage people and improve learning. Physical forms of art, if they are related to the topic, help to improve learning outcomes. Art can be used to engage medical students and make them experience how patients feel and live with their diagnoses.

From these findings and the theoretical framework surrounding them, a vision was built which proposes an approach to the integration of art into the medical curriculum. It builds upon the idea of art-based workshops which sets out to reach certain goals. Some goals can be achieved in the short term, while others are more suited for the long term, which should be taken into consideration when setting out goals for these art-based workshops. In order to mediate these goals pedagogical techniques are implemented in order to facilitate this process.

When these elements are all taken together, a recipe for the implementation of these art-based workshops into the medical curriculum falls into place. Many ingredients in this recipe are left open ended, this is an intentional act since there is not one single correct method, instead the approach should be adapted to the circumstances and what seems to work for that specific group of students and teachers. There is a perception that anatomical drawing is often a hard skill to learn, however, in the context of using art to improve anatomical learning it is more about this process than the actual product. The same goes for simply enjoying the creative activity of drawing. This perceived threshold is something that will need work in order to overcome these preconceived ideas in the minds of students. The value of this process lies within the acts which are performed by the mind in order to come to the product. More specifically the act of observing and attention to detail, the mental processing of the perceived image, the emotions which are summoned by this image and finally the mind having to put all this into a product, this is where the true value lies.

To close this chapter, we would like to revisit its essence, the synergy between art and science. It has led us to the appreciation that medicine should not be considered as an island, but rather as an element within a network. A network of things that touch our daily lives and make up our human experience. Both the physician and the patient are more than someone working in a medical institution and someone who is ill. Therefore, it makes perfect sense to integrate elements from different walks of life, such as art and the health humanities into the medical curriculum so that it can have its effects on and be integrated into the care of patients.

Acknowledgements We thank the following individuals for their expertise throughout all aspects of writing the manuscript. This chapter and the research behind it would not have been possible without the exceptional support of editor Mark Roughley at the Liverpool School of Art and Design. His enthusiasm, knowledge and exacting attention to detail have been an inspiration and kept our work on track.
Lieva Libiete, Head of the RSU Anatomy Museum in Riga and Elina Zunde, Teacher at The Art Academy of Latvia provided beautiful colour images for the manuscript. We thank Kitty Van Couillie, Mariela Devos and Geert Van Eeckhout, colleagues from the Royal Academy of Antwerp, Joanna Cameron, MAET Foundation Director and Luc Van Nassauw, Head Antwerp Surgical Training, Anatomy and Research Centre who provided insight and

expertise during a long history of art-based workshops on anatomy in Antwerp. Special thanks go to Joke Denekens, professor emeritus at the Faculty of Medicine and Health Sciences of the University of Antwerp for reviewing study interviews.
BIOMAB research is supported by the Royal Medical Association of Antwerp and the Culture Commission of the University of Antwerp. As for the ongoing research by BIOMAB and the Faculty of Medicine and Health Sciences of the University of Antwerp, it is our hope that the study is able to shine a more intimate light on how the average medical student views the integration of art in their curriculum and how this could benefit learning. Now already, it has proven to be beneficial by forming a theoretical framework which is now being tested and by analysing the interviews which have already been taken. In the future, this line of research may add to the 'why's' and 'how's' of this cooperation between the arts and medical education, which is something that excites us all.

References

Alsaid B, Bertrand M (2016) Students' memorization of anatomy, influence of drawing. Morphologie 100:2–6. https://doi.org/10.1016/J.MORPHO.2015.11.001

Anderson G, Dupré J, Wakefield JG (2019) Drawing and the dynamic nature of living systems. elife 8. https://doi.org/10.7554/ELIFE.46962

Bammes G (2017) The Complete guide to anatomy for artists & illustrators. Search Press; Illustrated edition. https://www.amazon.com/Complete-Guide-Anatomy-Artists-Illustrators/dp/1782213589#detailBullets_feature_div

Bell LTO, Evans DJR (2014) Art, anatomy, and medicine: is there a place for art in medical education? Anat Sci Educ 7:370–378. https://doi.org/10.1002/ASE.1435

Bendersky G (1997) The Olmec heart effigy: earliest image of the human heart. Perspect Biol Med 40: 348–361. https://doi.org/10.1353/PBM.1997.0052

Bronfenbrenner U, Cole M (1981) The ecology of human development, experiments by nature and design. Harvard University Press

Cohen RL (1981) On the generality of some memory laws. Scand J Psychol 22:267–281. https://doi.org/10.1111/j.1467-9450.1981.tb00402.x

Coss S (2017) Inktober Illness Series I–The art of Shawn Coss. https://www.shawncossart.com/inktober-illness-series-i. Accessed 26 Feb 2023

Dalia Y, Milam EC, Rieder EA (2020) Art in medical education: a review. J Grad Med Educ 12:686. https://doi.org/10.4300/JGME-D-20-00093.1

Delany C, Jones S, Sokol J, Gillam L, Prentice L (2021) Reflecting before, during, and after the heat of the moment: a review of four approaches for supporting health staff to manage stressful events. J Bioeth Inq 18 (4):573–587. https://doi.org/10.1007/S11673-021-10140-0/FIGURES/4

Daykin N, Orme J, Evans D et al (2008) The impact of participation in performing arts on adolescent health and behaviour: a systematic review of the literature. J Health Psychol 13:251–264. https://doi.org/10.1177/1359105307086699

Dhaliwal G (2011) Going with your gut. J Gen Intern Med 26:107. https://doi.org/10.1007/S11606-010-1578-4

Ebenstein J, Morbid AM (2016) The anatomical venus: wax, god, death & the ecstatic. D.A.P./Distributed Art Publishers. ISBN: 9780500252185

Frank J, Snell L, Sherbino J (2015) CanMEDS 2015

Gottlieb S (2001) Surgeons perform transatlantic operation using fibreoptics. BMJ Br Med J 323:713

Grogan K, Ferguson L (2018) Cutting deep: the transformative power of art in the anatomy lab. J Med Humanit 39:417–430. https://doi.org/10.1007/S10912-018-9532-2

Haanstra F (1996) Effects of art education on visual-spatial ability and aesthetic perception: a quantitative review. Stud Art Educ 37:197. https://doi.org/10.2307/1320854

Haidet P, Jarecke J, Adams NE et al (2016) A guiding framework to maximise the power of the arts in medical education: a systematic review and metasynthesis. Med Educ 50:320–331. https://doi.org/10.1111/medu.12925

Hanshumaker J (1980) The effects of arts education on intellectual and social development: a review of selected research. Bull Counc Res Music Educ 61: 10–28

Hardiman M, Rinne L, Yarmolinskaya J (2014) The effects of arts integration on long-term retention of academic content. Mind Brain Educ 8:144–148. https://doi.org/10.1111/mbe.12053

Harper D (2015) Early Chinese medical literature. Routledge

He B, Prasad S, Higashi RT, Goff HW (2019) The art of observation: a qualitative analysis of medical students' experiences. BMC Med Educ 19:1–6. https://doi.org/10.1186/S12909-019-1671-2/PEER-REVIEW

Hernández MM, Valiente C, Eisenberg N et al (2017) Elementary students' effortful control and academic achievement: the mediating role of teacher-student relationship quality. Early Child Res Q 40:98. https://doi.org/10.1016/J.ECRESQ.2016.10.004

Hoban G (2000) Using a reflective framework to study teaching-learning relationships. Reflective Pract 1: 165–182. https://doi.org/10.1080/713693148

Jones EK, Kittendorf AL, Kumagai AK (2017) Creative art and medical student development: a qualitative study. Med Educ 51:174–183. https://doi.org/10.1111/MEDU.13140

Lane T (2018) A short history of robotic surgery. Ann R Coll Surg Engl 100:5. https://doi.org/10.1308/RCSANN.SUPP1.5

Loftis C (2017) Mental flexibility. In: Kreutzer J, DeLuca J, Caplan B (eds) Encyclopedia of clinical neuropsychology. Springer International Publishing, Cham, p 1

Maslow AH (1943) A theory of human motivation. Psychol Rev 50:370–396. https://doi.org/10.1037/H0054346
McIlrath DA, Huitt WG (1995) The teaching-learning process: a discussion of models. Educ Psychol Interactive Vald State Univ
Merleau-Ponty M (2002) Phenomenology of perception. Routledge. https://www.google.is/books/edition/Phenomenology_of_Perception/oSgaSzvHbaoC?hl=nl&gbpv=0&kptab=editions
Mohamed M, Vymazalov H (2011) Ancient Egyptian royal circumcision from the pyramid complex of Djedkare. Anthropol 49:155–164
Older J (2004) Anatomy: a must for teaching the next generation. Surgeon 2:79–90. https://doi.org/10.1016/S1479-666X(04)80050-7
Passalacqua SA, Segrin C (2012) The effect of resident physician stress, burnout, and empathy on patient-centered communication during the long-call shift. Health Commun 27:449–456. https://doi.org/10.1080/10410236.2011.606527
Peppler K, Davis-Soylu HJ, Dahn M (2021) Artifact-oriented learning: a theoretical review of the impact of the arts on learning. Arts Educ Policy Rev 0:1–17. https://doi.org/10.1080/10632913.2021.1925609
Perry M, Maffulli N, Willson S, Morrissey D (2011) The effectiveness of arts-based interventions in medical education: a literature review. Med Educ 45:141–148. https://doi.org/10.1111/j.1365-2923.2010.03848.x
Pickering JD (2014) Taking human anatomy drawings for teaching outside the classroom. Surg Radiol Anat 36: 953–954. https://doi.org/10.1007/S00276-014-1282-X
Raab K (2014) Mindfulness, self-compassion, and empathy among health care professionals: a review of the literature. J Health Care Chaplain 20:95–108. https://doi.org/10.1080/08854726.2014.913876
Raddato C (2015) Fresco with Wounded Aeneas (Illustration)–world history encyclopedia. https://www.worldhistory.org/image/3854/fresco-with-wounded-aeneas/. Accessed 26 Feb 2023
Reid S, Shapiro L, Louw G (2019) How haptics and drawing enhance the learning of anatomy. Anat Sci Educ 12:164–172. https://doi.org/10.1002/ASE.1807
Shams L, Seitz AR (2008) Benefits of multisensory learning. Trends Cogn Sci 12:411–417. https://doi.org/10.1016/j.tics.2008.07.006
Spaner SJ, Warnock GL (1997) A brief history of endoscopy, laparoscopy, and laparoscopic surgery. J Laparoendosc Adv Surg Tech A 7:369–373. https://doi.org/10.1089/LAP.1997.7.369
Styn A, Scheiter K, Fischer MR et al (2023) Effects of tablet-based drawing and paper-based methods on medical students' learning of gross anatomy. Anat Sci Educ 16:266–279. https://doi.org/10.1002/ASE.2237
Tsafrir J, Ohry A (2001) Medical illustration: from caves to cyberspace‡. Heal Inf Libr J 18:99–109. https://doi.org/10.1046/J.1471-1842.2001.D01-16.X
Van Hee R, Biesbrouck M, Bols P et al (2020) In the shadow of Vesalius, 1st edn. Garant, Antwerp
Veelaert L, Santens P (2011) De fysiologie en pathologie van orthografische inputverwerking. Universiteit Gent
Vesalius A (1543) De humani corporis fabrica libri septum, 1st edn. Joannes Oporinus, Basel
Wammes JD, Meade ME, Fernandes MA (2018) Creating a recollection-based memory through drawing. J Exp Psychol Learn Mem Cogn 44:734–751. https://doi.org/10.1037/XLM0000445
Wammes JD, Meade ME, Fernandes MA (2016) The drawing effect: evidence for reliable and robust memory benefits in free recall. Q J Exp Psychol (Hove) 69: 1752–1776. https://doi.org/10.1080/17470218.2015.1094494
Wong JD, Kistemaker DA, Chin A, Gribble PL (2012) Can proprioceptive training improve motor learning? J Neurophysiol 108:3313–3321. https://doi.org/10.1152/JN.00122.2012/ASSET/IMAGES/LARGE/Z9K0231216670009.JPEG
Xie R, Jiang J (2022) Creativity: the effectiveness of teacher-student conflict. Int J Environ Res Public Health 19. https://doi.org/10.3390/IJERPH19031628

Using Medical Illustration to Improve Understanding of Endometriosis

2

Aimée Hutchinson and Caroline Erolin

Abstract

Endometriosis is a chronic and debilitating condition where cells similar to those found in the endometrium migrate elsewhere in the body. These rogue cells react to the menstrual cycle and bleed leaving the deposits to remain trapped. With nowhere to go, these endometriotic deposits go on to cause inflammation and scarring of healthy tissue and organs, as well as severe pain and discomfort. It is a condition that is said to affect 190 million women worldwide. Despite many theories regarding the epidemiology and pathogenesis of endometriosis, there is no conclusive evidence to state how, or why, the condition originates in only some individuals and because of this there is no cure.

Whilst it affects many women on a global scale, endometriosis is a condition that is still widely unheard of and lacking in research funding, further hindering diagnostic times, which currently stand at an average of 8 years in the United Kingdom; delaying the development of novel methods to stop the disease from progressing and impacting the quality of life for patients. Information surrounding endometriosis, particularly online, can be largely inaccurate. There are limited illustrations and educational tools available that explain and depict the disease or visualise its three highly individual subtypes.

This research project set out to discover if those living with the condition felt there was a need for more and better visual communications depicting endometriosis and the various methods used to manage and treat both it and its symptoms. Feedback, in the form of a questionnaire, was gathered from a group of participants who were formally diagnosed with, or were experiencing symptoms of, endometriosis. The questionnaire was shared via various social media platforms in July 2019. The respective findings from the questionnaire provided a foundational base for what was to be a website presenting various aspects of endometriosis, including a brief history of the condition along with digital illustrations of the three subtypes of endometriotic growth and where it can be typically found, as well as sites for extra-pelvic and rare endometriosis. The target audience was initially those with the condition, those who were awaiting an official diagnosis and doctors/healthcare providers. However, given the nature of endometriosis and the subsequent findings identified in the

A. Hutchinson (✉)
Independent Medical Artist and Instructional Design Illustrator, Belfast, UK

C. Erolin
Reader in Medical Art, School of Science and Engineering, University of Dundee, Dundee, UK
e-mail: c.d.erolin@dundee.ac.uk

M. Roughley (ed.), *Approaches for Science Illustration and Communication*, Biomedical Visualization 4,
https://doi.org/10.1007/978-3-031-41652-1_2

questionnaire which highlighted the lack of awareness regarding the condition, a wider audience that included academics and the general public were also deemed important to reach. In addition to creating an informative web resource for use by patients and healthcare providers, 3D digital and printed models were developed for public display.

Keywords

Endometriosis · Medical illustration · Gynaecology · Pelvic pain · Surgery · 3D models

2.1 Introduction: What Is Endometriosis?

Endometriosis, or 'endo' as it is often referred to in online forums discussing endometriosis, is considered one of the most common gynaecological conditions, second only to Polycystic Ovary Syndrome (PCOS). Whilst the cause of the condition is unknown and theories as to its origin are debatable, it is known that endometriosis occurs when cells like those found in the endometrium are found elsewhere in the body. Endometriotic cells are commonly found to affect areas within the pelvic region, such as the uterus and peritoneum (Kennedy et al. 2005), but they have also been discovered in extra-pelvic spaces like the navel, diaphragm, lungs, and even the pericardium (Horne and Pearson 2018).

The growth of endometriosis is fuelled by the hormone oestrogen, which is the same hormone that enables the monthly thickening of the lining of the uterus leading to menstruation. As the rogue endometriotic cells site themselves on, or within, various organs and tissues in the body, they too, react to oestrogen and follow the same menstrual cycle; bleeding and causing inflammation that goes on to create painful lesions and deep nodules of the disease. The bleeding can be so severe that it may lead to various organs adhering themselves to one another, and in some extreme cases, this may even cause complications such as a 'frozen pelvis' or lead to organ failure.

Of those who find themselves with endometriosis of the diaphragm or lungs, for example, they may have cyclical bouts of coughing up blood. If the endometriosis is within the navel, it too will expel blood, however, when the disease affects organs such as the bowels, endometriosis can mimic symptoms of other illnesses and chronic conditions, such as Irritable Bowel Syndrome (IBS) or even Pelvic Inflammatory Disease (PID) (Kennedy et al. 2005).

It is, in part, due to this mimicry that those living with endometriosis symptoms face such a lengthy diagnostic time, with it taking an average of 8 years in the United Kingdom to diagnose (Endometriosis UK 2011). For a condition that affects 10% of the female population, or 190 million worldwide, how can it be that women are continuing to be let down by the healthcare system with such a lengthy wait?

2.1.1 Is Endometriosis Older than we Think?

Endometriosis is noted as having been microscopically discovered in 1860 by Czech physician and pathologist, Carl von Rokitansky when he identified that, 'some fibrous tumors of the uterus contain gland-like structures that resemble endometrial glands' (Hudelist et al. 2009). However, the historically enigmatic subject of female anatomy and specifically, gynaecology, whilst often taboo, has always proved to be of great interest, with artworks from periods as early as the thirteenth century depicting representations of pelvic pain within women. During these times, pelvic disorders were known as the 'strangulation or suffocation of the womb', which scholars believe to have linguistically spawned from Hippocrates, whom in the fifth century BC first used the term *hysteria* to describe what was believed to be the 'wandering womb' (Tasca et al. 2012).

It was commonly believed that if a woman did not fulfil her duties, which were to marry and bear children, this would somehow deprive her womb of being used for its expected purpose and thus, the organ would begin to wander throughout the body, causing a myriad of issues for its host:

> ...if they have never been pregnant, the deranged state of menstruation is more common and more dangerous than when they have borne children... she will be released from this disease, when she is pregnant...—a Hippocratic author (Nezhat et al. 2020)

What is fascinating is that the components Hippocrates identified as being part of pelvic disorder in the fifth century BC, are strikingly similar to those linked to endometriosis today, with menstrual dysfunction, pain, and infertility as outcomes if untreated and, pregnancy may act as a possible 'cure', providing only temporary relief from the symptoms.

It was Claudius Galen (129–c.200 CE), one of the most prolific intellectuals of Western antiquity having written extensively on medicine and philosophy, and who is seen as one of the most important figures in medical history that uncovered a revolutionary finding in relation to gynaecological disorders (Singer 2021). He put forth the idea that violent contractions linked to the wandering womb were in fact caused by excessive menstrual blood enlarging the ligaments securing the uterus, forcing it to be twisted, pulled and contorted. This was a notable finding for academics tracing the history of the condition as it is highly suggestive of the adhesions and lesions which are intrinsically linked to severe, late-stage endometriosis. Whilst Galen presented us with what can be considered quite a revolutionary finding of its time, it was also he who gave fuel to the rhetoric of uterine distress being linked to a lack of sexual fulfilment and nulliparous women taking this cue from his predecessor, Soranus (c.98–138) (Tasca et al. 2012).

This notion continued in the years that followed, with the Middle Ages seeing physical violence unleashed on women (choking in particular) becoming a new form of therapy to counteract the internal strangulation of the disorderly womb (Tasca et al. 2012). The seventeenth century gave rise to the image of a woman as a witch who, if not able to conceive, it became assumed that 'magic' or 'evil' was used as a preventive measure and once more, women were subjected to demonisation (Mcculloch 1969).

However, the seventeenth century brought about more advanced findings that would discredit the belief that an ill woman was a 'witch'. The English physician, Thomas Sydenham, began to recognise that symptoms and pains were cyclical. He noted how menstruation could affect the stomach and with its onset cause vomiting and diarrhoea (Veith 1993). These novel discoveries continued throughout the seventeenth century with another physician, Daniel Schrön, taking the lead in identifying, what he described as, 'swelling globules' which were found in the uterus. The swelling globules Schrön refers to are said to be 'fibrous, glandular and membranous in part', a description that bears resemblance to the notes that Rokitansky would produce when identifying endometriosis microscopically for the first time (Nezhat et al. 2020).

The nineteenth century was a turning point in history with the industrial revolution bringing forth an era of great change as well as great advancements in female healthcare. The first successful abdominal oophorectomy was performed in 1809 to remove a large 22-pound tumour, followed by the first successful vaginal hysterectomy in 1815. Progress in medicine also gave rise to another success when in 1853 an abdominal hysterectomy was performed (Nezhat et al. 2020).

Whilst medical understanding significantly advanced throughout the nineteenth century, physicians needed cadavers to further their medical prowess. The Anatomy Act of 1832 was passed in Britain making it difficult to procure cadavers, however, in France there was no limit put on the number of cadavers that could be obtained. As a result, post-mortem studies flourished in a nineteenth century France and what followed were many revolutionary discoveries. French pathologists were amongst the first to macroscopically discover what would later be known as endometriosis, and it has been said that they did so without referencing the hysterical wandering womb or alluding to the individual as having manic tendencies, their findings were considered 'pure organic pathology' (Nezhat et al. 2020).

In the mid-to-late nineteenth century, we would meet the 'problematic young females' of the Salpêtrière whose symptoms then, were seen as psychological, however, they now share many similarities with what we know to be endometriosis, today. The Salpêtrière was an infamous Parisian insane asylum where French psychiatrist, Jean-Martin Charcot, began his famed studies on hysteria (Kumar et al. 2011). The young female patients were noted to have had frequent fits and convulsions akin to those associated with epilepsy. Charcot would map their 'hysterogenic zones', which were typically situated within the pelvic and abdominal regions, and with his fist, he would violently punch into the ovarian region of the patient in a bid to control the patient's convulsions.

An instrument known as an 'ovary compressor' (Fig. 2.1) was created by Charcot and intended to replicate the forceful pressure of his plunging fist. Further evidence reveals that Charcot would often make his hysterics 'perform' for large classes of male medical students, placing

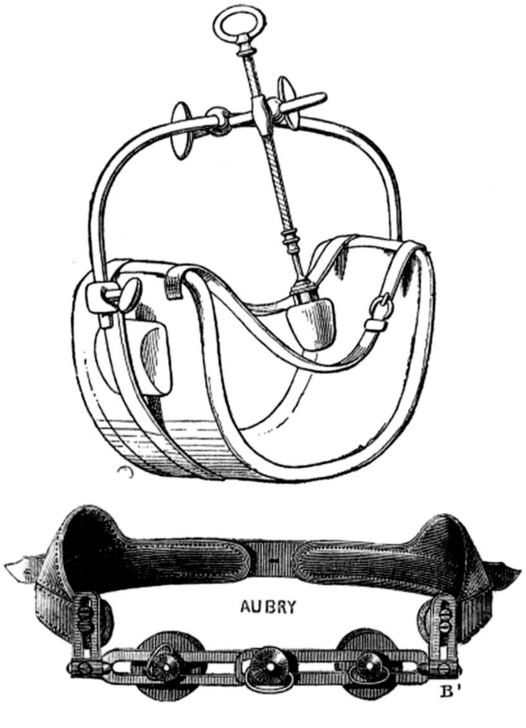

Fig. 2.1 Ovary compressor used by Jean-Martin Charcot. Image available at Bibliothèque numérique, Université René Descartes de Paris. This image is in the public domain

them under a state of hypnosis that would see the women crawl before them taking on the characteristics of an animal. Most disturbingly, whilst under this hypnosis, the students and their teachers would '. . .cut them, prick them, and burn them. . .' causing physical pain and leaving the female patients with lacerations and wounds (shown in Fig. 2.2) (Hustvedt 2011).

In establishing this connection between the ovaries and the hysterical female, Charcot undermined his own claims that hysteria was not a disease of the reproductive organs, and this notion was strengthened by the work of his predecessor, Philippe Pinel, who having also witnessed hysterical fits, noticed a trend which showed a synchronicity between the outbursts and the menstrual cycle (Nezhat et al. 2020).

Women would continue to endure the victimisation of the patriarchy as their illness continued to be an enigma, but the burgeoning twentieth century brought fresh discoveries and theories that would transform endometriosis into the condition that we know today. In 1927 the term 'endometriosis' was used for the first time in an article by gynaecologist, John Sampson, prior to this it was known only as 'diffuse adenomyoma' (Mann n.d., 1887–8). It was Sampson who discovered the viable, yet now heavily debated, theory that endometriosis was a direct result of retrograde menstruation. Within a mere two decades of Sampson's groundbreaking work and microscopic discoveries in which he found endometriotic tissues affecting the bowel and ovaries, endometriosis was being diagnosed throughout the whole body and was also found in teenagers. Animals had also been discovered to have endometriotic implants and lesions, and many began to realise that endometriosis was not as rare as previously thought (Hobbs and Bortnick 1940; Nezhat et al. 2020).

2.1.2 Societal Impact of Endometriosis

By the mid-twentieth century female healthcare was being transformed. Routine pelvic examinations and cervical cancer screenings

Fig. 2.2 'A Clinical Lesson at the Salpêtrière', André Brouillet, (1887). Image sourced from Université René Descartes de Paris. This image is in the public domain

were being performed, surgical standards were updated along with the utilisation of new technology, and the revolutionary hormonal contraceptive was introduced in 1957 to help with menstrual disorders and prevent pregnancy (Bailey 2013). By the 1970s, video-assisted laparoscopy was pioneered by Camran Nezhat and is still in use today, albeit with more modern equipment, however, the progress does not appear to have moved further when treating and managing endometriosis as a standalone condition. Diagnostic laparoscopy is still classed as the only way the condition can be confirmed and synthetic hormonal medications, such as the hormonal contraceptive, whilst helping to manage the symptoms and slow down further growth, do not cure endometriosis, but rather only mask the condition and provide a temporary release from the symptoms (Weisberg and Fraser 2015).

Symptoms of endometriosis are noted as dysmenorrhea, dyspareunia, chronic pelvic pain, and infertility, however, the symptoms are much more widespread, and each individual's list of symptoms can be as unique as the next. Perhaps one of the most frustrating aspects is the lengthy wait times women are faced with when seeking a diagnosis. In a world where women still face gender bias, particularly within the healthcare system, conditions like endometriosis are often under-prioritised and lack research funding (Ellis et al. 2022).

The delay in diagnosis may also be a result of women's pain being stigmatised and normalised for centuries. As young women mature and become attuned to their monthly menstrual cycles, they are often told that pain is merely a by-product of the nature at play and that it is to be expected, but when the pain intensifies and becomes so bad that it impedes upon daily activities, how should one know when to seek medical advice?

On average, women delay seeking help for 2.3 years from the onset of symptoms (J, 2019), and it may take up to seven visits to their doctor before they will be referred to a specialist. As diagnostic surgery is the only way endometriosis can be confirmed, this too slows down the process and proves to be an expensive option for the individual if private healthcare is sought. Simoens et al. (2012) state that 'annual healthcare costs for women with endometriosis treated in referral centres are similar to costs for other chronic diseases that receive more resources such as, diabetes, Crohn's disease and rheumatoid arthritis'. According to figures published by the United Kingdom's leading endometriosis charity, Endometriosis UK (2011), it costs the UK economy £8.2 billion annually through loss of earnings, healthcare costs, and purchasing sanitary products and medications. Of those who do seek medical help for pain related to endometriosis, three-quarters find their journeys are often plagued by misdiagnoses (As-Sanie et al. 2019). Irritable Bowel Syndrome (IBS) or an Inflammatory Bowel Disease (IBD), such as Crohn's Disease or Ulcerative Colitis, will too often be listed as the main perpetrator, as bouts of severe bloating and bowel issues varying from constipation to diarrhoea, and even bloody stools, appear to point to these disorders.

Misdiagnosis does not just affect women financially but also weighs heavily upon their emotional and physical well-being (De Graaff et al. 2013). The condition affects sleep, appetite, exercise, and relationships with romantic partners, along with family and friends, and often, patients are prescribed antidepressants as the first line of treatment. Sufferers also face further health issues in their later lives as endometriosis has been linked to an elevated risk of cancers such as breast and ovarian, as well as an increased risk of rheumatoid arthritis and cardiovascular disease (Kvaskoff n.d.). However, one of the most devastating consequences of the disease for individuals, is undoubtedly, infertility.

Infertility in women with endometriosis is said to be as high as 30–50% (Meuleman et al. 2009) and what is particularly devastating is that patients are often told that to cure their endometriosis, they must get pregnant. This harks back to the Hippocratic texts discovered 4000 years ago which allude to the wandering womb and how it would affect those who had not fulfilled their 'duties' to bear children and raise families. As a consequence of this, throughout the twentieth century, endometriosis was often referred to as the 'career woman's disease' because to fit into the societal norm, women are genetically predisposed to have children—not careers (Carpan 2003). With the introduction of hormonal contraception and women thriving academically and professionally, for some the prospects of motherhood are delayed or not desired at all, and this became a weak spot for those with endometriosis and infertility, it meant that by pursuing their studies or career, the individual was said to have brought the illness upon themselves.

The stigma of being a woman with a gynaecological disorder continues and unfortunately even in our progressive society, tradition often outweighs modernity, where we find medical professionals suggesting the same treatments and therapies that they did many years prior. With endometriosis affecting an increasing percentage of the female population, it is necessary that more efforts and funds go into further studies and research, and the topic itself needs to be highlighted and discussed.

2.1.3 Using Medical Illustration to Improve Communication Surrounding Endometriosis

The visualisation and communication of pain, and symptoms, related to endometriosis for both healthcare professionals and the public are generally lacking (Bullo and Weckesser 2021). Whilst researchers are working to create new and innovative ways to tackle the delay in diagnostic times using new methods, such as linguistics, there is still a noted gap in the realm of exploring the condition through visual communications, and perhaps this is due to the stigmatisation surrounding menstrual health (As-Sanie et al. 2019). New 2D medical illustrations and 3D representations of the condition may prove to be extremely valuable resources when aiding in the understanding of symptoms and pain dynamics.

Currently, textbooks are available, predominantly for academic use, which show a variety of endometriotic implants, lesions, and endometriomas, however, the growths themselves, do not all manifest in the same way (Tsamantioti and Mahdy 2023). The colours of these lesions make it difficult for surgeons to spot the disease and remove it (Schliep et al. 2017). Some growths can be black in colour which makes them easier to spot, whilst others can appear white or translucent, making identification difficult. There are three main subtypes that surgeons are aware of; superficial peritoneal endometriosis, which is black, or sometimes red in colour with a rash-like texture (Fig. 2.3), followed then by ovarian endometriosis, which is perhaps the easiest to see as it forms a fluid-filled cavity within the ovaries (Fig. 2.4). The

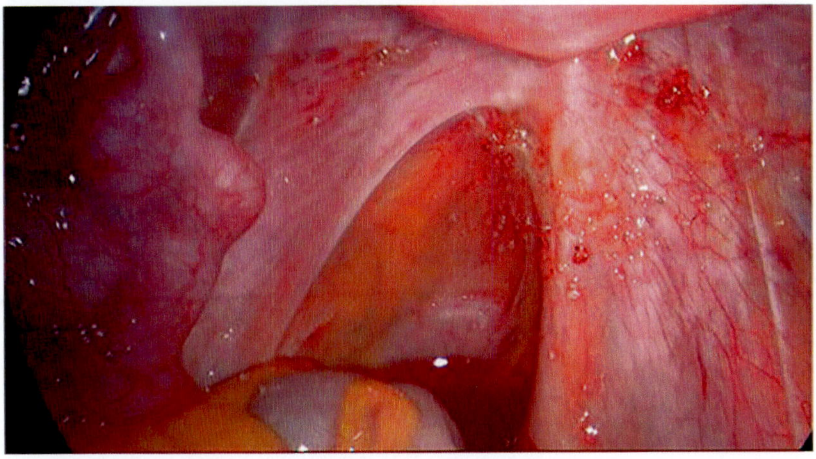

Fig. 2.3 Widespread superficial peritoneal endometriosis (Image provided by Professor Andrew Horne, July 2019)

final subtype is deep-infiltrating endometriosis (Fig. 2.5), and this may be the most problematic to spot, and remove, as this type of growth implants deep into the organ or tissue and takes on a white or clear appearance making it difficult to observe by an untrained eye, reaffirmed by Davidson (2005) who says, 'Most surgeons will look for endometriosis systematically on all the pelvic surfaces, but the lesions may be hidden and difficult to see'.

As endometriosis continues to baffle medical professionals and devastate the lives of the many who suffer from it, it would be beneficial to introduce visual illustrations that can be accessed by both. The advantage of using medical illustrations to visualise a condition or disease is that they are universal and inclusive. Medical illustrations allow the most experienced, or inexperienced, to view every part of the human body. Illustrations allow those who are not medically trained or have access to a cadaver or operating theatre, to understand the inner workings of the human body and they allow for the visualisation of certain parts that may have initially been difficult to observe (Oxner 2018). As Confucius is stated to have said, 'One picture is worth more than one thousand words' (Tsafrir and Ohry 2001). Medical illustrations also allow for call-out lines to be included which further aids the viewers understanding of the anatomy they are observing. The addition of this information can break down the barriers involved in anatomical education and greatly enhance the knowledge surrounding our own bodies (Harrison 2020, December 29).

Medical professionals can use such illustrations to their benefit by viewing a large collection of visual representations of the various

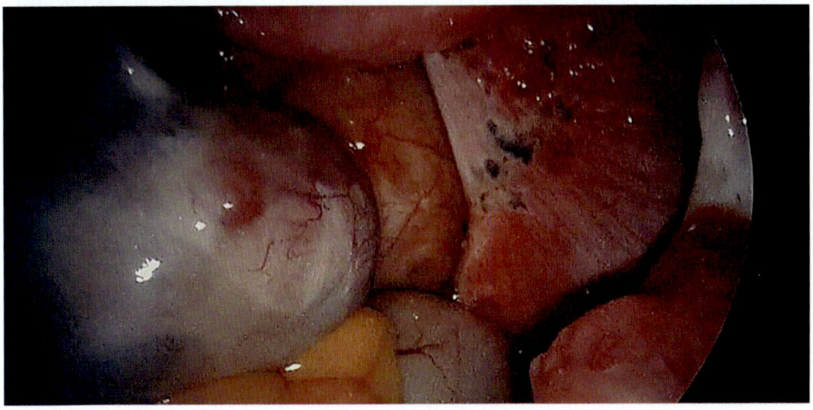

Fig. 2.4 Ovarian endometrioma (taken from researcher's own surgery, fifth July 2018)

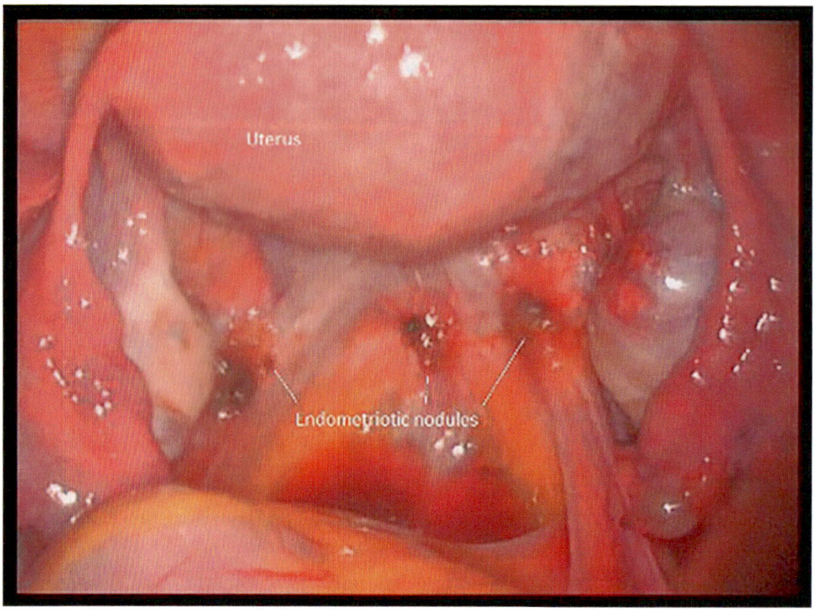

Fig. 2.5 Deep-infiltrating endometriosis (Image provided by Professor Andrew Horne, July 2019)

manifestations of the disease and the far-flung places it can be found. Whilst those living with the condition can begin to understand the concept of the disease and what is going on within their own bodies, giving the patient a sense of autonomy. Another aspect of medical illustration that can prove greatly beneficial for both classes of viewers is the depiction of surgical options available for diagnosing and treating endometriosis (Horne and Pearson 2018).

Whilst diagnosis is performed through a surgical procedure called a laparoscopy, there are two main surgical options available for treating the condition, ablation (Fig. 2.6) and excision (Fig. 2.7). Excision is performed by a specialist surgeon, who treats the endometriosis by dissecting it at the root, much like extracting a weed from a garden, whereas ablation uses diathermy to try to cauterise the growths with a heat laser. Diagnostic laparoscopy is an explorative surgery which sees the patient's abdomen filled with carbon dioxide and laparoscopic surgical tools placed in the abdomen via, what is usually, three incisions; one in the navel and two situated in the right and left-hand side of the lower pelvic region. Surgery can be the cause of severe anxiety for patients and observing medical illustrations as opposed to real footage or photographs of the procedure may appeal more to those who are squeamish.

3D visualisation is a highly transformative technology which is relatively new to healthcare and medical education, and it remains universally accessible, making it an area that is easily entered into by healthcare professionals and the public. 3D printing, in particular, is a cost-effective method that allows the observer to encounter something that is tangible and typically beyond their scope (Howard 2018). 3D-printed objects can now be created at home, and where once anatomical specimens were only to be viewed in preservation jars at museums, now they can be experienced from home and handled and studied for many years (Smith et al. 2018).

Whilst there are many benefits attached to 3D printing, the greatest flaw also lies therein in its accessibility. Online platforms, such as Sketchfab, allow individuals to liberally view, share, and buy and sell a wide range of 3D anatomical models, however, not all models found on this platform, and similar, can be deemed anatomically accurate due to the complex nature of the training involved in becoming, and working as, a qualified medical illustrator (Corl et al. 2000). Without having the 3D models examined by a healthcare professional or someone with

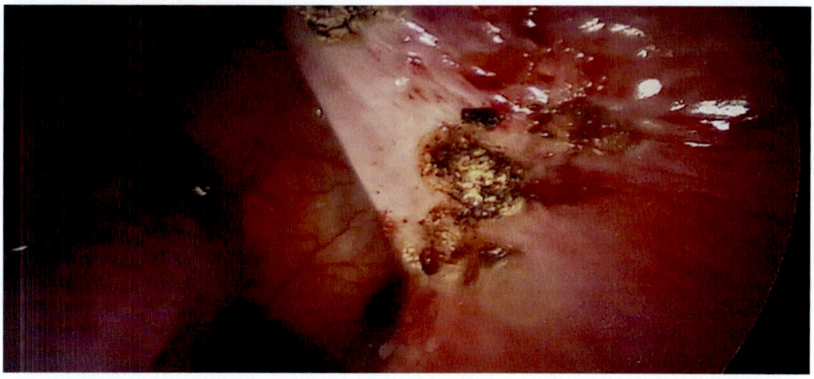

Fig. 2.6 Laparoscopic ablation technique for endometriosis (Image from researcher's own surgery, fifth July 2018)

an extensive knowledge of the subject matter in the first instance, it is very possible that misinformation can be spread thus enabling the widening of the cycle of confusion and doubt.

Although some negatives come with open-source learning and burgeoning technologies, the benefits and need for an overhaul in endometriosis education and awareness drastically outweigh this and need to be incorporated alongside patient care to ensure that successful treatments are met. By enlisting medical illustrators to create 2D and 3D visualisations, information can be accessed with ease, controlled, and viewed on a global scale, there is a real possibility to make this 'invisible' condition truly visible.

2.2 Methodology

Medical conditions can be easily researched online with search engines leading the way for at-home self-diagnosis, but unless an individual retrieves this from what can be deemed as a reliable and accurate source, the information they encounter may be wholly misleading (Hirsch et al. 2017). In recent years, endometriosis has been widely discussed online with a large amount of information being shared on social networking sites such as Facebook and Instagram, along with many online communities being established, too. Whilst these websites can provide much-needed support and friendship for some, they can also have a very detrimental effect, particularly when it involves discussions regarding treatment options and so-called 'cures'.

As endometriosis comes with such a broad spectrum of symptoms and can affect each individual differently, it was decided that before developing an online educational tool to visualise endometriosis, gathering feedback from a group of individuals with the condition would be essential. It was important to discover which aspects of

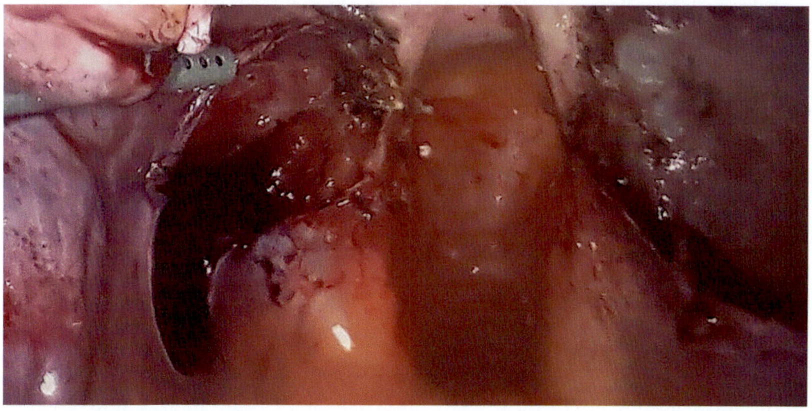

Fig. 2.7 Laparoscopic excision technique for endometriosis (Image from researcher's own surgery, 26th August 2020)

the condition sufferers felt required more emphasis. Ethical approval was obtained from the University of Dundee CAHID School Research Ethics Committee (approval number: UOD_SSEREC_CAHID_MSc_2019_021) to undertake these surveys.

2.2.1 An Online Resource for Endometriosis

To devise a list of questions, online health forums and social media platforms were scrutinised to discover what symptoms and topics were popular and discussed by the target audience. The questionnaire comprised three sections and 18 questions in total. 'Jisc Online Surveys' was designated for use by the University of Dundee for all projects involving surveying the public and was implemented in the research. Data collected remained anonymous and the questions consisted of a range of, 'Yes' or 'No' responses. Additional open-ended text boxes were anchored to each question, which gave the participant autonomy to voice their thoughts, opinions, and concerns regarding endometriosis.

Discussing endometriosis, or any health-related issue, can prove to be extremely distressing for the individual. As endometriosis deals significantly with issues around infertility, it was crucial that any potential participants were made aware of this prior to partaking in the questionnaire. To inform participants of the terms and conditions, and any upsetting aspects of the questionnaire, a Participant Information Sheet (PIS) was drafted detailing why the research was taking place, why it was important to collect feedback and how the information would then be stored and used. Following this information was a warning that some of the content may be upsetting or emotionally distressing, and if at any point the participant should feel any form of distress, they were free to close the webpage and cease participation. An additional inclusion to the PIS was a list of websites tailored to the individual's needs should they wish to seek further advice or guidance. As the questionnaire was distributed globally, information pertaining to international charities and support services was also gathered and input into the PIS.

To confirm participation in the research study and maintain anonymity, it was agreed between the primary investigator, Aimée Hutchinson, and their academic supervisor, Dr. Caroline Erolin that implied consent would take place by clicking the instruction 'Next', following the participant having read the PIS. Having done so, this would then lead the participant into the questionnaire where it was estimated that it would take no longer than 30 min to complete.

To gather sufficient feedback, following ethical approval, it was agreed that social media websites such as Facebook, Twitter, and Instagram could be implemented as a means of seeking active participation. Various call-outs appropriate to each social platform were created and approved by the ethics committee. The call-outs were also put forward to the moderators of the targeted groups to gain their permission to post within their forum. The call-outs were then rolled out to audiences on endometriosis-specific support group pages on Facebook, whilst hashtags were used to garner attention on the photo-sharing platform, Instagram, and subsequently the microblogging site, Twitter (Fig. 2.8).

Individuals wishing to partake in the research were informed to e-mail the researcher using the academic e-mail address provided. Once an e-mail was received stating interest, a private weblink to the questionnaire was then sent, and the participant's contact details were then deleted. The questionnaire opened on the June 26th, 2019, with the final response submitted on July 15th, 2019. A total of 136 individuals actively completed the questionnaire. The data from the questionnaires was downloaded as a.CSV document from Jisc and entered into Microsoft Excel where the findings were then visualised through the use of simple bar and pie charts. As there were a large number of responses to some questions, it was vital that the information provided be output as an easy-to-read resource.

It is to be noted that with some open-ended text boxes, participants wrote lengthy responses detailing large sections of the individual's medical history and current health situation. As the

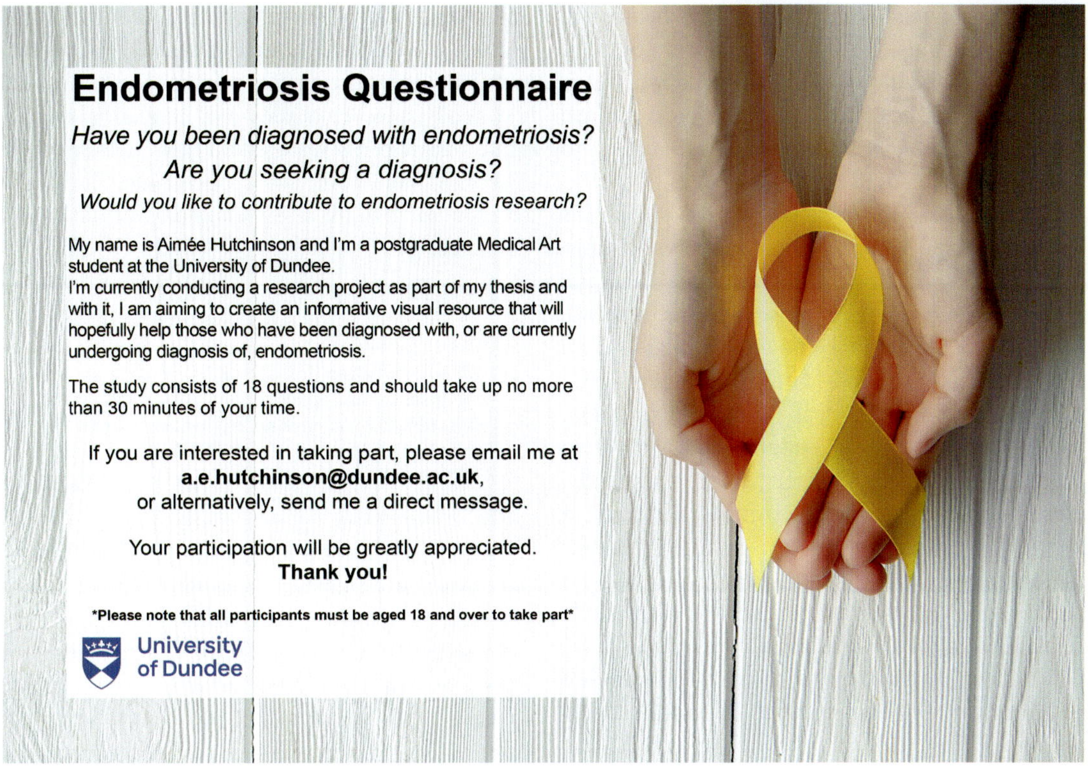

Fig. 2.8 Participant call-out for endometriosis questionnaire. By Aimée Hutchinson

research topic is of a sensitive nature and within an area that is stigmatised and not widely spoken about, having created an environment where the individual feels a willingness and ability to share, and has the opportunity to have their voice heard, it was to be expected that some responses may be of a more delicate nature than others.

The questionnaire was intended to gather feedback before any creative work was completed, this was to ensure that those living, and those working with, endometriosis felt there was a need for visual communications and illustrations, and to discover what areas they felt were particularly underrepresented. There were 136 participants over the course of 3 weeks and 2 days. All participants identified as female, and the majority of respondents were aged between 25 and 34 years and from Europe.

The age of 17 years was noted as having been the age when most respondents began to primarily notice their symptoms, however, there were some whose symptoms only appeared in their 40s and shockingly one participant said she was 7 years old when her symptoms began to affect her daily life. Whilst the ages are not reflective of when endometriosis was confirmed via diagnostic methods, it must be taken into consideration that most medical texts on the condition allude to it affecting females predominantly in their 30s and 40s. The most common age for diagnosis from the participant questionnaire was 25 years old, which is 2 years less than stated in the various facts and figures relating to endometriosis, perhaps this shows that healthcare regarding the condition is improving, however, it still conveys the drastically long wait times between the average onset of symptoms and the age at which diagnosis occurs.

Of the participants, 86% had received an official diagnosis of endometriosis, whilst the remaining 14% were undergoing diagnosis, or suspected they had the condition. Including those undergoing diagnosis was intended, as it was assumed that it was these participants who would be actively going through the healthcare system to attain an official diagnosis for their symptoms, therefore the information they provided would allow us to explore the approach healthcare professionals were actively taking to diagnose and treat the condition.

A positive aspect to be taken from the data is that 67% of the 136 participants involved admitted to having heard of endometriosis before an official diagnosis via laparoscopy was reached, with much of their prior knowledge arising from conversations had with family members who, they themselves, also dealt with the condition. As the pathogenesis of endometriosis is yet to be discovered, there are theories alluding to the element of genetic predisposition. Perhaps, if research was carried out on a much broader scale and significantly more data collected, we could further explore the genetic link as a viable starting point for the condition.

Of those involved in the data, 80% were not aware of the visual appearance of endometriosis and out of 134 participants who answered, 92% felt it would be helpful to have illustrations and visual communications depicting endometriosis. A further 80% of participants had not received any further information from their healthcare provider following the discussion, or diagnosis, of endometriosis. Of those who did receive further information 58% deemed it to be unhelpful.

When asked if illustrations showing the surgical procedures involved would be useful, 134 participants out of 136 agreed by saying 'Yes', however, when Professor Andrew Horne (MB ChB, PhD, FRCOG, FRCP Edin), Chair of Academic Board, Royal College of Obstetricians and Gynaecologists, and Co-Director of EXPPECT Centre for Pelvic Pain and Endometriosis, was asked the same question he felt that emphasis should not be placed on this area of the condition as there was yet to be a confirmed option for which surgical technique is better as this depended wholly on the individual and their symptoms, as well as lifestyle factors involved.

It was important to ascertain how severe of an effect endometriosis has on the lives of those who live with it daily. Pain and a lack of understanding were the two answers that appeared frequently, with the latter putting into perspective how great the need is for resources and education on the topic. Participants were asked how, if they could visualise their pain, what appearance it would take on. Below are a few examples of what was collected:

> A web suffocating me.
>
> A lot of the time I see it like a cancer, a cancer that can't kill you but will cause you to suffer just as much in silence.
>
> It looked like cigarette burns—black and lumpy.
>
> A rubber band around my insides squeezing tighter and tighter.
>
> Figuratively, as shards of glass or flaming knives.

Additional feedback was sought from Professor Andrew Horne. A telephone interview between Hutchinson and Horne took place on Wednesday July 10th, 2019, in which a series of questions, similar to those presented in the questionnaire were put forward. Professor Horne has been a triumphant voice in endometriosis research and raising awareness for the condition as well as being an advocate for the UK-based charity, Endometriosis UK. Extracts from the telephone call can be found below (reproduced with permission):

> *Question:* How would you describe the visual appearance of endometriosis?
>
> *Answer:* It comes in three subtypes; superficial peritoneal endometriosis, ovarian endometriosis (endometriomas), and deep-infiltrating endometriosis. They are all variable. The first subtype is similar to an angry rash in appearance and severe endometriosis can look like cancer in a very advanced stage. Endometriomas can look blueish and their inner contents look like chocolate, which is why they're called chocolate cysts (Tactile, nodular).
>
> *Q.* Do you think it is important to have more visual references available for those with endometriosis?
>
> *A.* Leaflets would be helpful, however, some patients may be squeamish, so it is important to take this into consideration. Some patients also might see their photos following surgery. There is definitely a role for imagery though.

Q. Do you think it is important to have visual depictions of surgical approaches to endometriosis, such as ablation and excision?

A. There is debate around ablation and excision and which is the better option. As a surgeon, we try to use our own judgement to determine which will be the better option for the individual. It would be helpful to illustrate the difference in approaches, however, and to show that laser can both ablate and cut out the damaged tissue.

Q. Are there any aspects of endometriosis that you feel are underrepresented?

A. I feel that pain is very much underrepresented. Fatigue is another thing that is not recognised, and it is extremely symptomatic of endometriosis. Fatigue also has a severe impact on things like relationships; intimate or platonic, career progression, and life attainments. Our research centre aims to ensure that women with the disease can do what those without the disease can do, too.

2.2.2 Creating the 3D Uterus: Conceptual Idea

What is often mentioned when discussing endometriosis with someone who has the condition, is the retort that they are often told—*but you don't look sick*. There are many chronic conditions that are classed as invisible but when searching for research funding available for endometriosis, it is one area that pales in comparison (Horne et al. 2017). Following the initial survey, and prior to the development of an online resource for endometriosis, a 3D digital model of a uterus was developed with the intention for it to be hosted on the online resource. An accompanying 3D printed model would also be produced to tackle the notion that endometriosis is largely classed as an 'invisible illness' and to enable another vehicle for visualisation.

Whilst endometriosis affects many parts of the body, the uterus was chosen as it is synonymous with the condition, and presumably, the first thing people visualise when they hear of gynaecological disorders. An average healthy uterus is 7.6 cm long and 4.5 cm wide, with a weight of 60 grams (Christiano and Biggers 2017). For a 3D-printed model to communicate information about a gynaecological disorder, this size was discounted as it would prove to be too small and as a result provide an inadequate foundation for portraying the information desired. It was decided that to fully convey the condition, as well as creating a uterus complete with ovaries and fallopian tubes, there would be an accompanying coronal cross-section. The cross-section model would depict the inside of the uterus and ovaries.

The various subtypes of endometriosis depicting the superficial lesions, an ovarian endometrioma and deep-infiltrating endometriosis, were to be depicted in the 3D digital model and 3D printed model. These details would be created using the 3D modelling software, ZBrush. The 3D-printed model would contain LED lights that would slowly light up to reveal the damaged and impacted areas of the uterus. These models aimed to effectively convey the message of the condition, to focus on visibility and perhaps provoke a discussion.

The 3D sculpting process took place in Pixologic ZBrush 2019.1.2. ZBrush is a digital sculpting tool that uses 'pixol' technology to combine 3D modelling, painting, and texturing. To begin, a sphere was loaded onto the canvas and manipulated using the 'Move' and 'Move Topolgy' brushes within ZBrush. The 'Smooth' brush was also used in between the manipulation of the shape and by doing this, it ensured that the uterus would be kept neat. When the main body of the uterus was complete, a cylinder was then added to the canvas which was to represent the area of the cervix and vagina. Both the 'Move' and 'Move Topology' brushes were used again in conjunction with the 'Smooth' brush to achieve an accurate shape. The cylinder shape was added to the bottom of the newly sculpted uterus and both were then merged together to create one 'polygroup' (Fig. 2.9). In doing this, it allowed for further manipulation and perfecting of the sculpture. The 'Flatten' brush was utilised to remove any visible seams where the two 'polygroups' had been connected.

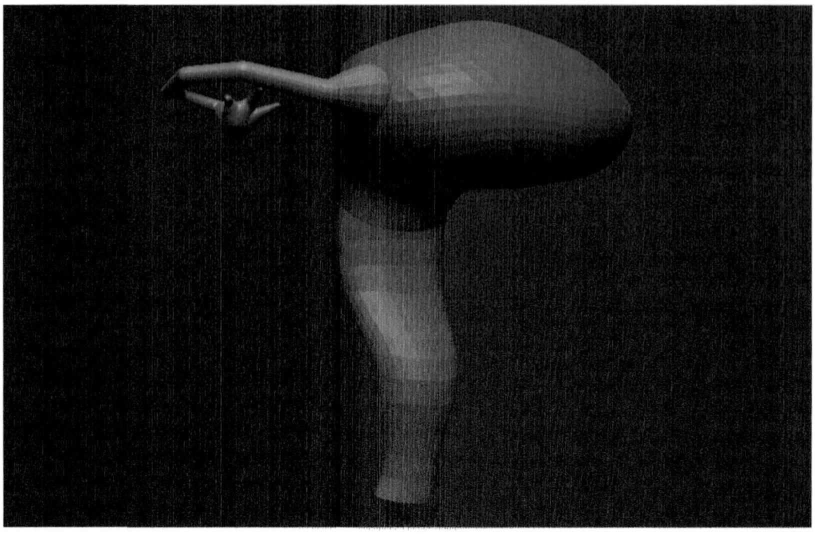

Fig. 2.9 Initial modelling of the uterus in ZBrush. Image by Aimée Hutchinson

When sculpting the fallopian tubes and fimbriae, 'ZSpheres' were used (Fig. 2.10). This allowed for the creation of quite an organic tubular shape and gave full control as to the direction of the fallopian tubes. To capture the fimbriae, once content with the shape and flow of the fallopian tubes, a large 'ZSphere' was added and on top of this several smaller 'ZSpheres' were placed to depict the finger-like projections. To ensure the symmetry was correct on both sides of the uterus, the initial 'ZSpheres' portraying the fallopian tube and fimbriae were 'Copied' and 'Mirrored'. An 'Adaptive Skin' was created for the 'ZSpheres', and this allowed for further sculpting and texturing work.

In creating the ovaries, two spheres were loaded onto the canvas and once again, they were manipulated using a similar process as

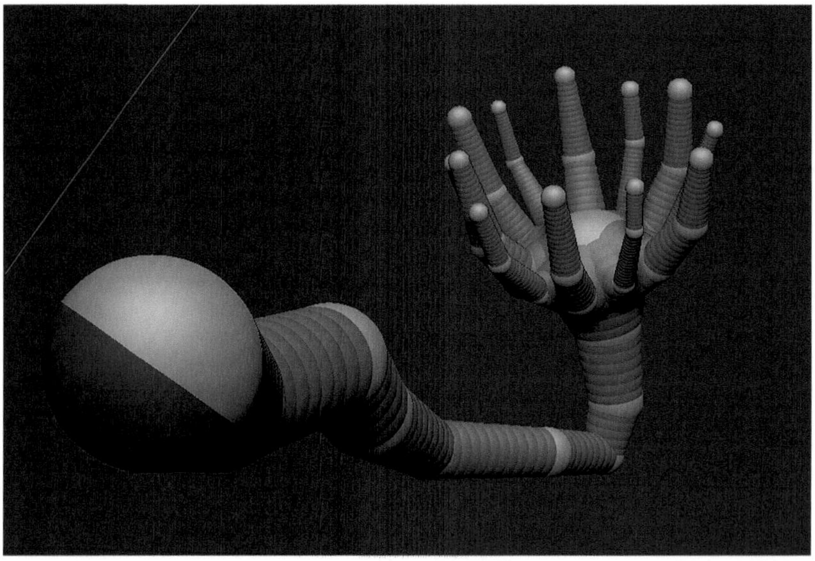

Fig. 2.10 Modelling of the fallopian tubes and fimbriae in ZBrush. Image by Aimée Hutchinson

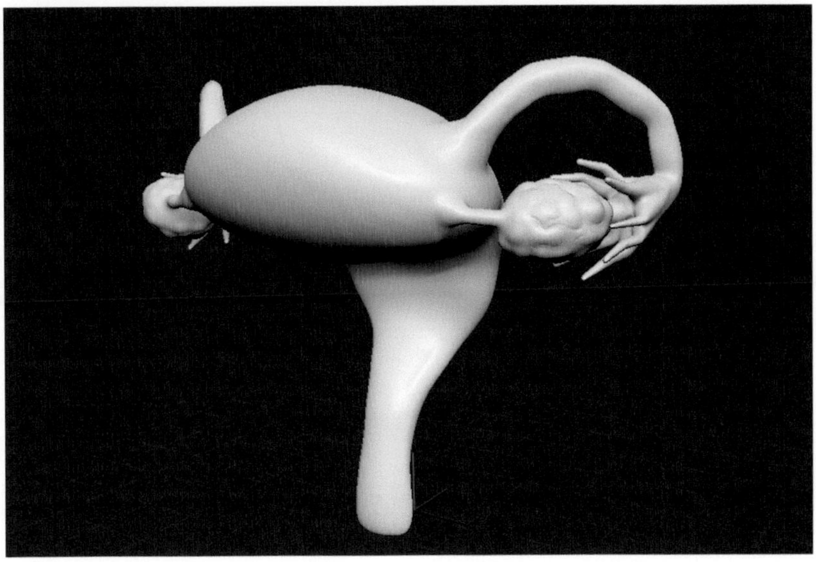

Fig. 2.11 Rendering of the uterus, fallopian tubes, and ovaries in ZBrush. Image by Aimée Hutchinson

aforementioned. As the ovaries take on an oval, egg-like shape, as opposed to a perfectly round sphere, it was important to ensure this was adequately captured. The ovarian ligament was sculpted using 'ZSpheres' which were then placed in an 'Adaptive Skin' and merged together to achieve a seamless look (Fig. 2.11).

Pixologic ZBrush is an appropriate tool for sculpting life-like anatomical models. Its many functions allow the user to embellish models with a variety of textures and colours, all of which can be dynamically changed in a few small steps. ZBrush comes loaded with a broad range of 'Alphas', texturing brushes to add or subtract textures from your digital clay and can also be used in conjunction with 'Material' and 'PolyPaint' elements.

'Alphas' were used to add texture to the 3D uterus. Using the material 'Skinshade 4', Alpha brushes 24 and 07 gave texture to the ovaries and some minor embellishments were also added to the fundus of the uterus. 'PolyPaint' was then used to capture the colour and further details, including vessels and veins (Fig. 2.12).

When it came to creating the coronal cross-section of the uterus, the same initial process that was used for the first model was implemented. A sphere was loaded and manipulated to achieve the correct shape. As the coronal cross-section would be detailing the uterus and ovaries only, once both components were on the canvas and positioned, the 'Slice Curve' brush was used to cut the model in half. Using the 'DamStandard' brush set to a larger size, symmetry was turned on and the carving out of the inside of the uterus began. Symmetry was an extremely helpful tool for this process, as it ensured consistency throughout the sculpture. 'ZSpheres' were again used to model the fallopian tubes and fimbriae, and as before, the initial sculpt was copied and mirrored on the opposite side (Fig. 2.13).

The 'Masking' tool was used to block out areas where there would be no texture, for instance, on the endometrium and exterior of the uterus. Striations were added to the myometrium using 'Alpha 58'. When the first set of striations was placed, the 'smooth' function was operated, and the process repeated until the desired look had been achieved. A similar method was carried out on the ovaries; however, the symmetry tool was turned off for this particular action (Fig. 2.14).

As the left ovary on the 3D model was to depict an endometrioma, the 'Inflate' brush added volume to the area, so that this ovary

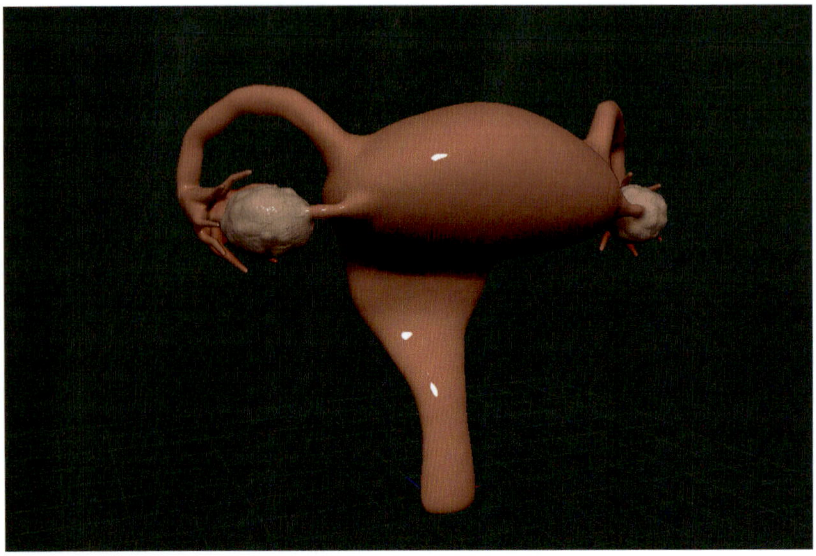

Fig. 2.12 Development of painting process of uterus, fallopian tubes, and ovaries in ZBrush. Image by Aimée Hutchinson

would appear much larger in comparison to its counterpart (Fig. 2.15). The outside surface of the ovaries was textured in a similar manner with brush alphas working veins and organic nuances onto the surface. Several gynaecology books retrieved from the Medical Library at Ninewells Hospital Dundee, were consulted as references for the cross-section details of the uterus and ovaries.

2.2.2.1 But What Is an Ovarian Endometrioma?

Endometriomas are known as 'deep infiltrating endometriosis' and affect 17–40% of women with the condition (Jayaprakasan et al. 2018). They are benign, fluid-filled cavities made up of ectopic endometrial tissue and are often referred to as 'chocolate cysts' due to the darkened appearance of the blood they contain.

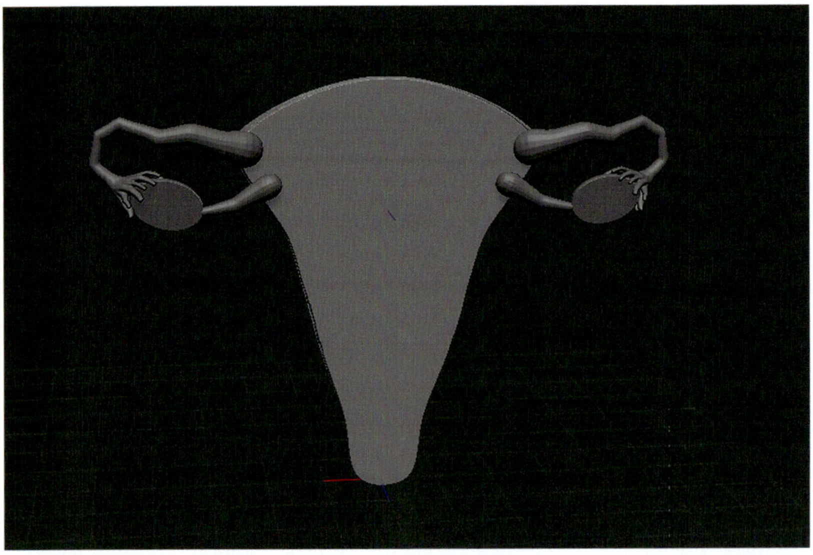

Fig. 2.13 Initial modelling of coronal cross-section of the uterus, fallopian tubes, and ovaries in ZBrush. Image by Aimée Hutchinson

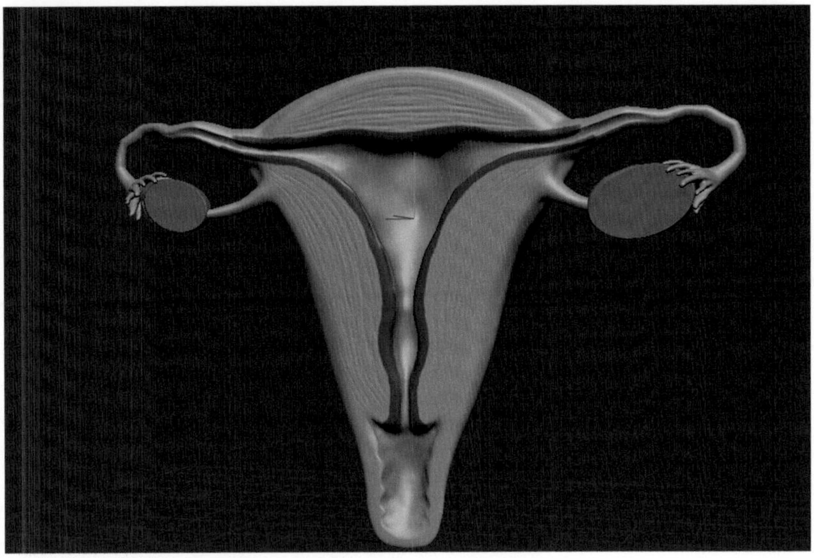

Fig. 2.14 Adding textures and rendering to coronal cross-section of the uterus in ZBrush. Image by Aimée Hutchinson

Endometriomas come in varying sizes and whilst they are monitored via regular transvaginal or pelvic ultrasound scans, they can be extremely painful and surgical intervention is put forward as an option when chronic pain begins to greatly impact upon an individual's life. Endometriomas larger than 1 cm do not go away on their own (Endometriosis: diagnosis and management 2017) and without treatment to impede the growth, or surgically removing the endometrioma completely, they will continue to grow.

Endometriomas can also have a detrimental effect on the fertility of an individual. Whilst ovarian function can continue as the endometrioma proceeds to grow, it is a surgical intervention that will ultimately impinge upon a woman's ovarian reserves. Medical professionals are directed to consult with the patient and take into consideration their age and whether or not they intend to have children before making a final decision on how to treat the endometrioma (Demircivi Bor et al.

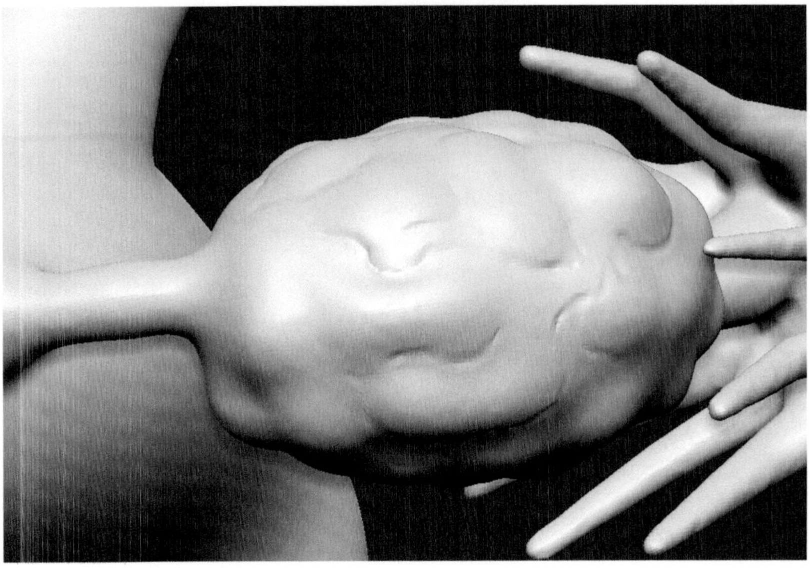

Fig. 2.15 Detail of texture added to the ovarian endometrioma in ZBrush. Image by Aimée Hutchinson

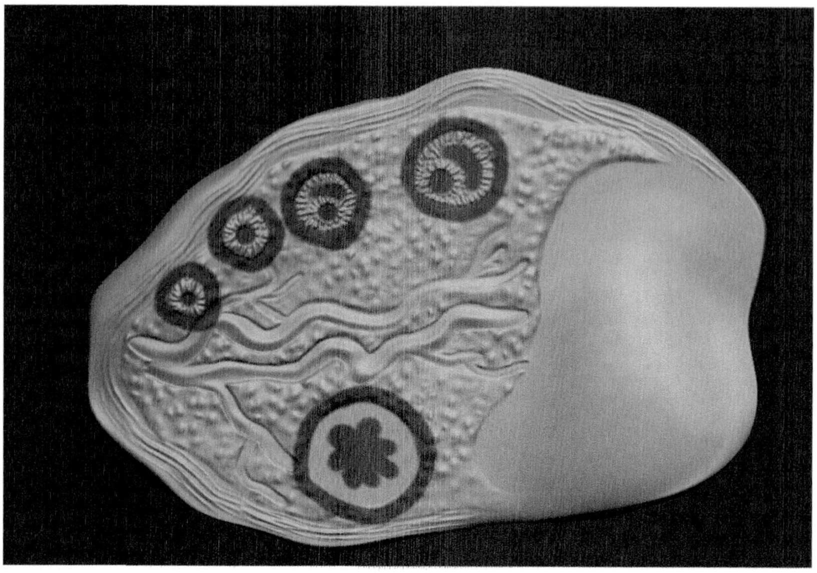

Fig. 2.16 Detail of masking and texturing process of the coronal cross-section of the ovarian endometrioma in ZBrush. Image by Aimée Hutchinson

2013; Endometriosis: diagnosis and management 2017).

To achieve the appearance of the ovaries and the endometrioma, online resources were consulted and Professor Andrew Horne of the EXPPECT Clinic in Edinburgh, provided further reference images retrieved during laparoscopic excision. Whilst the initial model depicts a full uterus, the coronal cross-section model was inspired by traditional anatomical models. Contrast is shown between the healthy right ovary and the cystic left ovary. The ovarian cycle was depicted in both ovaries to add an additional element to the piece, showing the maturing follicle, the shedding of the ovum, and the development of the corpus luteum (Figs. 2.16 and 2.17).

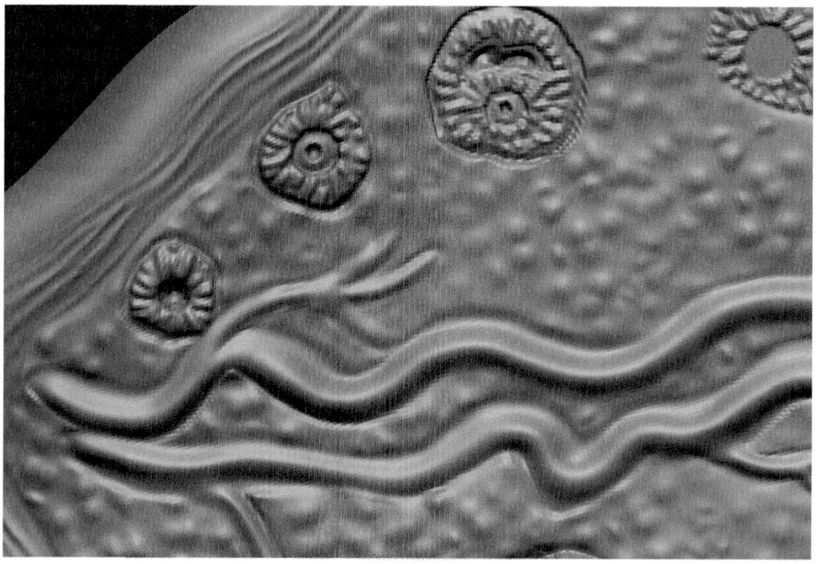

Fig. 2.17 Detail of coronal cross-section of ovarian endometrioma depicting the ovarian cycle in ZBrush. Image by Aimée Hutchinson

2.2.3 Rendering Process and Sketchfab

The finished 3D models were exported into Keyshot for rendering. Keyshot is a real-time ray tracing and global illumination programme used to create 3D renderings and animations. ZBrush to Keyshot Bridge was used and this piece of software brought the models, along with their textures and 'polypaint' properties, into Keyshot8 (version 8.2.80). Slight adjustments were made to the environment and lighting within Keyshot, however, a clean output was desired, so background changes were kept minimal, and a 'Hard Shiny Plastic' material was added to the models to create a high gloss finish. Rendered images were captured of the models from various angles and then processed in Adobe Photoshop to improve the vivacity of the colour and enhance sharpness (Figs. 2.18, 2.19, 2.20, 2.21, 2.22, 2.23, 2.24, and 2.25).

To add an interactive element and improve accessibility, the 3D models were uploaded to Sketchfab directly from ZBrush. Sketchfab (www.sketchfab.com) is a platform that allows its members to upload and host 3D models as well as virtual and augmented reality content. By creating an account on this website, the user can share 3D creations with a global audience. Annotations can also be added to virtual models, making them a valuable tool for medical communications and education (Figs. 2.26 and 2.27).

2.2.4 Printing the 3D Model

The printing of the 3D model was carried out on a 3D Systems SLA 5000 machine using Accura Xtreme material. Accura Xtreme is a tough material that is resistant to breakage and was specifically chosen to protect against any damage to the delicate fallopian tubes and fimbriae. To prepare the model for building, supports, technical holes for drainage and electrics, and the hollowing of the insides were created using Magics software. The duration of the build took approximately 24 h after which it was washed in an isopropanol bath, the supports removed and cured in a UV oven. 120–240 grit sandpapers were used to smooth build lines before the models were bead blasted with aluminium oxide 120-grit. Bead-blasting helped to even out the surface and remove shine whilst also accessing and smoothing smaller, hard-to-reach crevices.

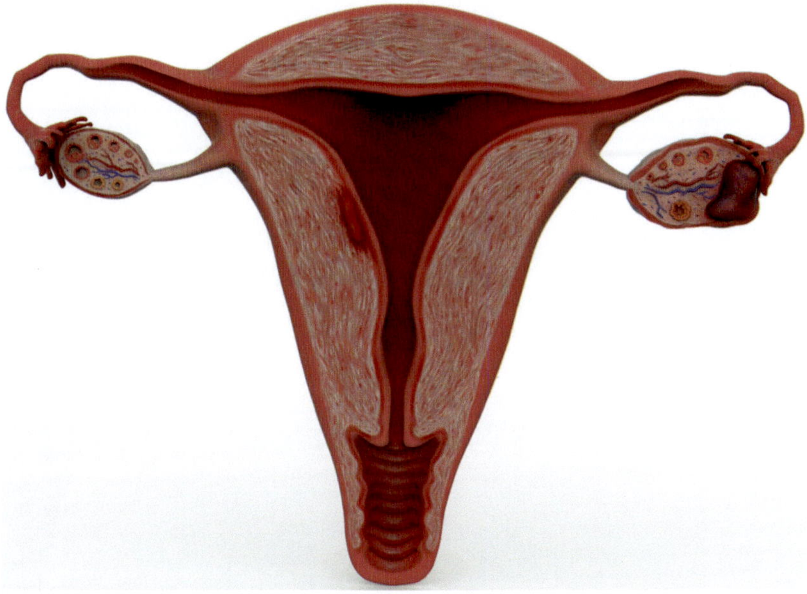

Fig. 2.18 Final rendering of coronal cross-section of uterus depicting ovarian endometrioma and adenomyosis. Image by Aimée Hutchinson

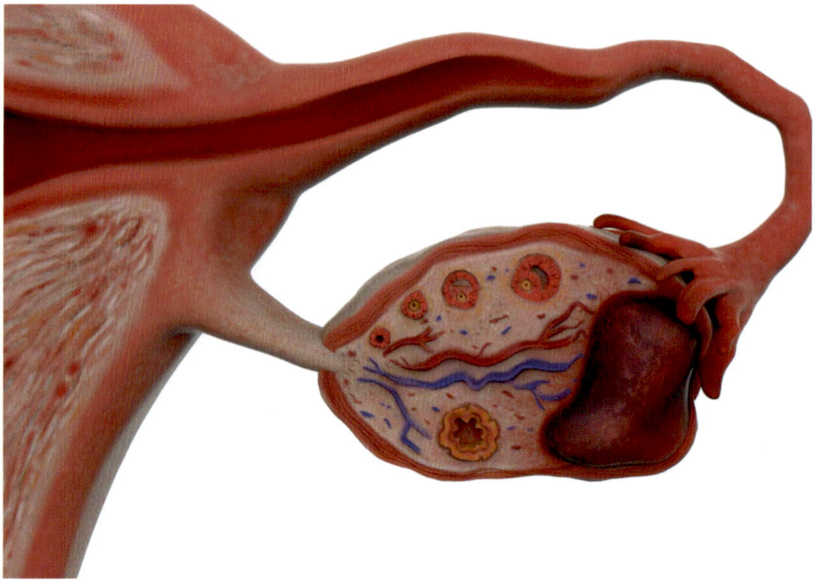

Fig. 2.19 Detail of final rendering of ovarian endometrioma. Image by Aimée Hutchinson

Before any further aesthetic work was completed, USB-powered LEDs, which were fully programmable and purchased online, were glued into the areas that would later be painted to resemble endometriotic lesions. The technical holes on the model were closed and any remaining gaps were filled with U-Pol Dolphin Glaze—a standard, self-levelling polyester body filler typically used in the automotive industry. Further sanding was done to the prints to ensure a smooth, even finish was achieved before being primed with 2 K Primer.

Both models were finished in the same way, however, to have the cross-section model resting at a slight diagonal angle, a special base and support was crafted. A 12-mm piece of MDF

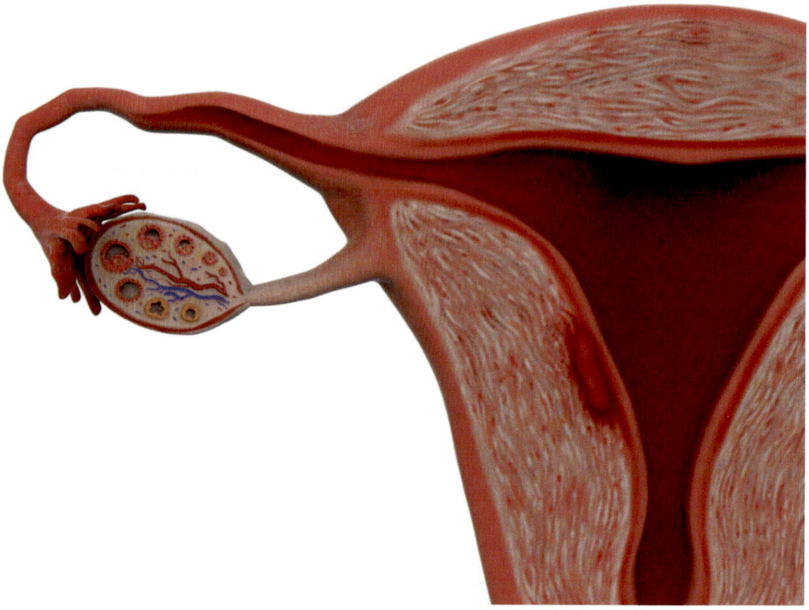

Fig. 2.20 Detail of final rendering of healthy ovary & adenomyosis. Image by Aimée Hutchinson

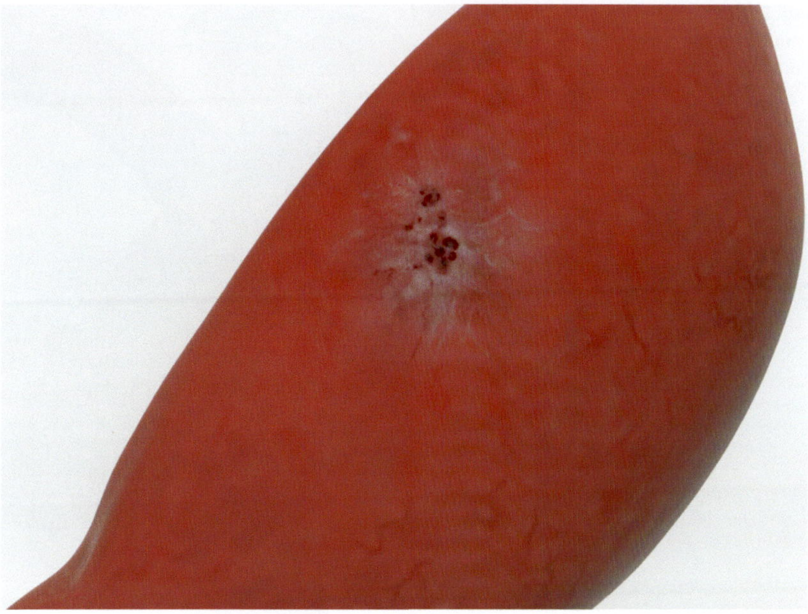

Fig. 2.21 Detail of superficial endometriosis lesions on the fundus of the uterus. Image by Aimée Hutchinson

was cut and shaped into a circle to match the base of the accompanying model. A hole was drilled into the back of the cross-section and a plastic rod was inserted for support. A metal pin was used to connect the model and the MDF base before finally sculpting more support out of body filler.

Fig. 2.22 Final rendering of the uterus depicting superficial endometriosis lesions of fundus, outer vagina and ovarian endometrioma. Image by Aimée Hutchinson

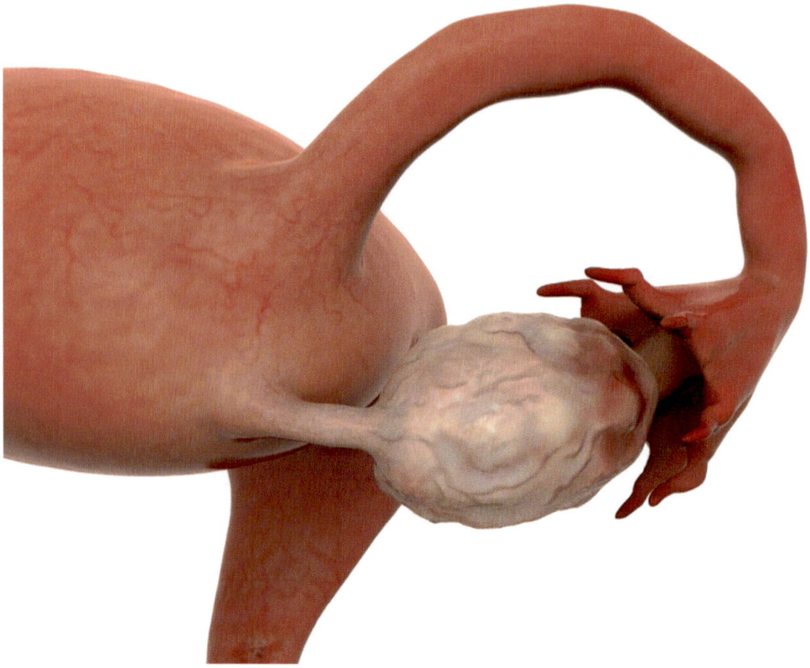

Fig. 2.23 Detail of ovarian endometrioma. Image by Aimée Hutchinson

2.2.5 Painting the 3D-Printed Models

The painting process proved to be the most challenging aspect of the project as the acrylic paint chosen did not adhere to the 3D-printed material well. Having created the digital model, textured, and painted it within ZBrush, when it came to capturing the likeness with traditional painting methods it proved to be complex. Many layers had to be added and each had to be left to dry for a total of 24 h before applying further detail. This proved to be time-consuming. Once the base

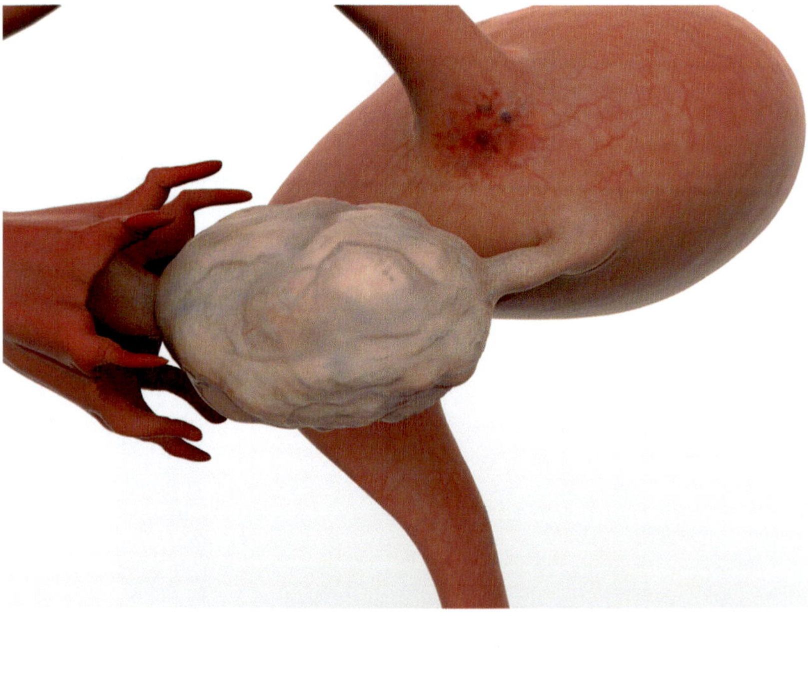

Fig. 2.24 Detail of deep-infiltrating endometriosis of the fallopian tube, along with healthy ovary. Image by Aimée Hutchinson

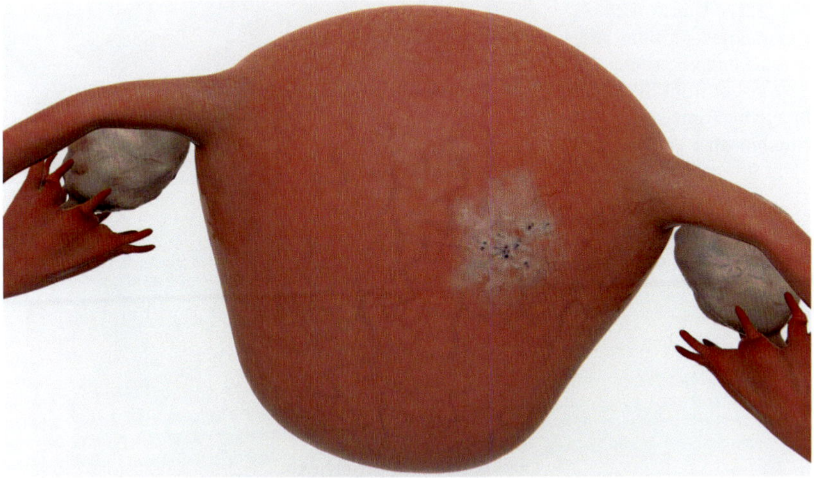

Fig. 2.25 Detail of 'powder-burn' superficial endometriosis lesion of the fundus. Image by Aimée Hutchinson

colour was completed in acrylic, details were added using a sponge in which colour was gently dabbed onto areas and then softened with a clean, wet brush (Fig. 2.28).

To depict the blood vessels of the uterus, coloured pencils were used (Fig. 2.29). The models were predominantly painted in the evening as this made the lights inside them easier to see, and the amount of paint added to areas that would 'light-up' using lights was strongly considered before being applied. Acrylic paint was used on the 'Accura Xtreme' material and as this paint was difficult to thin down, the opacity of the paint tended to dull the lights of the model. To counteract the fading of the lights, a thin wash of paint was added to the model and coloured pencils were used over the top to capture finer details. The models were then varnished using an acrylic spray varnish, and this gave them a more visceral, realistic appearance akin to human organs. The base of each model was also painted white again to cover any flaws that may have occurred during the painting process (Figs. 2.30 and 2.31).

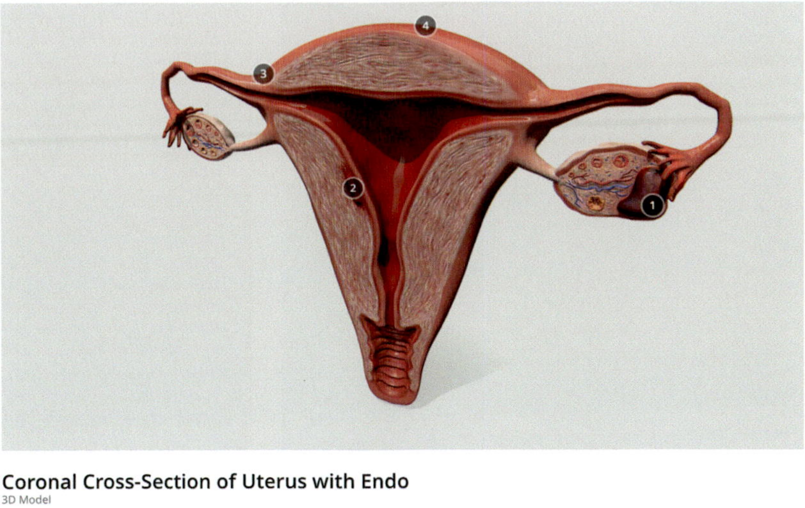

Fig. 2.26 Coronal cross-section of the uterus with endometriosis. Screenshot from Sketchfab. Image by Aimée Hutchinson

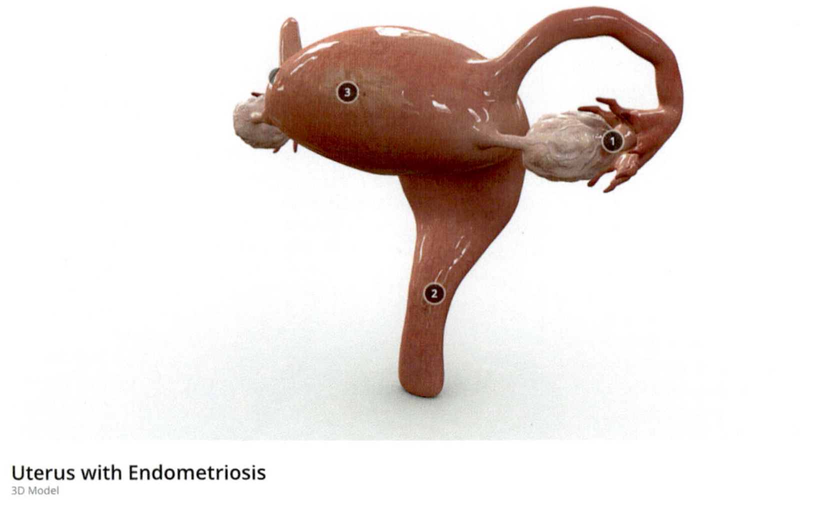

Fig. 2.27 Uterus with endometriosis lesions and ovarian endometrioma. Screenshot from Sketchfab. Image by Aimée Hutchinson

2.2.6 Creating the Online Resource for Endometriosis

To begin building the online resource for endometriosis, the cloud-based web development service *Wix* was used as it does not require the user to possess any coding skills to create the foundation of the website, and it also allows for flexibility when developing a design and layout. As the website would predominantly host illustrations and 3D digital models, when it came to choosing a name for the web page and an appropriate link, 'Illustrating Endo' was deemed to be an appropriate option (Fig. 2.32).

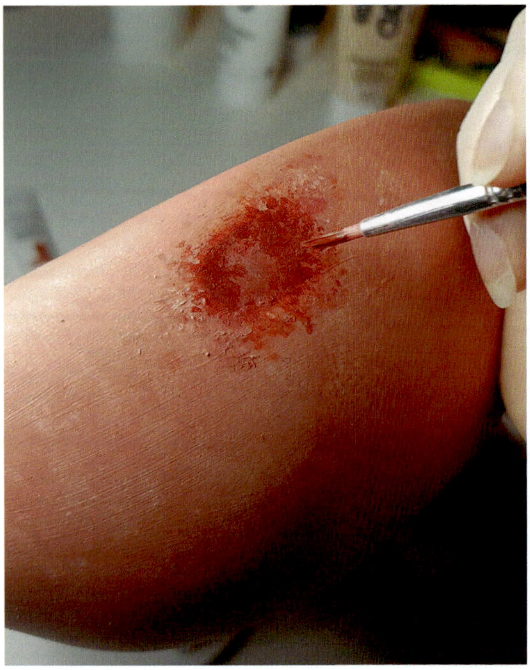

Fig. 2.28 Detail of the painting process of the 3D printed coronal cross-section of the uterus. Image by Aimée Hutchinson

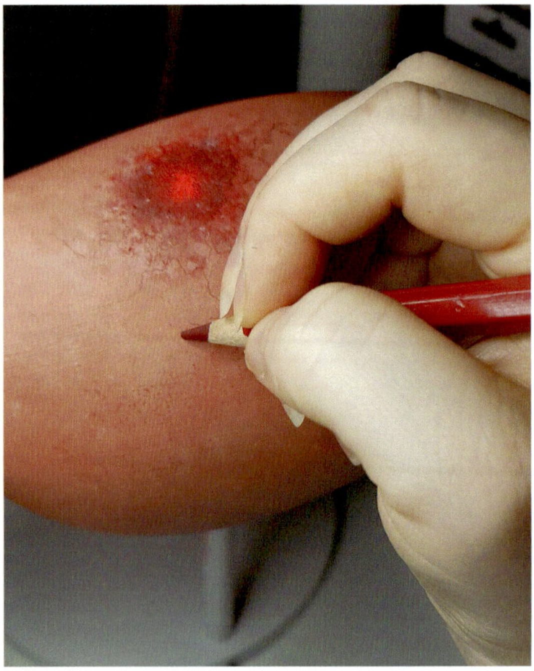

Fig. 2.29 Capturing smaller details of endometriosis lesions on the 3D-printed coronal cross-section of the uterus. Image by Aimée Hutchinson

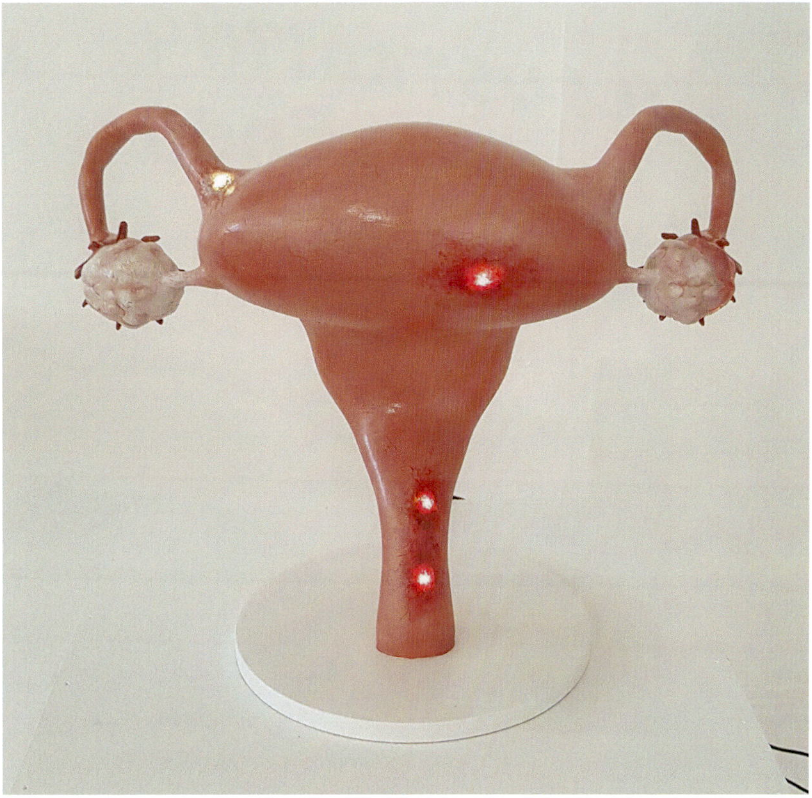

Fig. 2.30 Image of 3D printed uterus model depicting endometriosis lesions and ovarian endometrioma. Here you can see the LED lights which were installed to highlight the deposits. Image by Aimée Hutchinson

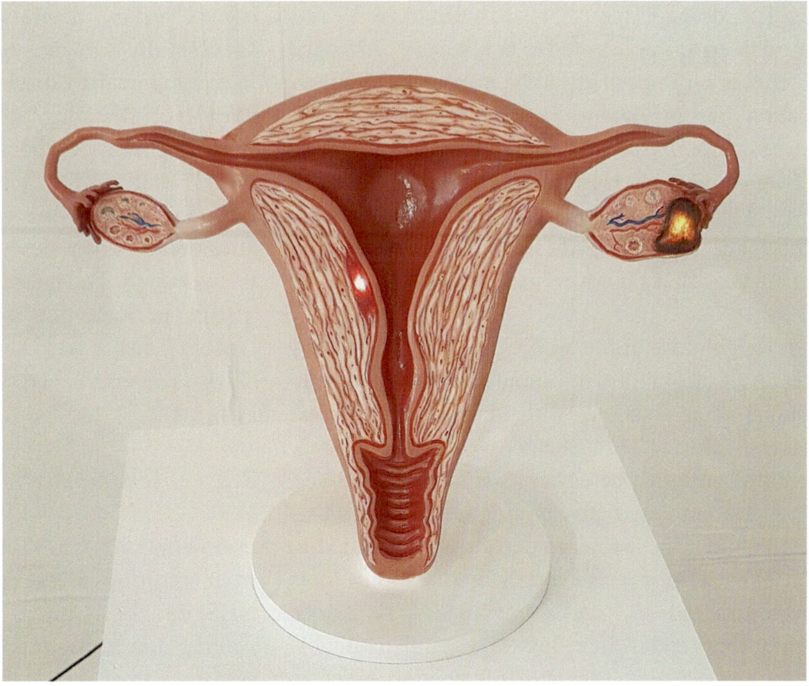

Fig. 2.31 Image of a 3D-printed coronal cross-section of the uterus depicting endometriosis lesions and ovarian endometrioma. Again, the LED lights can be seen which highlighted the deposits of growth. Image by Aimée Hutchinson

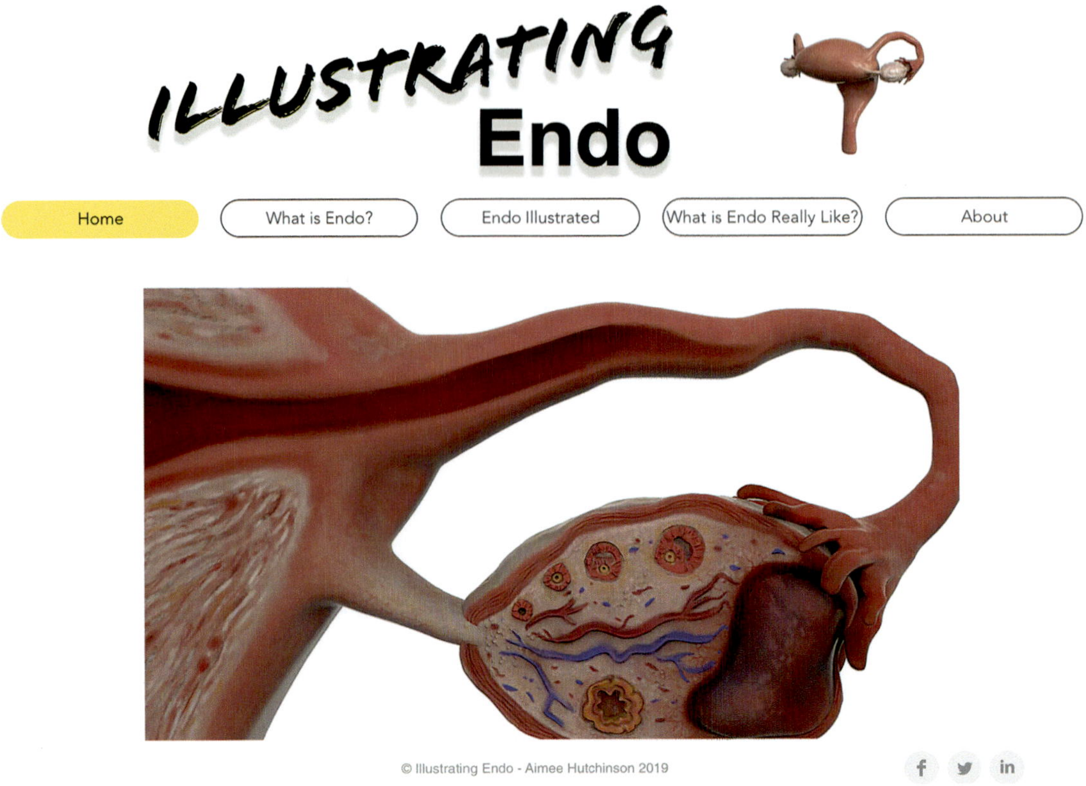

Fig. 2.32 Image of the landing page for IllustratingEndo.com website. Image by Aimée Hutchinson

The domain for 'IllustratingEndo.com' was purchased and connected to the website. As social media is an important factor in social communication, an Instagram and Twitter handle was secured, both of which are: @illustratingendo (Instagram) and @EndoIllustrated (Twitter). It would have been preferable for both to remain the same, however, there was a limit to the characters that could be used as a username on Twitter.

The content of the webpage was informed by feedback from the questionnaire and the telephone conversation with Professor Andrew Horne. The colour scheme was predominantly yellow, with this being representative of the yellow ribbon used to highlight endometriosis awareness. The webpage design was intended to be eye-catching and modern and various fonts were trialled in programmes such as Adobe Photoshop and Adobe InDesign until a suitable font was chosen. Medical information tends to come from formal sources that, whilst successfully carrying out their intended purpose in communicating to a broad range of individuals, may feel daunting and not very welcoming (Topac 2013). Medical terminologies may also be off-putting for some (Baker et al. 1996; Osborne 2006; Tong et al. 2020) and this was considered when devising the text content for the webpage. The intention was to create an informative resource that will educate, advise, and create awareness without condescension (Figs. 2.33 and 2.34).

2.3 Discussion

Following the COVID-19 pandemic there has been insurmountable pressures placed on the NHS in the United Kingdom with hospital waiting times reaching all-time highs. Patients seeking treatment are now facing greater

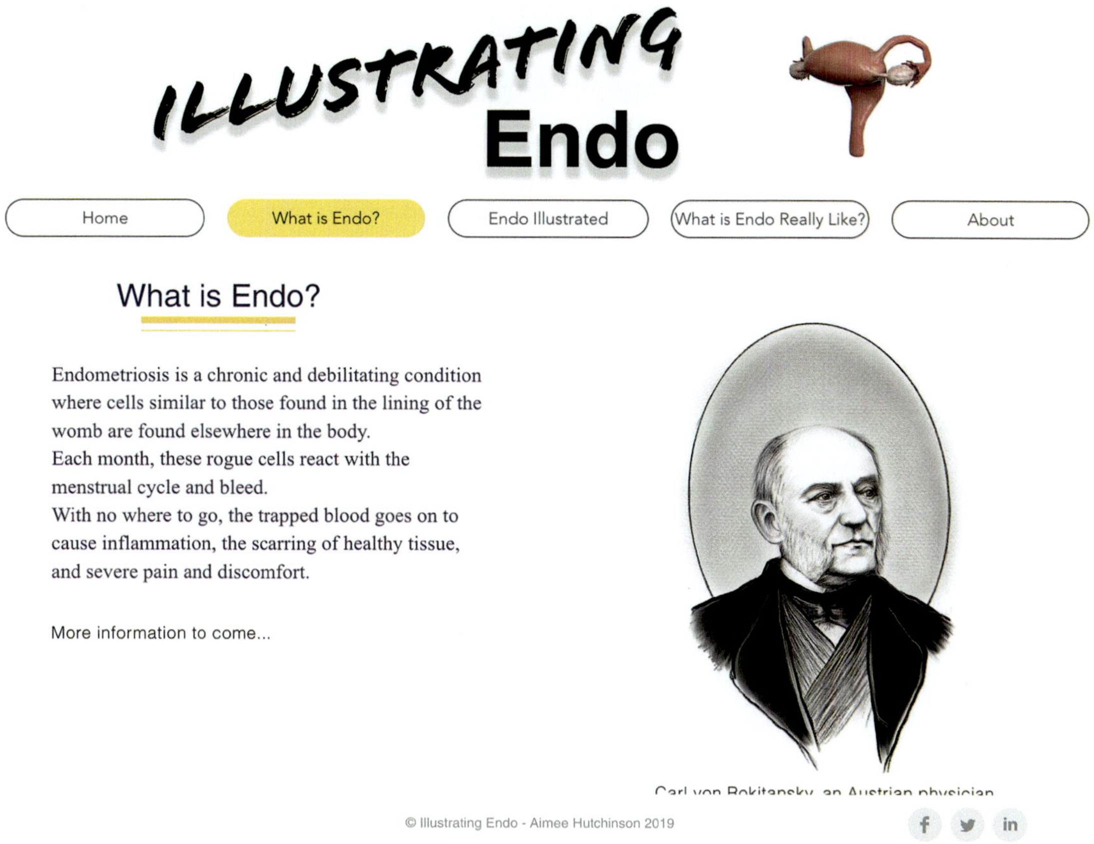

Fig. 2.33 Image of the 'What is Endo?' section from IllustratingEndo.com website. This section gives a brief rundown of the history of the disease and includes digital drawings by the researcher. Image by Aimée Hutchinson

challenges, and healthcare staff are battling extreme workloads and stresses. Patients living with endometriosis and its symptoms are enduring the failings surrounding gynaecological healthcare. Diagnostic surgeries have been delayed, and in some instances, cancelled altogether. Hospital consultations to discuss further treatments face the same fate. This has meant that some women are opting to seek treatment abroad (health tourism) or remain in the United Kingdom and potentially put their lives on hold. A brief search on social media will direct you to many GoFundMe pages set up by those living with the condition who are attempting to regain control of their lives and free themselves of the burdens that come with endometriosis. Anxieties surrounding the preservation of fertility and the financial impact that endometriosis has on an individual's life is predominantly what is driving those to seek treatment abroad. The lack of BGSE-accredited endometriosis centres in the United Kingdom and Ireland also plays a huge factor.

The questionnaire outcomes found that a staggering 80% of the 136 participants were not given further information by their GP or Gynaecologist and this fuelled the development of IllustratingEndo.com to provide illustrations of various aspects of endometriosis and most importantly, information sourced from trusted academic leads—many of which have been referenced in this research. The responses to the questionnaire demonstrated that many people are ready and willing to speak out and raise awareness of endometriosis and the impact it has on many individuals. They also suggested that a series of illustrations available in the form of an online

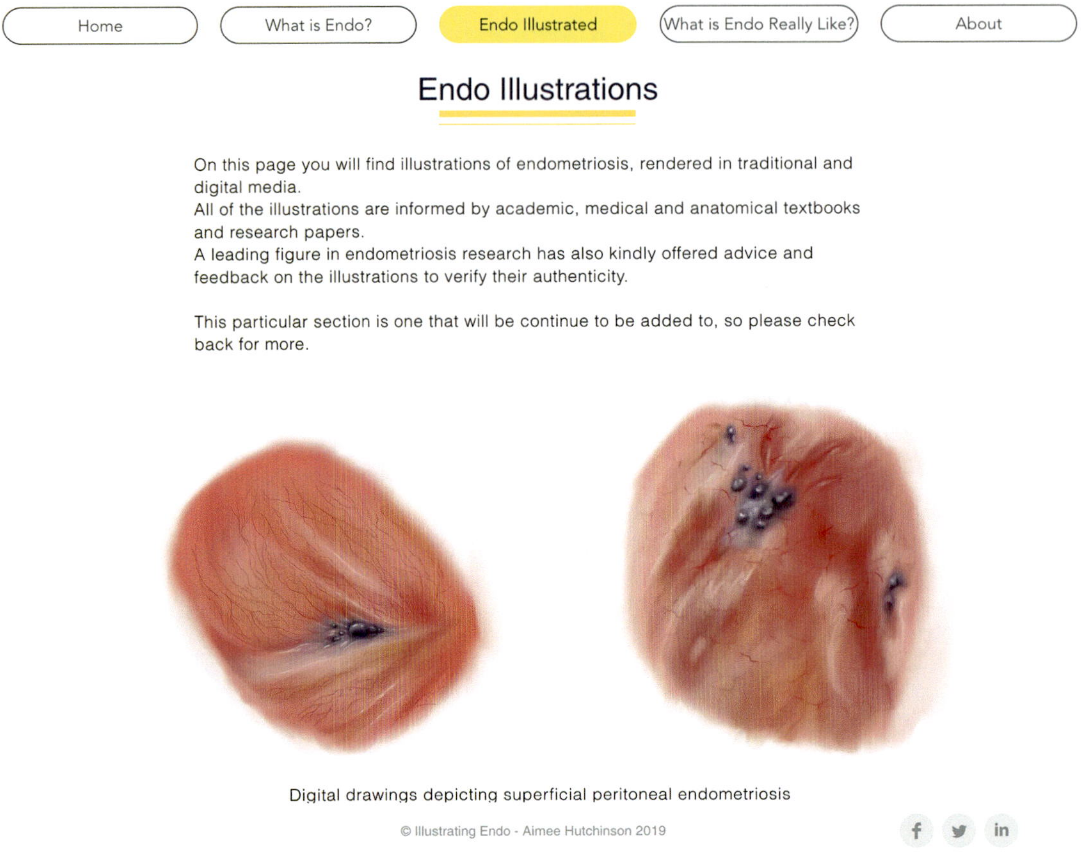

Fig. 2.34 Image of 'Endo Illustrated' from IllustratingEndo.com showing examples of 2D media depicting Endometriosis. Image by Aimée Hutchinson

educational resource is an absolute necessity, and the majority of participants said they had never encountered anything of its kind before. Whilst gynaecology and anatomy textbooks are readily available for academics, not everyone has access to these texts and so, creating something for the public realm would be extremely advantageous.

When speaking with Professor Horne and Endometriosis UK, they too, believed a visual resource was needed. Endometriosis UK has a number of information leaflets that tell the reader what signs and symptoms they should look out for, however, these are not available in all doctors' surgeries or hospitals. Working alongside patients and professionals to create an online visual resource available to all for free can be an effective way to educate and give a voice to those with endometriosis.

In the 3 years since the launch of the online resource, it is reassuring to note that there has been an influx of Internet activity and publicity surrounding the condition. Using 'hashtags' related to endometriosis on social media will bring you directly to those advocating for better healthcare. You will find personal stories from those living with the condition describing how they battle their daily pain and symptoms, and some users, having experienced the gap within this section of healthcare, have begun to take it upon themselves to create informative guides to aid and help others. The online resource created for this project seeks to provide an informed multi-media alternative to some of these patient-created resources.

2.3.1 Limitations and Future Considerations

Given the vast amount of information that should be illustrated, there are possibilities to expand the content of IllustratingEndo.com, however, the website that was created did not gain as much traffic as originally intended. As of 2023, the website has been taken offline. The researchers still maintain the domain name and are intent on producing further work and information to add to the website, and along with a revamp of the design; there is hope that a new online resource can be developed in the very near future.

It would have been beneficial to create a second questionnaire to gather feedback from the same set of participants following the creation of the illustrations and web resources, but as the initial questionnaire was completed anonymously, this would not have been feasible given the limited timeframe of this Master's research project.

The 3D-printed models and online resources were exhibited at the University of Dundee's Master's Show in August 2019. They were also presented at the annual National Endometriosis Network Meeting in Edinburgh in September 2019, and the Institute of Medical Illustrator's Annual Conference in Belfast in October 2019 with attendees comprising academics, researchers, and medical illustrators. Anecdotal feedback received on the project at these events was positive, however, more interest was shown towards the 3D digital and printed models than the online resource itself. This might suggest that there is an interest in developing more 3D digital models depicting the various states of endometriosis, and as a full-body disease showing how it can have a devastating impact on the body. To date, the two digital models pictured and discussed within this chapter have been purchased numerous times via the Sketchfab store by a wide range of individuals working for academic institutes and within a healthcare setting. This supports their success and highlights the limited number of models of this kind available.

2.4 Conclusion

The intended purpose of this project was to create an educational, visual resource which was inspired by the thoughts and feelings of those living with endometriosis and would be available as an educational resource by the public, those with the condition, and healthcare professionals. Comments from those who used the online resource and viewed the 3D printed models are positive, with many of the questionnaire participants expressing their gratitude for highlighting and raising further awareness for endometriosis. In the years that have elapsed since the creation of IllustratingEndo.com, there is still the utmost need for work of this nature. Until endometriosis loses its title of the 'invisible disease', medical visualisations of this kind will be needed and welcomed.

References

As-Sanie S, Black R, Giudice LC, Gray Valbrun T, Gupta J, Jones B, Laufer MR, Milspaw AT, Missmer SA, Norman A, Taylor RN, Wallace K, Williams Z, Yong PJ, Nebel RA (2019) Assessing research gaps and unmet needs in endometriosis. Am J Obstet Gynecol 221(2):86–94. https://doi.org/10.1016/J.AJOG.2019.02.033

Baker DW, Parker RM, Williams MV, Pitkin K, Parikh NS, Coates W, Imara M (1996) The health care experience of patients with low literacy. Archives Fam Med 5(6):329

Bailey MJ (2013) Fifty years of family planning: new evidence on the long-run effects of increasing access to contraception (No. w19493). Nat Bureau Econ Res 2013:341. https://doi.org/10.1353/eca.2013.0001

Bor ED, Oncul M, Atakul N, Yilmaz H, Tekirdag AI, Sahmay S (2013) Does the endometrioma effect the ovarian reserve? Fertil Steril 100(3):S161. https://doi.org/10.1016/j.fertnstert.2013.07.1526

Bullo S, Weckesser A (2021) Addressing challenges in Endometriosis pain communication between patients and doctors: the role of language. Frontiers in Global

Women's Health 2. https://doi.org/10.3389/fgwh.2021.764693
Carpan C (2003) Representations of Endometriosis in the popular press: "the career Woman's disease". Atlantis 27(2):32–40. Available at: http://journals.msvu.ca/index.php/atlantis/article/download/1317/1189/0
Christiano D, Biggers AMD (2017) What causes an enlarged uterus and how is it treated? Healthline. Available at: https://www.healthline.com/health/womens-health/enlarged-uterus. Accessed 5 June 2019
Corl FM, Garland MR, Fishman EK (2000) Role of computer technology in medical illustration. Am J Roentgenol 175(6):1519–1524. https://doi.org/10.2214/ajr.175.6.1751519
Davidson K (2005) Ndometriosis: information from the internet. J Consumer Health On the Internet 9:43–53. https://doi.org/10.1300/J381v09n03_05
De Graaff AA, D'hooghe TM, Dunselman GAJ, Dirksen CD, Hummelshoj L, WERF EndoCost Consortium, Simoens S, Bokor A, Brandes I, Brodszky V, Canis M (2013) The significant effect of endometriosis on physical, mental and social wellbeing: results from an international cross-sectional survey. Hum Reprod 28(10):2677–2685. https://doi.org/10.1093/humrep/det284
Ellis K, Munro D, Clarke J (2022) Endometriosis is undervalued: a call to action. Front Glob Womens Health 3:902371. https://doi.org/10.3389/fgwh.2022.902371
Endometriosis: diagnosis and management (2017) NICE guideline: short version
Endometriosis UK (2011) Endometriosis UK facts and figures, facts and figures. Available at: https://www.endometriosis-uk.org/endometriosis-facts-and-figures#5. Accessed 25 May 2019
Harrison S (2020) In a pandemic, medical illustrators made science accessible. Wired. Available at: https://www.wired.com/story/in-a-pandemic-medical-illustrators-made-science-accessible/
Hirsch M, Aggarwal S, Barker C, Davis CJ, Duffy JM (2017) Googling endometriosis: a systematic review of information available on the internet. Am J Obstet Gynecol 216(5):451–458. https://doi.org/10.1016/J.AJOG.2016.11.1007
Hobbs JE, Bortnick AR (1940) Endometriosis of the lungs: an experimental and clinical study. Am J Obstet Gynecol 40(5):832–843
Horne AP, Pearson C (2018) Endometriosis: the experts' guide to treat, manage and live well with your symptoms. Vermilion
Horne AW, Saunders PT, Abokhrais IM, Hogg L (2017) Top ten endometriosis research priorities in the UK and Ireland. Lancet 389(10085):2191–2192. https://doi.org/10.1016/S0140-6736(17)31344-2
Howard C (2018) Scientific models: practical uses of 3D printing in science classes, British Columbia science teachers' association. Available at: https://bcscta.com/2018/05/30/3d-models-in-science-classes/. Accessed 31 May 2019
Hudelist G, Keckstein J, Wright JT (2009) The migrating adenomyoma: past views on the etiology of adenomyosis and endometriosis. Fertil Steril 92(5):1536–1543. https://doi.org/10.1016/j.fertnstert.2008.08.086
Hustvedt A (2011) Catalepsy: statues and automatons. In: Medical muses, pp 63–71
Jayaprakasan K, Becker C, Mittal M (2018) The effect of surgery for endometriomas on fertility: scientific impact paper no. 55. BJOG-AN Int J Obstetrics Gynaecol 125(6):E19–E28. https://doi.org/10.1111/1471-0528.14834
Kennedy S, Bergqvist A, Chapron C, D'Hooghe T, Dunselman G, Greb R, Hummelshoj L, Prentice A, Saridogan E (2005) ESHRE guideline for the diagnosis and treatment of endometriosis. Hum Reprod 20(10):2698–2704. https://doi.org/10.1093/humrep/dei135
Kumar DR, Aslinia F, Yale SH, Mazza JJ (2011) Jean-Martin Charcot: the father of neurology. Clin Med Res 9(1):46–49
Kvaskoff M (n.d.) Endometriosis and co-morbidities. endometriosis.org
Mann MD (n.d.) A system of gynecology, vol 2. Lea Brothers, Philadelphia, pp 1887–1888
Mcculloch TAH (1969) Theories of hysteria. Can Psychiatr Assoc J 14(6):635–637. https://doi.org/10.1177/070674376901400614
Meuleman C, Vandenabeele B, Fieuws S, Spiessens C, Timmerman D, D'Hooghe T (2009) High prevalence of endometriosis in infertile women with normal ovulation and normospermic partners. Fertil Steril 92(1):68–74. https://doi.org/10.1016/j.fertnstert.2008.04.056
Nezhat C, Nezhat F, Nezhat CH (2020) Endometriosis: ancient disease, ancient treatments. Springer International Publishing, pp 13–127
Osborne H (2006) Health literacy: how visuals can help tell the healthcare story. J Visual Comm Med 29(1):28–32. https://doi.org/10.1080/01405110600772830
Oxner E (2018) Chapter 14–a picture is worth a thousand words: the benefit of medical illustration in medical publishing. In: Markovac J, Kleinman M, Englesbe M, B. T.-M. and S. P (eds) Medical and scientific publishing. Academic Press, pp 131–136. https://doi.org/10.1016/B978-0-12-809969-8.00014-0
Schliep KC, Chen Z, Stanford JB, Xie Y, Mumford SL, Hammoud AO, Boiman Johnstone E, Dorais JK, Varner MW, Buck Louis GM, Peterson CM (2017) Endometriosis diagnosis and staging by operating surgeon and expert review using multiple diagnostic tools: an inter-rater agreement study. BJOG Int J Obstet Gynaecol 124(2):220–229. https://doi.org/10.1111/1471-0528.13711
Simoens S, Dunselman G, Dirksen C, Hummelshoj L, Bokor A, Brandes I, Brodszky V, Canis M, Colombo GL, DeLeire T, Falcone T (2012) The burden of endometriosis: costs and quality of life of women with endometriosis and treated in referral centres. Hum Reprod 27(5):1292–1299. https://doi.org/10.1093/humrep/des073

Singer PN (2021) Galen. The Stanford encyclopedia of philosophy. https://plato.stanford.edu/archives/win2021/entries/galen/

Smith CF, Tollemache N, Covill D, Johnston M (2018) Take away body parts! An investigation into the use of 3D-printed anatomical models in undergraduate anatomy education. Anat Sci Educ 11(1):44–53. https://doi.org/10.1002/ase.1718

Tasca C, Rapetti M, Carta MG, Fadda B (2012) Women and hysteria in the history of mental health. Clin Pract Epidemiol Ment Health 8:110. https://doi.org/10.2174/1745017901208010110

Tong A, Levey AS, Eckardt K-U, Anumudu S, Arce CM, Baumgart A, Dunn L, Gutman T, Harris T, Lightstone L, Scholes-Robertson N, Shen JI, Wheeler DC, White DM, Wilkie M, Craig JC, Jadoul M, Winkelmayer WC (2020) Patient and caregiver perspectives on terms used to describe kidney health. Clin J Am Soc Nephrol 15(7) https://journals.lww.com/cjasn/Fulltext/2020/07000/Patient_and_Caregiver_Perspectives_on_Terms_Used.9.aspx:937

Topac V, Stoicu-Tivadar (2013) Patient empowerment by increasing the understanding of medical language for lay users. Methods Inf Med 52(05):454–462. https://doi.org/10.3414/ME12-02-0006

Tsamantioti ES, Mahdy H (2023) Endometriosis. https://www.ncbi.nlm.nih.gov/books/NBK567777/

Tsafrir J, Ohry A (2001) Medical illustration: from caves to cyberspace‡. Health Info Libr J 18(2):99–109. https://doi.org/10.1046/j.1471-1842.2001.d01-16.x

Veith I (1993) Hysteria: the history of a disease. Jason Aronson (The Master work series). Available at: https://books.google.co.uk/books?id=n9ISAQAAMAAJ

Weisberg E, Fraser IS (2015) Contraception and endometriosis: challenges, efficacy, and therapeutic importance. Open Access J Contracept 6:105–115. https://doi.org/10.2147/OAJC.S56400

Editorial Medical and Scientific Illustration

3

Jennifer E. Fairman and Tim Phelps

Abstract

Editorial illustration has become a widely utilized form of medical illustration with a rich history of trailblazers whose art paved the way for editorial visuals to become what they are today. This chapter is intended to cover how editorial illustration has evolved, is used, and continues to be a genre of visual communication in science and medicine that is highly visible to both consumer and trade-based audiences alike.

Keywords

Editorial illustration · Medical illustration · Storytelling · Conceptual illustration · Visual communication · Cover illustration

3.1 Introduction

In the field of biomedical visualization, editorial work is commonly defined as, "illustrations that summarize the content of a publication or article by calling attention to the subject matter of the same, either as cover or inside art" (Association of Medical Illustrators 2022). Cover art, whether for a journal, magazine, or book, is often the most recognizable and effective use of editorial illustration. Editorial illustrations span a variety of uses in consumer-facing publications such as those for health, fitness, science, and nature magazines or newspapers, intended for a public/lay audience. Many of these publications, including Popular Science, National Geographic, and Scientific American, will employ editorial illustrations as visual metaphors to capture the essence of a featured article (Goldberg 2021). Similarly, such illustrations are also often used in medical or science academic journals and magazines that are intended for a physician, basic science researcher, or health provider audience. Other professional trade publications (e.g., Contemporary Urology and American Family Physician) are peer-reviewed and use editorial illustrations that are more literal in style. However, many scientific journals, like Nature, Cell, Science, and PNAS, have recently featured less realistic and more fantastical styles on their covers.

In this chapter, we focus on the use of editorial and conceptual illustrations in biomedical visualization, the genres of editorial illustration, and describe the creative process supported by case studies. As Medical Illustrators, we both teach and practice editorial illustration in the Medical and Biological Illustration Graduate Program at Johns Hopkins University School of Medicine,

Supplementary Information: The online version contains supplementary material available at https://doi.org/10.1007/978-3-031-41652-1_3.

J. E. Fairman (✉) · T. Phelps
Art as Applied to Medicine, Johns Hopkins University School of Medicine, Baltimore, MD, USA
e-mail: fairman@jhmi.edu; tphelps1@jhmi.edu

M. Roughley (ed.), *Approaches for Science Illustration and Communication*, Biomedical Visualization 4,
https://doi.org/10.1007/978-3-031-41652-1_3

USA. In the following sections, we discuss the process of creating various types and styles of editorial illustrations, primarily within the literal, metaphorical, and fantastical genres with many nuanced combinations in between. Finally, we will walk through the practices of formulating ideas with research and thumbnailing, and the iterative process of idea formulation and storytelling. First, various examples and case studies of historic and modern artists are what follow.

3.1.1 Telling a Well-Thought, Engaging Story

Illustration serves as a universal language that can be understood by people from different cultures and backgrounds, making them a powerful tool for communication. They can convey information in a way that words alone cannot and help to make complex ideas or concepts more accessible and understandable. This is especially true in medical illustration.

Editorial illustration refers to the use of conceptual illustrations in magazines, newspapers, and other forms of print and/or digital media to support and enhance written content (Association of Illustrators 2022). These illustrations can take wide-ranging forms of drawings, highly rendered digital paintings, cartoons, or montages, and are used to visually explain, reinforce, or comment on the accompanying text. The purpose of editorial illustration is to provide a visual representation of the ideas and concepts being discussed, making the content more engaging and accessible to readers.

As academics at the Johns Hopkins University School of Medicine, we teach editorial and conceptual illustration at the graduate level to students who are training to become medical illustrators. Their education in this area includes many facets beginning with learning the history and evolution of the field, as well as the various techniques used in creating illustrations for print and digital media. The students also study visual storytelling and learn how to create images that effectively convey information and ideas. In addition to traditional illustration techniques such as pencil drawing, watercolor, and oil painting, students also learn about digital illustration tools and techniques, such as using vector and raster graphics as well as 3D software to create images. They discover how to effectively use color, composition, and typography to create illustrations that are visually appealing and engaging. Lastly, students learn about the business side of illustration, including how to market their work, how to negotiate contracts, and how to build a professional portfolio. They can have opportunities to work on real-world projects and receive feedback from content experts. The goal is to provide students with the skills and knowledge they need to succeed as professional illustrators and to create illustrations that effectively communicate information and ideas in print and digital media.

3.1.2 Early Influencers

Editorial illustration has a long and rich history, with roots dating back to the earliest forms of printed media. During the Middle Ages, illustrations were used in books and manuscripts to provide visual representations of religious stories and historical events (Norman Rockwell Museum 2023; Department of Medieval Art and The Cloisters 2001). In the seventeenth and eighteenth centuries, political cartoons, and illustrations became popular in newspapers and magazines as a means of commenting on current events and societal issues (First Amendment Museum 2023). With the advent of mass printing in the nineteenth century, illustrations became an integral part of printed media, used to tell stories, explain complex ideas, and entertain readers (Ives 2004).

In the twentieth century, the rise of illustrated magazines and newspapers, such as *The New Yorker* and *The Saturday Evening Post*, led to a golden age of editorial illustration. During this period, many of the most renowned artists of the time, such as Norman Rockwell, J.C. Leyendecker, N.C. Wyeth, James Montgomery Flagg, Howard Pyle, and many others, produced work for these publications

(Levin 1980; Normal Rockwell Museum 2023). In the decades that followed, the rise of mass media and digital technology led to changes in the production and use of editorial illustrations. Today, editorial illustration continues to play an important role in print and online media, serving as a visual complement to written content.

Within the scientific community, the visual display of statistical information had a rich history within the editorial illustration landscape. Data visualization has seen a growing need for the layperson to understand data typically presented in scientific journals, such as graphs and charts, in a more accessible and visually appealing format. Edward Tufte is one of many influential contributors to changing the way our field has presented this type of information in an engaging way for the lay public. His work and subsequent publications provided a foundation of principles that are often used in today's data visualization and its use in scientific editorial visuals (Greenbaum Kasson 2016).

Before data visualization became the recognized method to creatively interpret and represent scientific information, Friedrich "Fritz" Kahn (1888–1968) was creating whimsical and surreal ways of metaphorically depicting the human body and many of its physiological processes (von Debschitz and von Debschitz 2017). Kahn was a German gynecologist and medical writer, known for his innovative illustrations that aimed to explain complex medical concepts in a simple and engaging way. Little known as somewhat of a father of the infographics movement, his illustrations combined elements of scientific accuracy with humor and imaginative storytelling, making them popular in both scientific and popular culture. His works were influential in shaping the field of medical and editorial illustration, especially in the early twentieth century. One of his most famous works includes "Man as Industrial Palace" (1926) as seen in Fig. 3.1, which visualizes the human body as a factory busily managed by numerous homunculi of varying disciplines. Another of his most well-known works is "The Living Machine" (1927), which depicts the human body as a complex machine composed of recognizable everyday objects of the industry representing various organs and systems. These illustrations are still widely admired today in medical and educational contexts.

Kahn was a master of metaphor—using the imagery of engineering and industry as visual analogies (e.g., the bellows of a camera for the eyes, copper pipes representing vasculature, books inside of the brain for thinking and ideas, a switchboard with telephone operators directing signals of the nervous system up and down the spine, and a pipe organ for illustrating the production of sound coming from the vocal cords) for the systems he wanted to convey to his viewer. Kahn might be seen as a pioneer in what is now termed *information visualization* and a master of graphic design.

3.2 Modern Editorial Illustration

Today, editorial illustration is widely used in the medical world to communicate complex medical concepts like innovations in surgery, cellular processes, or discovery of molecular mechanisms in an engaging way (May 2020, see Fig. 3.2). For example, editorial illustrations can be used to explain the spread of diseases or to show the effects of medical treatments. Illustrations can also be used in medical textbooks, scientific journals, and patient education materials to help clarify complex medical information and make it more accessible to a wider audience.

Editorial illustrations can also be used to convey information about health and wellness, such as the benefits of exercise and healthy eating, the dangers of smoking, or bring an awareness to a public health crisis like Human African Trypanosomiasis (HAT). Figure 3.3 shows an illustration created for inclusion within the report on Neglected Tropical Diseases (NTDs), disseminated Médecins Sans Frontières (MSF)/ Doctors Without Borders. Human African Trypanosomiasis (HAT, aka Sleeping Sickness) is one of six topics illustrated for inclusion in the report. Tsetse fly is the vector, typical environment is near water and dense vegetation (i.e., Congo). The afflicted often cannot sleep at

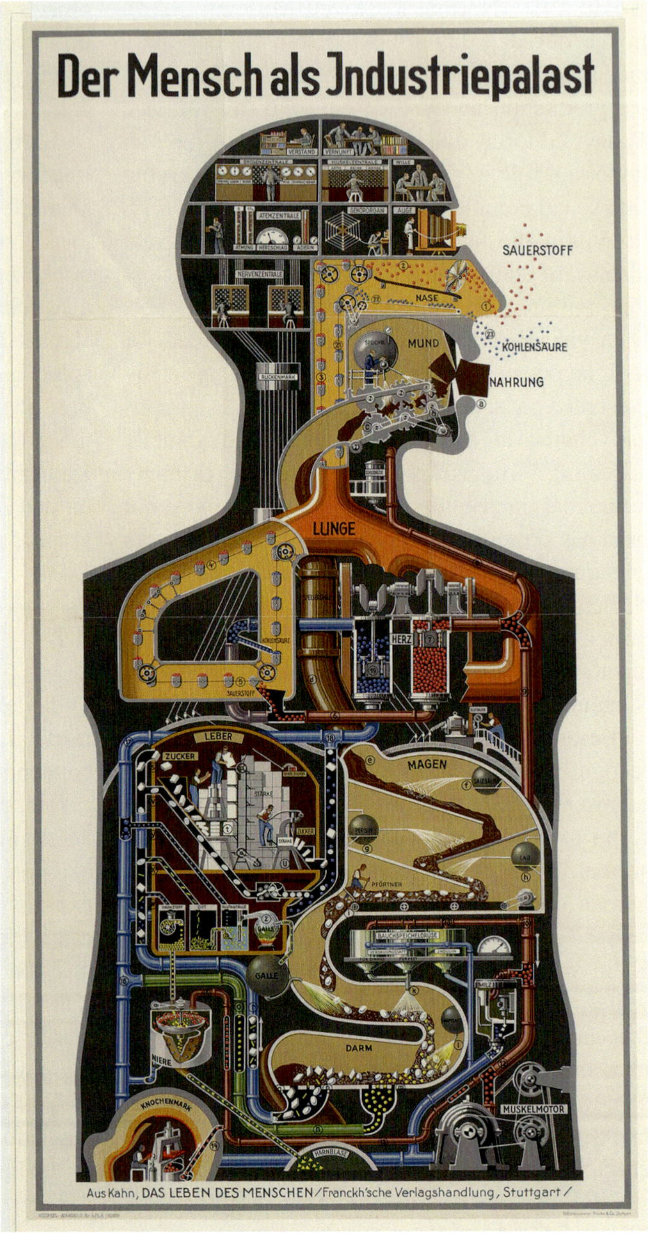

Fig. 3.1 *Der Mensch als Industriepalast (Man as Industrial Palace),* Fritz Kahn, 1926, public domain, Credit: The William H. Helfand Collection, 2010, Philadelphia Museum of Art

night, but then sleep during day wherever they are. The illustrations are used for patient education, to raise public awareness of NTDs, and to raise funding for research and treatments.

Editorial illustrations can be a valuable tool in health promotion and disease prevention. They can be created using a variety of techniques including hand-drawn sketches, digital 2D vector or raster techniques, and 3D modeling. Regardless of the technique used, the goal of editorial illustration in the medical world is to provide a visual representation of medical information that is accurate, engaging, and accessible to a wide audience (Errea 2017; Rengen 2020).

In addition, the use of animation can make a big impact on bringing a story to life, adding a dynamic and engaging element to a visual. Animated illustrations can be used to demonstrate a

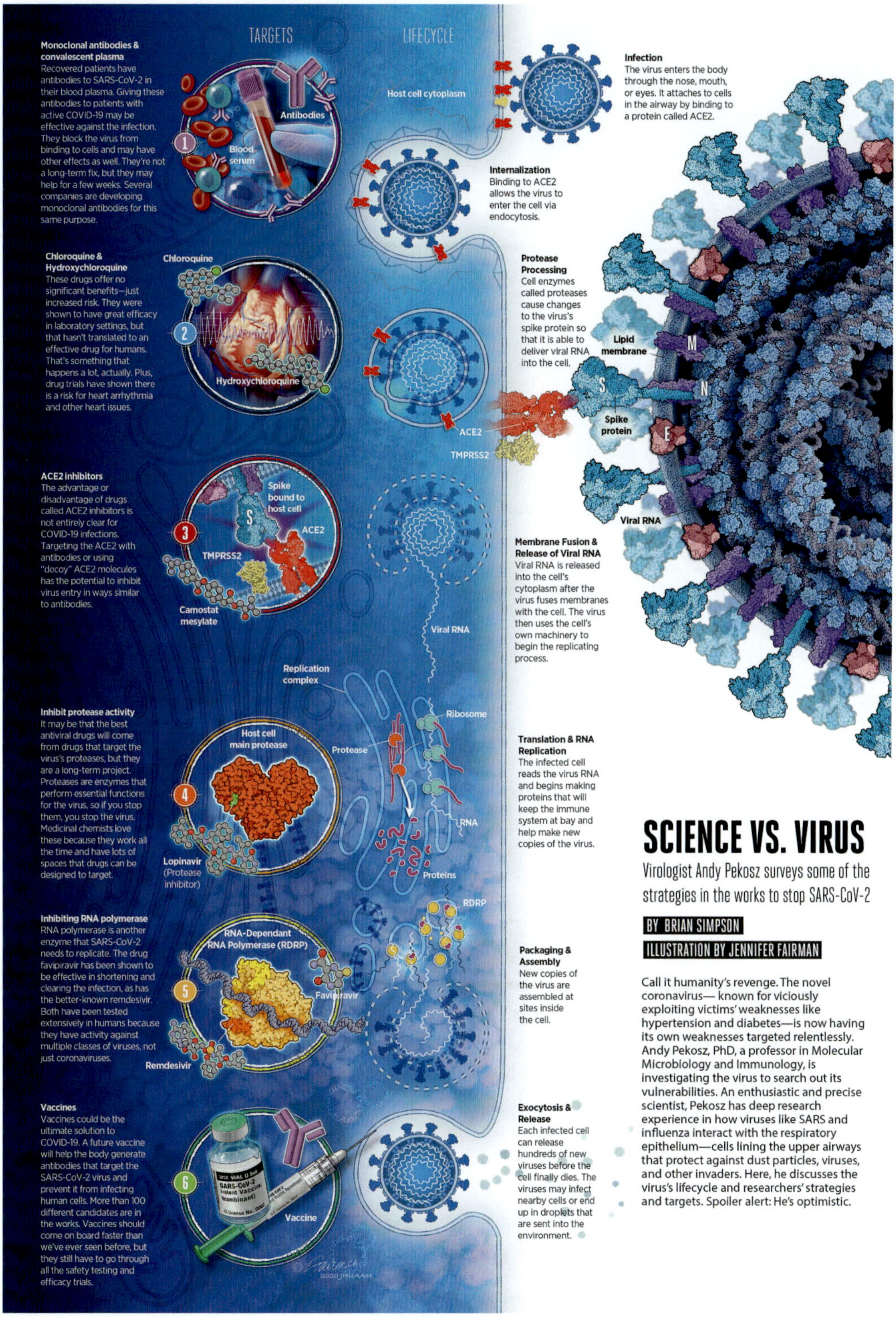

Fig. 3.2 *Science* Vs. *Virus*, Bloomberg School of Public Health Magazine, Jennifer E. Fairman 2020, © Johns Hopkins University

Fig. 3.3 *Human African Trypanosomiasis (HAT),* with permission from Todd Buck, © 2021, Todd Buck

process, highlight a change over time, or simply bring a static illustration to life, encouraging a viewer to click and learn more. Animated GIFs are frequently used as editorial illustrations because they provide a simple and engaging way to convey information and ideas in a more dynamic and interactive visual format than a still image.

Animated GIFs are also commonly used to add humor, personality or alleviate tension to editorial illustrations. Such usage can illustrate a light-hearted story or article, bringing a touch of levity to the content. These simple web animations can be created using a variety of tools, including image editing software and online tools, and can be optimized for different platforms and devices. They are also relatively small in file size, making them easy to share and view online (see Figs. 3.4 and 3.5 for example).

Unlike simple diagrams, editorial illustrations also can evoke emotions and create a connection with the viewer. This can be especially powerful in editorial illustration, where the goal is to engage the viewer and to convey complex information in a way that is memorable and impactful, encouraging them to learn more about a particular subject. With these advantages, editorial illustration will complement the words of a feature article, making the information more rich and complete. The use of symbolism, personification, palette and style, and space, are techniques incorporated into editorial illustration to aid in visual storytelling.

3.2.1 Use of Symbolism

Widely recognized symbolic imagery throughout society is often used to convey ideas or emotions in a way that is more powerful and memorable than words. Figure 3.6 illustrates in an eye-catching manner, the major molecular components involved in cellular balance and stability portrayed by intertwined ribosomes,

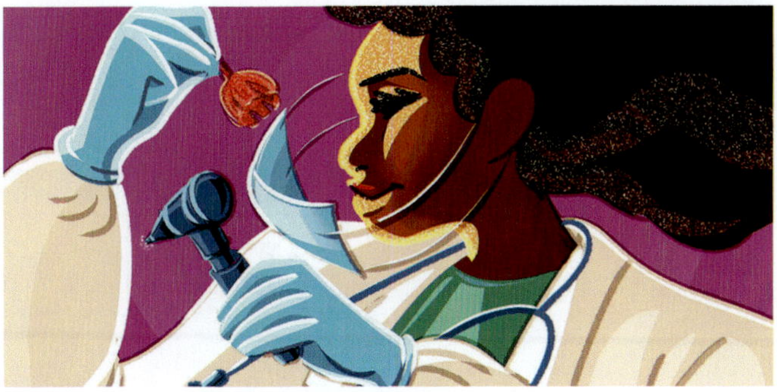

Fig. 3.4 *How to Wear Glitter to the Operating Room*, Animated GIF, with permission from April Brust and Doximity, © 2019, Doximity. https://opmed.doximity.com/articles/how-to-wear-glitter-to-the-operating-room

tRNAs, protein chains, mRNA strands, and alpha helix ribbon polypeptides involved in the quality control of gene expression. A mandala, often recognized as a symbol of balance, can be used as a visual metaphor to represent balance within a cellular system. Here, the mandala is a visual metaphor representing balance to the cell provided by the quality control of gene expression.

The illustration in Fig. 3.7 metaphorically uses an analogy of calm vs. stormy waters to depict the difference between biochemical wave activity exhibited on normal and cancer epithelial cell surfaces. Signal transduction waves propagate as coordinated waves (like rolling ocean surf) on the cell cortex and delineate molecular feedback causing turbulence and excitability. As normal cells transform into cancer cells, they display more waves on their surfaces suggesting cancer signal transduction activity is increased in metastatic cancer cells. An analogy juxtaposes calmer basal waves on the left, and turbulent cancerous waves on the right. A splashing crab is shown as an additional symbolic nod to cancer—the illustration of a crab on a tumultuous seashore can symbolize cancer and the unpredictable nature of disease progression. These added touches are considered "easter eggs" that add to the meaning behind the visual message.

3.2.2 Personification

Personification is the incorporation of attribution of a personal nature or human characteristics to something nonhuman, or the representation of an abstract quality in human form. For example, a heart, though living tissue, can be transformed

Fig. 3.5 *The Distinction Between Patient and Customer*, Animated GIF, with permission from April Brust and Doximity, © 2022, Doximity. https://opmed.doximity.com/articles/the-distinction-between-patient-and-customer (For the animated GIF, please see supplementary file 1)

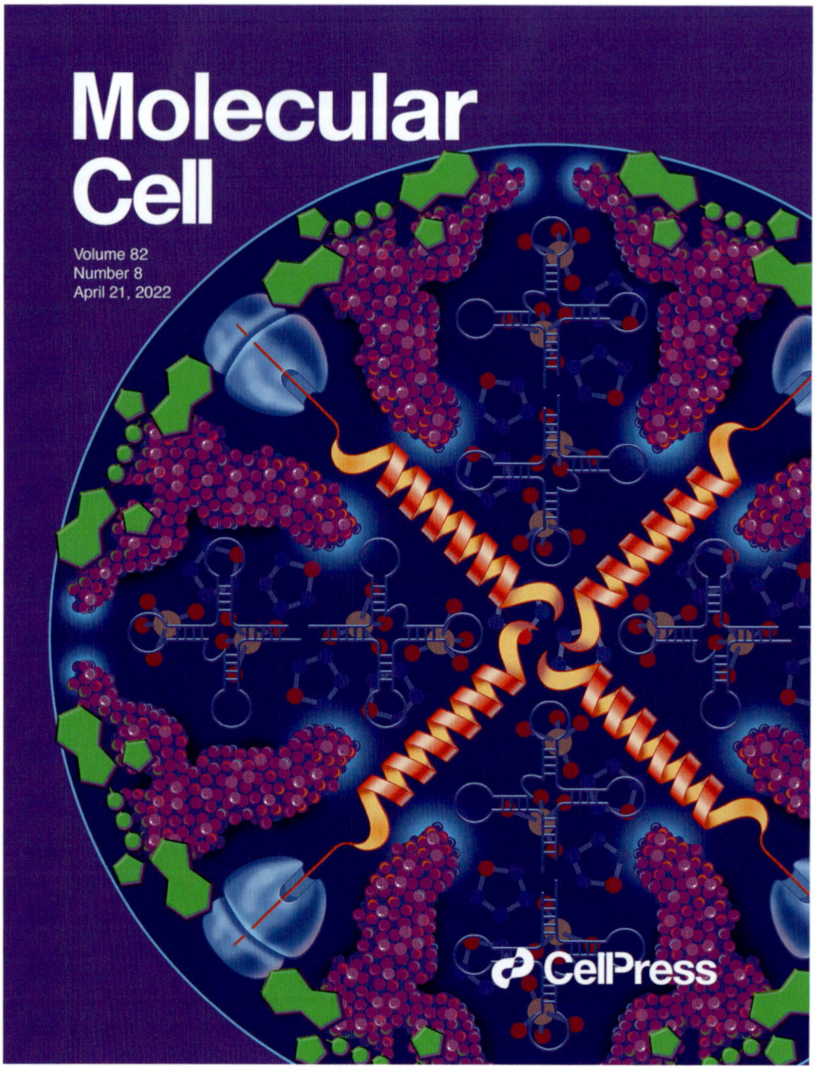

Fig. 3.6 *Quality control of gene expression provides balance to the cell,* Tim Phelps, © 2022, Johns Hopkins University

into a character with deep emotional stress as seen in Gilbert Chen's *Let your Heart Rest* cover concept (Fig. 3.8). Here the heart lies restless and awake. Chen conveys to his viewer that while they may sleep at night, their heart is affected by too much exposure to light, thereby affecting cardiometabolic function while sleeping.

In Fig. 3.9, Emily Cheng uses personification in *Midnight Mechanic: Immune cells rewire and repair brain cells while we sleep* (see Fig. 3.9). In this image, Cheng created a large anamorphic immune cell wearing a conspicuous hard hat and is shown intently working at a workbench with a variety of tools as it repairs a mechanical brain late at night.

3.2.3 Palette and Style

The use of color can be instructional, evoke emotions and create a visual hierarchy. For example, using bright or contrasting colors to highlight important elements in an illustration can draw the viewer's attention to those elements (Figs. 3.10 and 3.11). With the advent of digital media, many styles can be incorporated into a conceptual piece which can range from the hyperrealistic to more

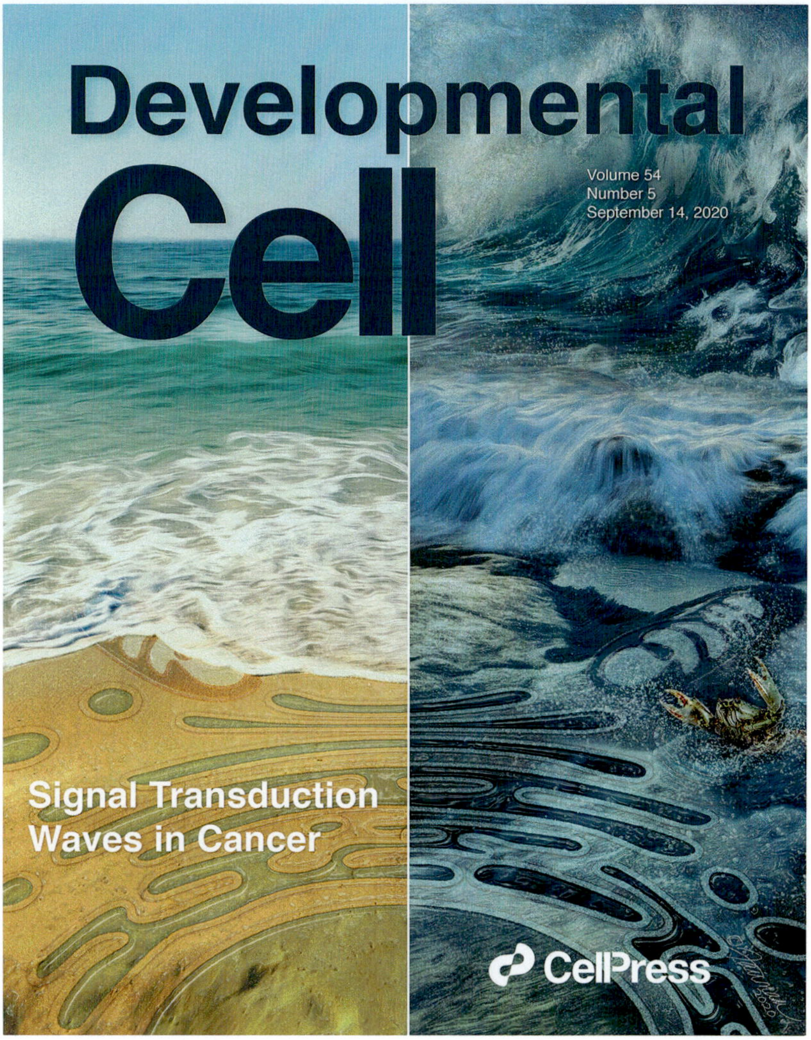

Fig. 3.7 *Signal transduction waves in cancer,* Jennifer E. Fairman, © 2021, Johns Hopkins University

cartoon-like or caricature style. Images can possess a flat info-graphic motif (see Fig. 3.12) or a more robust and three-dimensional quality (Aparaschivei 2022).

3.2.4 Space

The use of spatial imagery can create an immersive experience for the viewer. For example, an editorial illustration of a cityscape can transport the viewer to a new place and create a sense of atmosphere, space, or mood. An *enclosed* space can give the viewer a sense of discomfort or *claustrophobia,* as shown in Fig. 3.13.

3.3 Editorial Illustration Genres

Fritz Kahn's use of the visual metaphor is but one of the many genres that can be used in editorial illustration. Three of the most used categories in medical illustration are *realistic/literal*,

Fig. 3.8 *Let your heart rest*, with permission from Gilbert Chen, © 2022, Gilbert Chen

metaphorical/analogies, and *fantasy/surrealism*. Each of these visuals requires a thorough understanding of the underlying theme of the manuscript being illustrated.

The first is a *literal summary approach* that requires a complete understanding of the scientific subject matter. The editorial illustration will need to contain accurate depictions of the subject matter in correct proportions and descriptive visual textures and/or accepted scientific field images and icons. The second approach is through *metaphoric visual analogies and symbolism* that are used to tell a story. It is important to select appropriate metaphors that relate directly to and are universally understood by your intended audience. Slight deviations can be considered and depicted by employing visual puns, plays on words and object context twists such as portraying a subject out of its familiar context and placing it into a new setting making the strange familiar and the familiar strange. The third category places the subject into *a fantasy or pop culture* setting. A mood is captured through a vivid use of color and shape. The images may appear surrealistic, ethereal, or atmospheric to encourage a desired emotion or gut reaction.

Fig. 3.9 *Midnight Mechanic: Immune cells rewire and repair brain cells while we sleep*, with permission from Emily Cheng, © 2020, Emily Cheng

3.3.1 Realistic/Literal Scientific Summary Genre

After a full understanding of the science and processes within a given topic, this genre requires the illustrator to tell a clear story that uses realism. Realism and accuracy are necessary for surgical, anatomical, cellular, and molecular imagery to gain the trust of an audience. Spatial relationships, textural details, and the relationship of objects to one another need to be believable for them to be effective if the illustrator chooses this genre. The use of specimen dissections, datasets, X-ray crystallography, and medical imaging are a few examples of tools used to create images that are as close to reality as possible, and the outcomes of a realistic scientific summary approach are shown in Fig. 3.14.

Even at the cellular level, an illustrator can tell a vivid story of what one might envision is happening at the microscopic level in a more realistic sense. A good example is David Rini's *Antigen driven proliferation of HIV-infected T-cells* cover for the *Journal of Clinical Investigation* (Fig. 3.15). Here, he uses two sweeping cascades of viral particles and immune system cells to depict the proliferation of antigen-responsive HIV-infected T cells isolated from HIV-infected individuals on antiretroviral therapy. Antigen-driven clonal selection influences the persistence

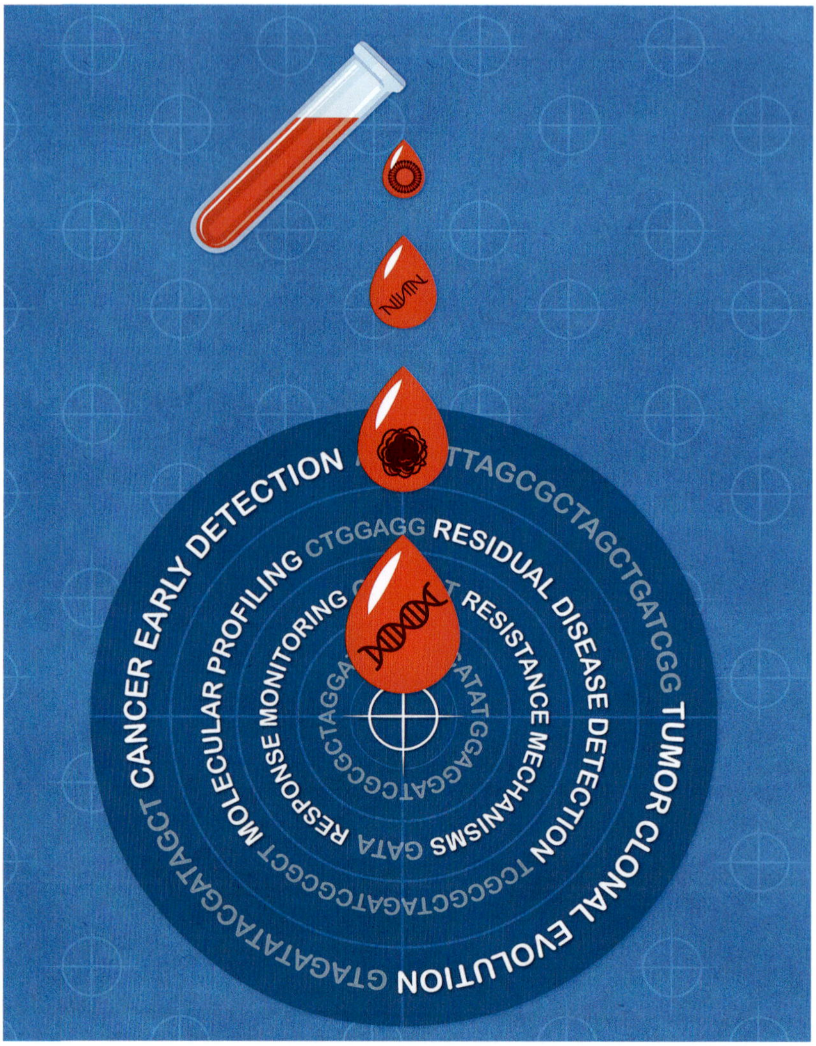

Fig. 3.10 *Targeted immunotherapy,* Jennifer E. Fairman, © 2023, Johns Hopkins University

of HIV-1–infected CD4+ T cells. The clonal expansion of infected CD4+ T cells is a major mechanism of HIV-1 persistence and a barrier to achieving a cure. His image depicts the potential role of recurrent antigenic exposure in maintaining the HIV-1 reservoir. Proviral populations in cytomegalovirus (CMV) antigen-responding cells are shown in large clones, including some harboring replication-competent proviruses. Infected cells under antigenic stimulation are depicted as proliferating. Analysis in the featured article shows that infection can occur early or late during a clone's response and can generate infected cell populations too large to be explained solely by homeostatic proliferation. These findings implicate antigen-driven clonal selection as a major factor in HIV-1 persistence, a finding that may prove a difficult challenge to eradication efforts.

3.3.2 Conceptual/Design Style

A more conceptual approach allows an illustrator to depict real quantitative data using a more artistic style as seen in Valeria Altounian's *Synthetic Chromosomes* cover image for the March 2017 cover of Science Magazine (Fig. 3.16). The image

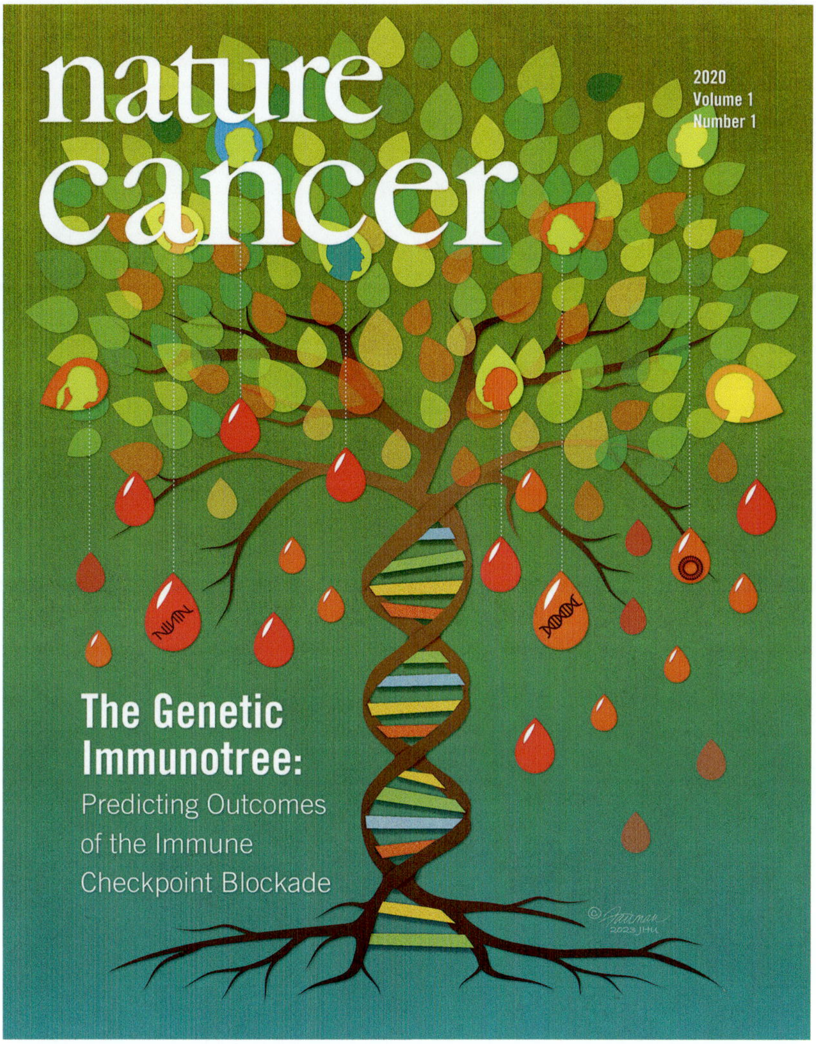

Fig. 3.11 *The genetic immunotree,* Jennifer E. Fairman, © 2023, Johns Hopkins University

is an illustrative data visualization of a yeast cell genome containing all the chromosomes currently synthesizable in a lab. The goal of this cover was to illustrate the progress made toward a fully synthetic yeast genome. The author used chromosomal contact data in the nucleus to reconstruct their conformation in 3D space to compare the synthetic versions against their wild-type counterparts. Wild-type chromosomes fade into the background while gold calls attention to the 6.5 synthetic arms. Chromatin fiber is visible through one end of each synthetic arm. By conceptualizing the chromosome structures, her approach simplifies the complexities of science while connecting with the audience using engaging imagery. As Jay Perlman describes, "conceptual illustration is a way of visually communicating an idea in a concise but thought-provoking way . . .The end result of a quality conceptual illustration is an image that encourages the user to find a deeper meaning" (Perlman 2022).

Another great example can be found in Satyen Tripathi's *Real. Genuine. Heart* (Fig. 3.17). This piece commemorates a series of philanthropic gifts from Genuine Parts Company (GPC) to the

Fig. 3.12 *Filling the gaps: Closing in on the human genome, Science Magazine,* Valerie Altounian, © 2022, American Association for the Advancement of Science. From Zahn, L.M., 2022. *Filling the gaps.* Science, 376 (6588), pp. 42–43. Reprinted with permission from AAAS

Fig. 3.13 *The Dark Secret Behind New Hospital Designs*, with permission from April Brust and Doximity, © 2020, Doximity

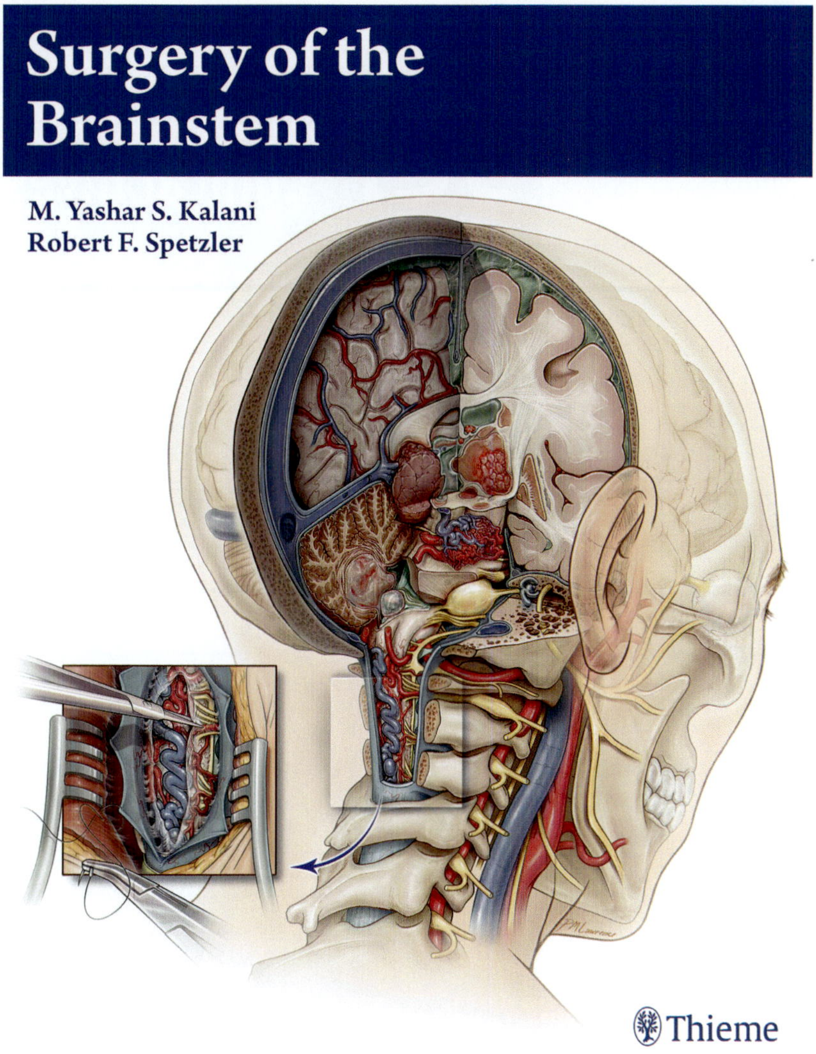

Fig. 3.14 *Surgery of the Brainstem, Controversies in Skull Base Surgery.* With permission from Peter Lawrence, © 2017, Barrow Neurological Institute, Department of Neurosurgery

Carlyle Fraser Heart Center of Emory University. The illustration depicts a heart composed of automotive parts and machinery in a nod to the donor, GPC. GPC is an organization engaged in the distribution of automotive replacement parts and materials. Their gifts were used to serve the mission of the Heart Center. Additionally, the title was chosen to reference both the spirit of the gifts and the name of the donor, Genuine Parts. This clean graphic style allows for the final presentation to be letterpress printed to be displayed in the corporate offices of the donor.

3.3.3 Metaphoric Genre

This genre uses common/everyday cultural ideas and analogies to help the audience 'get' what the creator is trying to convey. Illustrators can use the same conventions of realism as the literal or scientific genre, but a lot of artistic license and creativity can be used to visually tell a story. Visual metaphors are a powerful tool in illustration that can be used to convey complex ideas and concepts in a simple and memorable way. A visual metaphor is a comparison between two

Fig. 3.15 *Antigen-driven proliferation of HIV-infected T-cells*, Journal of Clinical Investigation, David Rini, © 2021, Johns Hopkins University

unlike things that are used to explain or describe one of the things in a creative and imaginative way.

For example, in the case of *Snip* Vs. *Shred* (Fig. 3.18), Jennifer Fairman uses the metaphor of a scissor and a shredder to depict two CRISPR-Cas systems, the very particular scissor-snipping Cas9 and imprecise shredder-like Cascade/Cas3 complexes, highlighting differences between their structure and function. Both complexes are genetically engineered tools based on one of several naturally occurring bacterial immune systems that fight bacteriophages by targeting viral DNA sequences. The spots on the cow in the sidebar also utilize the molecular structures as an added element to the visual story.

Visual metaphors are created by combining elements from different domains and using them to represent a new idea. For instance, an illustration of a sprouting flower bulb in the form of a heart can be used to represent an idea such as growth or healing (Fig. 3.19), and a dart board can be used to represent a pin-point target of accuracy (Fig. 3.20). Emily Taylor, who created *Clinical Trials of Cancer Drugs Fail Due to Missed Target Protein* says this image was created to reveal that, “the reason for some cancer drugs failing during clinical trials was that they

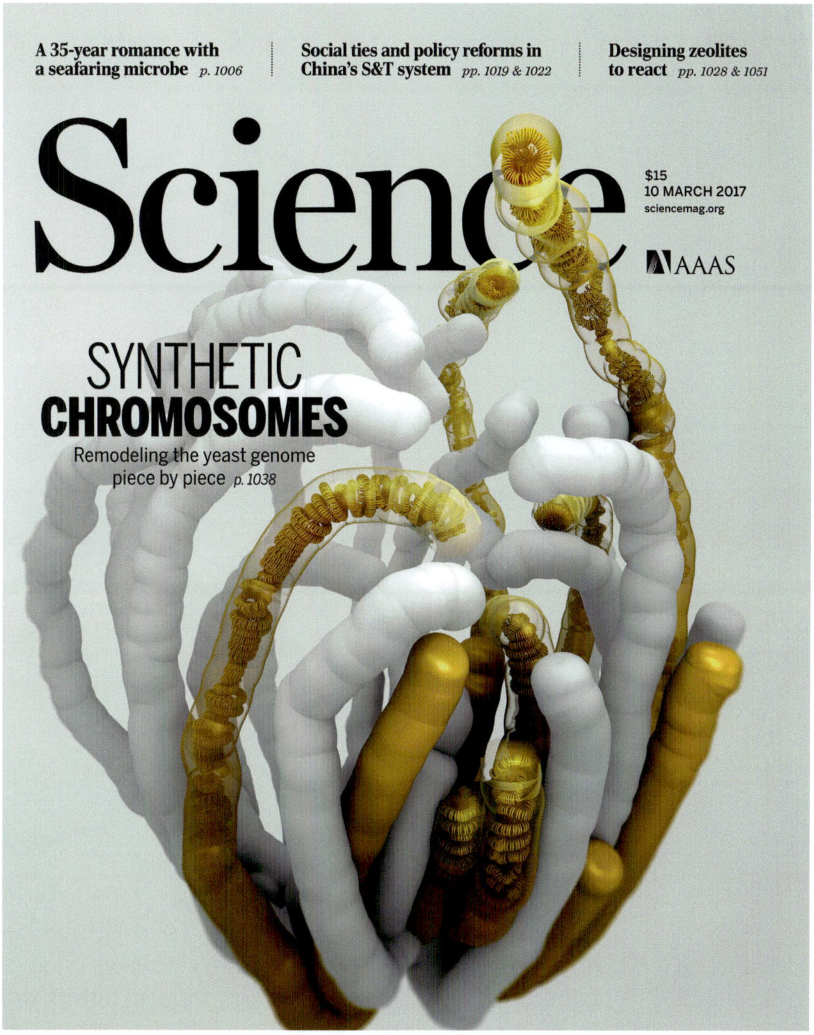

Fig. 3.16 *Synthetic Chromosomes: Remodeling the yeast genome piece by piece, Science Magazine,* Valerie Altounian, 2017, © American Association for the Advancement of Science. From Zhang, W., Zhao, G., Luo, Z., Lin, Y., Wang, L., Guo, Y., Wang, A., Jiang, S., Jiang, Q., Gong, J. and Wang, Y., 2017. Engineering the ribosomal DNA in a megabase synthetic chromosome. Science, 355(6329), p.eaaf3981. Reprinted with permission from AAAS

were not targeting the protein that the researchers had intended to target. As well, some drugs that had worked in clinical trials were still found to target a different protein than intended. This piece serves as a visual metaphor for the study… The piece uses a "cellular dart board" to represent the cancer cell being targeted. The target drug is the bulls-eye, however I chose to represent the case where the drug is still effective despite hitting the wrong protein, in this case, I put that protein in the triple-20 spot of the dartboard."

By using familiar objects and concepts to represent new ideas, the audience can quickly understand and connect with the message being conveyed.

Metaphors provide a way to simplify complex concepts and make them more accessible to a wide audience. When used effectively, visual metaphors can be both memorable and impactful, helping to convey the intended message in a way that is easy to understand and remember. The purpose of *Recreational Waterborne Illness* by Jennifer E. Fairman, (2017) was to visually summarize the journal's featured article, which highlights the causes of recreational waterborne illnesses (Fig. 3.21). These illnesses are most common in the summer months due to increased recreational exposure. Depicted are the environment and the causes of recreational waterborne illnesses. These pathogenic agents, including

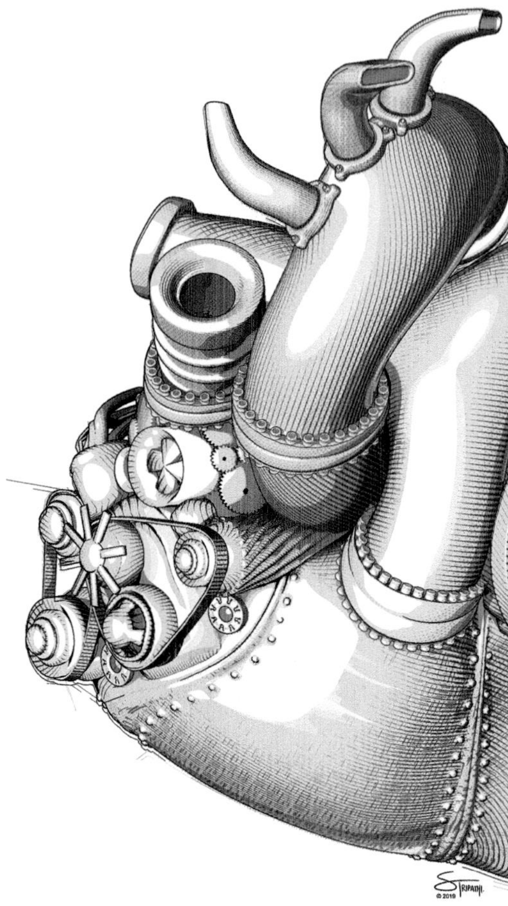

Fig. 3.17 *Real. Genuine. Heart*. With permission from Satyen Tripathi, © 2019, Emory University

bacteria (e.g., *E. coli* and *Legionella*), viruses (Adenovirus, Rotovirus, and Norovirus), parasites (*Giardia intestinalis*), and algae (Cyanobacteria), are represented metaphorically by inflatable pool floats and toys. The CDC has recently begun tracking multi-person outbreaks associated with recreational waterborne activities. Although many people believe that untreated ponds and rivers are the culprits, a public pool or the hotel spa is more often the source.

3.3.4 Fantasy or Surrealist Genre

This genre allows the illustrator to make up fantastical environments, personify inanimate objects and create near impossible imagery, engaging the viewer to want to learn more. Influence of commercial artists like Frank Frazetta and Marvel Comics. Some illustrations may even borrow from art history with a suggestive nod toward surrealistic artists such as Salvador Dali, Hieronymus Bosch, and Rene Magritte. In Fig. 3.22, Fairman pays a homage to Magritte's 1928 painting *The False Mirror* where the eye is all-seeing and so is the immune system. The iris is alive with the chromosomal plot of a persistently mutating cancer cell that is the target within the pupil. Persistent mutations and encoded neoantigens cannot be edited under immune pressure and drive sustained immunologic tumor control. In other works that pay tribute to Dali, the illustrator might use cellular landscapes and scientific imagery that is placed in a visually distorted perspective and may display unfamiliar interiors or exteriors. Pop culture references to superheroes or folk heroes can provide inspiration for a subjects' descriptive looks or actions or

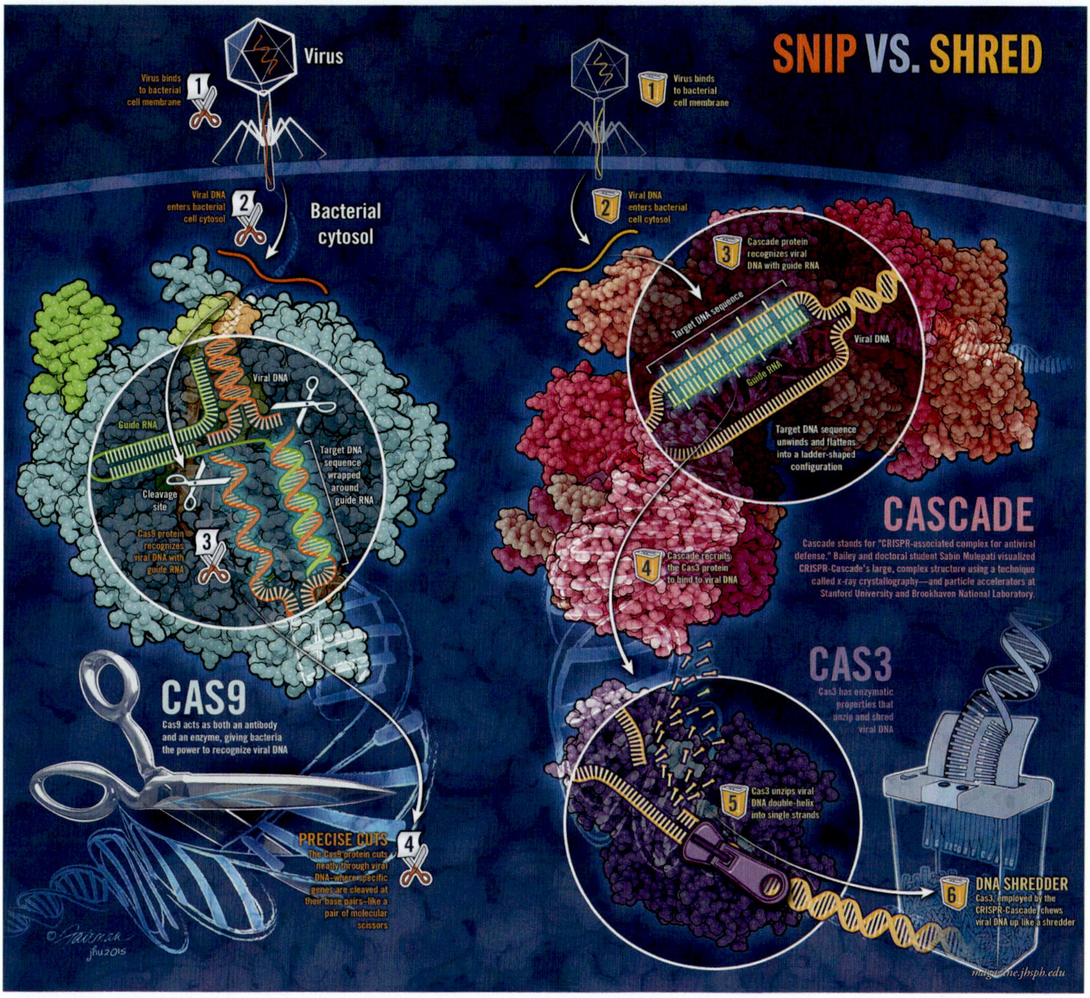

Fig. 3.18 *Snip vs. Shred, Johns Hopkins Bloomberg School of Public Health Magazine,* Jennifer E. Fairman, © 2015, Johns Hopkins University

subjects may be treated in an anamorphic fashion. Fantasy illustrations often depict mythical creatures, enchanted environments, and otherworldly landscapes, providing a rich visual canvas for storytelling (Aparaschivei 2022).

In editorial illustration, fantasy can be used to create imaginative illustrations for journal covers, scientific publications specifically for children, or *Graphic Medicine* comic books. These illustrations can help bring the story to life and engage the reader in a fantastical world. They can also be used in advertising and marketing to evoke a sense of whimsy and magic, and to create an emotional connection with the audience.

Figure 3.23 shows a fantastical dramatization of the flowering of gastrointestinal microbiota fertilized by the seething gastric environment. The flowering is depicted by petals of *Bacteroidetes, Bifidobacterium*, and *Enterococcus faecalis*, pistils of *Lactobacillus* and stamens of *Enterococcus faecalis* perched on stalks of *Escherichia coli*. The flower itself emerges on stalks of *Clostridium difficile*. Free-chaining *Streptococcus* mutans floats by under the flower.

Fig. 3.19 *Breaking New Ground*, with permission from Audra Geras, © Audra Geras

Bacillus cereus, long rod-shaped bacteria in their arthomitus stage attach by fibers to the intestinal epithelium, grow filamentously, and sporulate from their distal ends.

Fantasy illustrations often rely on symbolism and allegory to convey deeper meanings and messages. To create effective fantasy illustrations, illustrators need to have a strong sense of visual storytelling and the ability to create believable and engaging worlds. They need to be able to use color, composition, and line to create an imaginative and fantastical visual narrative that captures the reader's imagination. The use of fantasy as a visual narrative, like the depiction of bacteria in a floral pattern may convey growth within the gastrointestinal microbiota fertilized by the seething gastric environment, provides a unique and imaginative way to convey a unique idea, engaging the reader in a world beyond their own.

The literal, metaphorical, and fantastical genres combined with a framework and methodology for understanding the subject as well as the audience allow the illustrator to structure and conceptualize an image that tells a story (Heavner 2019).

Fig. 3.20 *Clinical Trials of Cancer Drugs Fail Due to Missed Target Protein,* with permission from Emily Taylor, © 2020, Emily Taylor

3.4 The Editorial Illustration Process

Editorial illustration is generally meant to be an engaging yet educational medium (Bossert 1996). For the student eager to take on this genre of biomedical visualization, understanding the science is an essential first step for a medical cover or spread to be effective. Inaccurate or misleading illustrations can have serious consequences, as they can spread misinformation, perpetuating incorrect ideas about a topic. Understanding the complex science of the malaria life cycle, for example, can also help the illustrator to create a more engaging and effective illustration that is more imaginative, creative, and visually appealing (Fig. 3.24). This is a process the authors teach in the Medical and Biological Illustration Graduate Program in the Department of Art as Applied to Medicine at the Johns Hopkins University School of Medicine. Here, we explore the path that students take when learning to create an editorial illustration for a medical or scientific journal.

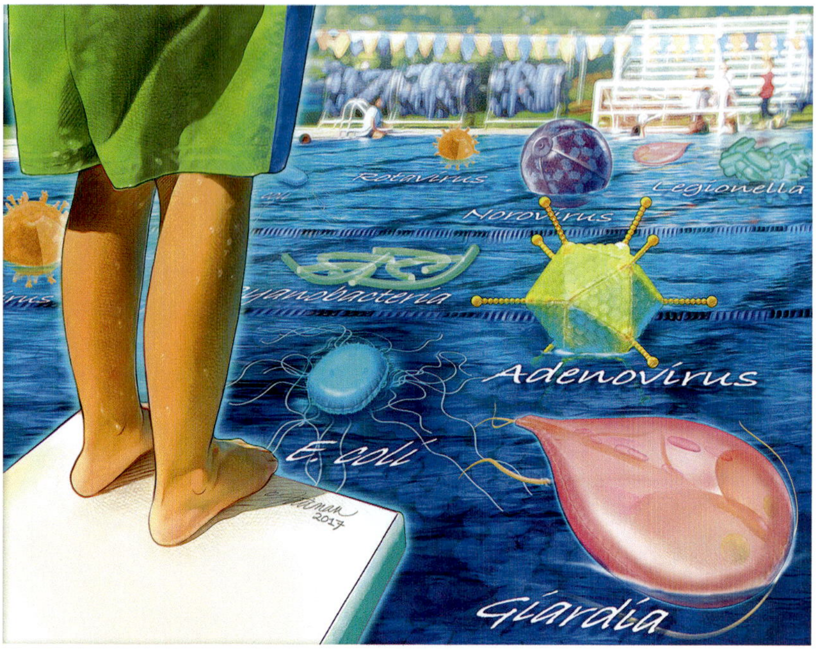

Fig. 3.21 *Recreational Waterborne Illness, American Family Physician,* Jennifer E. Fairman, © 2017, Fairman Studios, LLC

Fig. 3.22 Above, *The False Mirror,* Rene Magritte, © 1928, Rene Magritte, fair use; *R*, *Cancer's False Mirror*, Nature Medicine, Jennifer E. Fairman, © 2022, Johns Hopkins University

Fig. 3.23 *Flora,* with permission from Cynthia Turner, © 2019, Cynthia Turner

3.4.1 Thumbnailing and Mood Boards

One of the first steps to developing an effective editorial illustration is to begin brainstorming visual ideas through quick thumbnail sketches. Thumbnailing rough concepts that come to mind is a useful method to record ideas early in the process. Nurturing creativity is an iterative process and many tools, like a sketchbook or mood board, can be utilized to find creativity, let ideas and concepts simmer, and acknowledge that no idea is bad. The artist must be uninhibited and not judge too soon.

Figure 3.25 shows a concept development through fast sketches and thumbnailing. The left side of the image shows lists and highlighted terms accompanied by sketches produced as-fast-as-possible, followed by thumbnailing of ideas on the right. Cecelia Johnson describes this

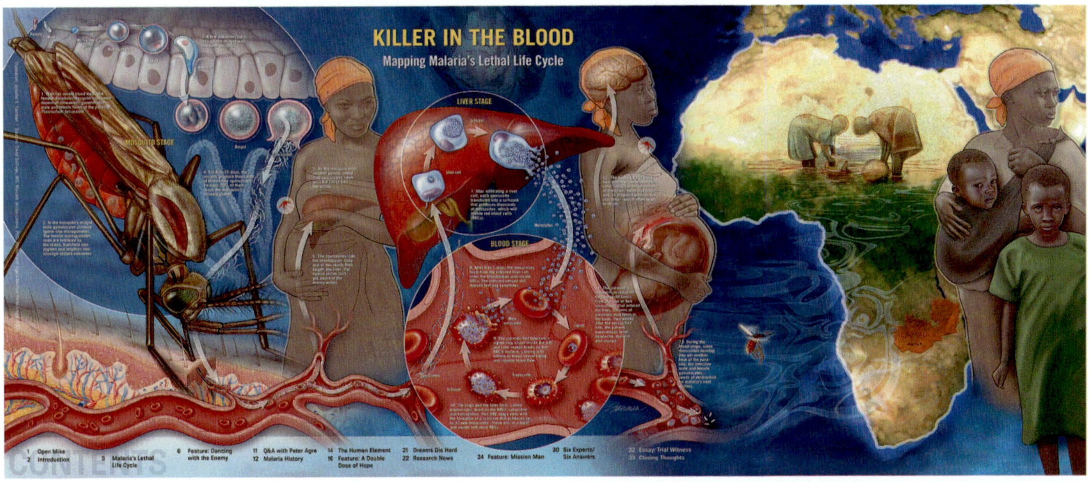

Fig. 3.24 *Killer in the Blood*, Johns Hopkins School of Public Health Magazine, Jennifer Fairman, © 2011, Johns Hopkins University

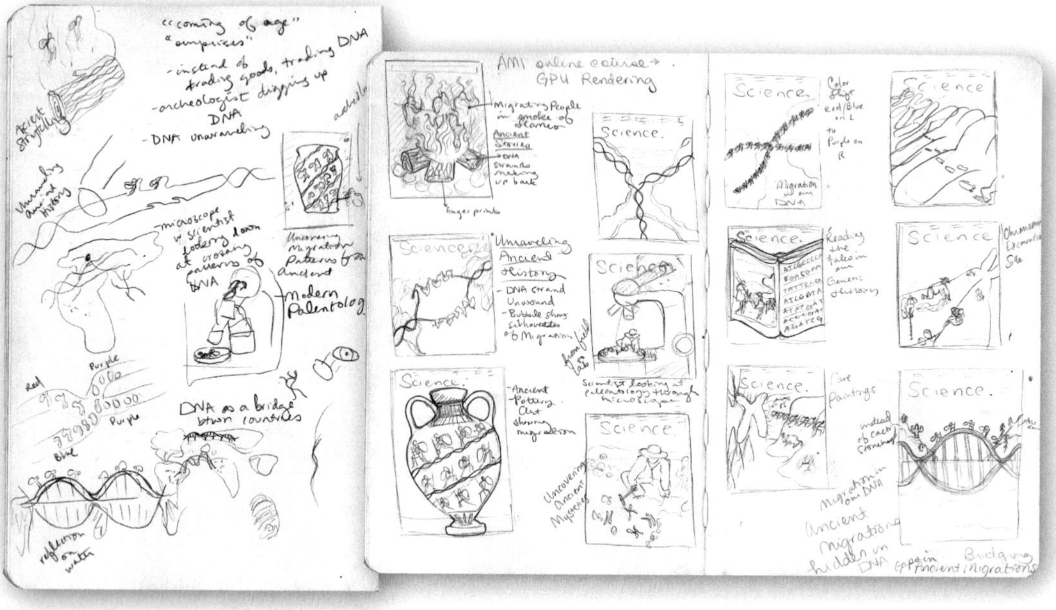

Fig. 3.25 Concept sketches and thumbnails for *Piecing Together Ancient Migration*, with permission from Cecelia Johnson, © 2018, Cecelia Johnson

process as "getting out my ideas as fast as possible - some while I'm reading the article, some afterwards while I was 'digesting' the material." These fast facts and thumbnails spur the conceptualization process.

The use of an illustration mood board can help illustrators understand their subject matter further. Illustration mood boards are visual collections of images, including photographs and other artwork, color palettes, typography samples, textures, and other elements used to communicate the visual style, tone, and atmosphere of a project. Figure 3.26 shows a mood board that accompanies Fig. 3.25. The mood board includes supplemental visual research on the topic to be illustrated. In this case, the Bell Beaker People, an archaeological culture named after the inverted-bell beaker drinking vessel used at the very beginning of the European Bronze Age. Questions emerge when generating concept sketches and mood boards can help to pose answers. Here, the mood board helps to posit, "what are Bell Beakers and what do they look like? Who were the Bell Beaker people, and what do they look like?"

The purpose of a mood board is to establish a visual direction for a project and to ensure that everyone involved in the project, from the client to the illustrator, is on the same page in terms of the desired aesthetic. By providing a visual representation of the desired style and tone, mood boards can help ensure that the final illustration accurately reflects the desired vision for the project. They can also be used to present a project to clients, as well as to showcase the designer's or illustrator's vision and style to potential clients or employers. A few things to consider when creating a mood board include:

1. Exploring sources of inspiration: not just looking at what others have done, but also, looking outside of your normal environment, liberal arts approach, tap into anything that inspires you, look outside your normal comfort zone for inspiration.
2. Make connections with what may appear to be disparate ideas.

Fig. 3.26 Mood board for *Piecing Together Ancient Migration*, with permission from Cecelia Johnson; fair use

3. Science behind creativity—you may not be born with it, and you need to nurture it into practice.
4. Tools, references, brainstorming, thumbnails of the ideation process—same story told in different ways.
5. Diversity of techniques, use of multimedia, what can be used to enhance your message, software, hardware, or even raw materials.
6. Audience—patient, medical, bench science, and lay public.

3.4.2 Audience and Context

It is important to consider the intended audience when creating an editorial illustration given that the purpose of the illustration is to communicate information to a specific viewer. Understanding the audience's needs, preferences, and perspectives is key to creating an illustration that effectively communicates its intended message. Think about things like age, gender, culture, education, and socioeconomic background. Considering these things can also help to ensure that the illustration is culturally sensitive and appropriate. For example, it may be important to consider the cultural representations and stereotypes used in the illustration, and to avoid any imagery that might be offensive or inappropriate. This information can help to inform decisions about the visual style, tone, and content of the illustration; an illustration intended for a young audience may be more colorful and playful, while an illustration intended for a scientific audience may be more technical and detail oriented.

It is also important to consider the context in which the illustration will be viewed, such as in a print publication, on a website, or on social media as well as the physical boundaries of the art, journal format, or page layout. This can impact the size and format of the illustration, as well as the types of visual elements that are used. Ask, "What is the style and primary readership of the journal?" These considerations help to ensure that the illustration effectively communicates its intended message to the right people in the right way.

3.4.3 Conceptualization and Progression Toward a First Draft

Students with a given assignment will research topics of interest in the fields of medicine, healthcare, anatomy, or biology. Topics can be narrowly confined centering on a single defined topic or may be developed and combined from multiple readings of similar manuscripts into a single topic. The student may already have scientific knowledge and experience in a particular field of study and is encouraged to select a topic that takes advantage of their special attributes.

Students are introduced to multiple publications and online sources. They will spend a half day researching and viewing online professional journals for covers and possible article ideas by searching such journals as *Nature, Scientific American, Cell, Cell Biology, Neuron, Discover, JAMA, Science, Scientist, Howard Hughes Medical, Genome Research, American Family Physician,* and many more. Another recommended resource can be found on Ella Maru Studio's website where great examples of covers can be found (Ella Maru Studio 2023). Students will also Review examples of commercial illustration and medical illustration online in sourcebook searches and the AMI Gallery of Professional Artists (Association of Medical Illustrators 2022). They also will review past AMI Salon members' work on the AMI Archives. Our dept. library has multiple Communication Arts Illustration and Photography Annuals (Coyne & Blanchard, Menlo Park CA) and spiral-bound yearly Medical Illustration and Animation Sourcebooks (Bajema et al. 2017). Upon completion of their searches, students are tasked with sharing two successful covers or images that communicate the desired message and two poorly executed covers/images describing in detail issues of effective uses of visual communication and message, tagline, color, mood, and format for publication for each.

Upon completing a first reading of the text, the student will begin to understand, imagine, and develop an overall concept. This is achieved by determining the basic theme or story line. At this stage, personal questions and inquiries begin such as "What is the main message?" "Which important features stand out?" "What are the key elements in the main theme?" "Which elements provide the best visual opportunities for storytelling" "Who or what are the main players?" and "Are there members of the secondary or supporting "cast" that can be identified?" With a better idea of scope and content, it is advised that the student read the manuscript a second time to further connect to the content that they believe is the central concept and can be the most visually compelling to the audience.

After a second or third reading and review of the manuscript an important active stage commences. The student should begin making lists of the subject matter attributes or cataloging their ideas. They can create mood boards of similar topics, color co-ordinations, and styles of illustration they would like to immolate or develop proficiencies. They may create a visual word ranked "TAG cloud," a concentrated hierarchical diagram with varying sizes of type with the largest words delineating the most important words and ideas with secondary and lesser words being smaller in size. Students are encouraged to think about adjectives, descriptive slogans, and nouns, and, about action verbs that accurately and elegantly describe the actions of the subject and theme. Each of these tactics assists in organizing information to establish a hierarchy and supports the intended main concept.

These efforts are tailored around understanding and targeting your readership through the use of more refined and detailed sketches that become communicating visuals for effective storytelling. This draws directed attention to the specific theme, creating art that is eye-catching and

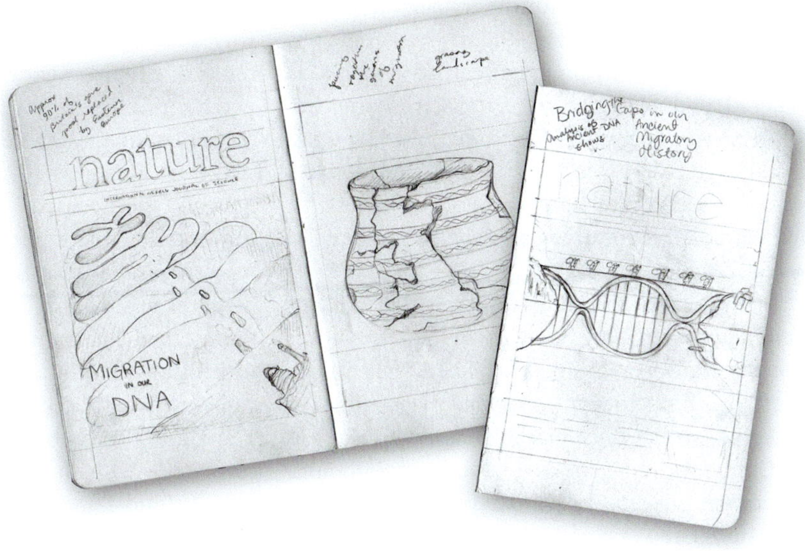

Fig. 3.27 Concept and composition for *Piecing Together Ancient Migration*, with permission from Cecelia Johnson, © 2018, Cecelia Johnson

attractive, and visually communicating one of the more important concepts (Fig. 3.27).

Discussing ideas with peers can assist in pushing concepts further. In relation to the Bell Beaker People illustration, reviews by peers highlighted a desire for more context on the 'migration' aspect of these people, so it was suggested, since this is a museum artifact, that it could be shown as placed in a natural history museum setting, perhaps in front of a mural of migrating people with the natural history painting feel blurred in the background in addition to using symbolism subtly placed as 3D texture maps (Fig. 3.28). Thumbnails, mood boards, and peer-feedback function together to further develop concepts in the sketchbook.

Fig. 3.28 3D Asset development: Efficient workflow: displacement maps and Boole within Cinema4D; Final stage: background and compositing for *Piecing Together Ancient Migration*, with permission from Cecelia Johnson, © 2018, Cecelia Johnson

In the act of brainstorming, take ideas and thumbnails and scribble them very quickly. There should be no prejudgment, as you get as many ideas down on paper as quickly as you can. Remember—There are *no bad ideas*. Students are encouraged to prepare 15–20 quick sketches. As your ideas come to light each one will lead to another possible solution and may take an artist to unrealized places.

In the sketch preparation stage, the student should consider *scale, size, placement, direction and mood, hierarchy,* major players, and minor roles. It is important to consider the amount of available space the central image will reside in. Consider the positions and sizes of adjoining supporting images. A preliminary color composition is prepared that considers hue and tint of the color portrayed in the central figure and in important areas of the image with an effective use of light and shadow to portray mood and emotion. Direct focus on the central idea or message through decision-making will guide the amount of textural detail portrayed in the central figure and the limiting of such detail in the peripheral areas of the presentation image. Each of these approaches will inform and guide the reader to the central theme in the presentation of the final image on the printed page (Fig. 3.29).

Directives to the student include preparing rough sketches and continuing to further refine the student's approach. In thinking of the intended image, provide a working imaginative summary short sentence caption. Consider a tagline that "plays" and highlights the subject matter either quickly identifying the concept in just a few distinct and descriptive words or drawing attention to the topic.

3.5 Diversity

Diversity plays a critical role in medical and editorial illustration, as it reflects the reality of the world we live in and helps to ensure that all members of society are represented and included. In medical illustration, diversity is important to ensure that medical information is presented in a way that is accessible and relatable to a wide range of people, regardless of their race, ethnicity, gender, or other factors.

For example, illustrations that depict a diverse range of people and body types can help to increase the accuracy and relevance of medical information, making it more accessible to people who may otherwise feel excluded (Fig. 3.30). It helps to ensure that all members of society are represented and promotes a more inclusive and respectful world. In editorial illustration, diversity can also help to challenge stereotypes and prejudices, promoting equity, equality, and respect for all members of society. It is important for illustrators and publishers to actively seek out diverse perspectives, work with a diverse range of models, artists, and other professionals, as well as carefully consider the visual elements used in illustrations, such as skin tone, hair, and body type. Health and healthcare affect everyone. American Family Physician has done a consistent job of contracting medical illustrators to consider a wide range of subjects from diverse backgrounds depicted monthly in their featured articles (Fig. 3.31).

A great example of a publication that incorporates diversity is National Geographic (NatGeo). NatGeo has become a leader in publication illustrations due to its commitment to high-quality storytelling and visually stunning imagery. The publication has always placed a strong emphasis on visual storytelling, using photographs, illustrations, and graphics to bring its stories to life. National Geographic's illustrators and photographers are known for their skill, creativity, and attention to detail, and their work has won numerous awards and accolades over the years.

National Geographic's commitment to visual storytelling has also been reflected in its investment in new technologies and techniques, such as aerial and underwater photography, and digital illustration. This commitment has helped the publication to stay at the forefront of the industry and

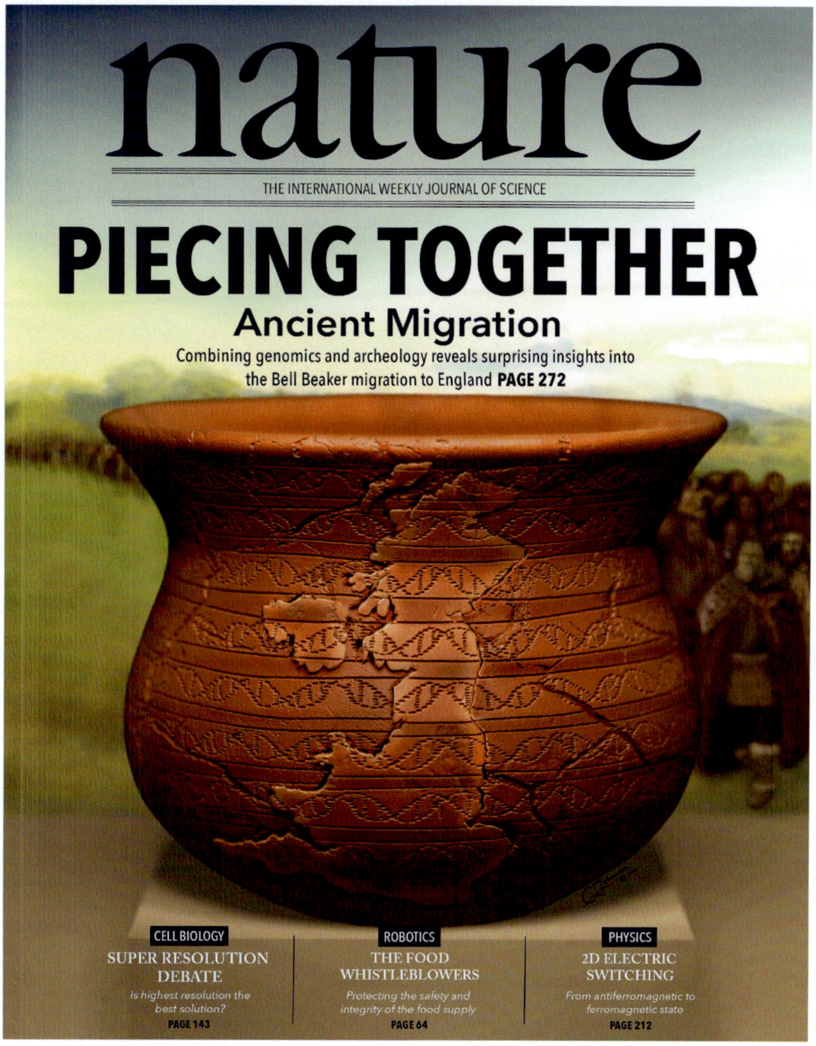

Fig. 3.29 *Piecing Together Ancient Migration*, with permission from Cecelia Johnson, © 2018, Cecelia Johnson

maintain its reputation as a leader in publication illustration.

Another reason why National Geographic has become a leader in publication illustrations is its commitment to diversity and representation. The publication has worked to ensure that its illustrations and photographs accurately reflect the world's diverse populations and cultures, promoting cultural understanding and promoting a more inclusive and equitable world.

3.6 Beyond the Cover: A Look at the Future

The emerging field of *Artificial intelligence (AI)-generated* images in communication publishing has come to the forefront in many creative fields, especially in editorial illustration. It behooves the authors to address the idea of AI and its place in editorial; image creation and use in ideation. AI is increasingly being used as a tool to create

Fig. 3.30 *Overturning Roe* Vs. *Wade: Reproductive Rights and Rheumatology,* Jennifer Fairman © 2023, Johns Hopkins University

editorial illustrations in various fields, including advertising, publishing, and design. AI can be used to automate certain aspects of the illustration process, such as generating concepts and sketches, creating digital illustrations, and even colorizing and compositing illustrations or scenes with which to use as backdrops.

One example of AI-generated editorial illustration is the use of generative adversarial networks (GANs), which can create new images based on a set of existing illustrations (Fig. 3.32). Illustrators can train these networks on their own styles and preferences, and then use the AI to generate new illustrations that are similar in style and content. Another example is the use of AI-powered illustration software, which can assist illustrators in the creation process by suggesting colors, lines, and shapes based on a set of rules and algorithms. AI can be used to analyze the visual elements of an illustration, such as color, line, and form, and provide recommendations for improvement, providing insights into their style and content. This can save time and help streamline the illustration process, allowing illustrators to focus on the creative aspects of their work.

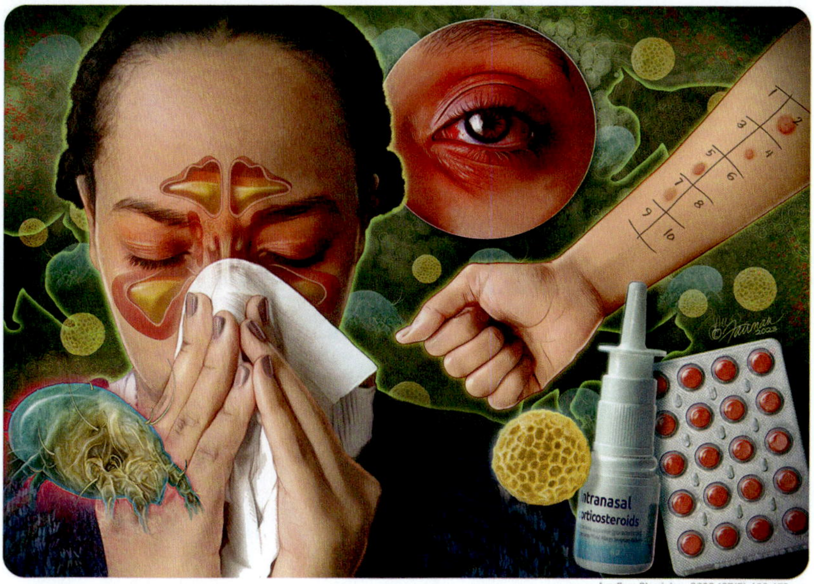

Am Fam Physician. 2023;107(5):466-473

Fig. 3.31 *Allergic Rhinitis: Rapid Evidence Review,* American Family Physician, Jennifer Fairman, © 2023, Johns Hopkins University

The use of AI *as a tool* in conjunction with the process of creating editorial illustrations has the potential to increase efficiency and productivity, while still allowing illustrators to retain control over the creative aspects of their work. However, it is important to note that while AI can be a helpful tool, it is not a replacement for the creativity, skill, and expertise of human illustrators.

With this new technology, new questions arise. Will artificial intelligence replace editorial artists? It is unlikely that AI will completely replace editorial artists in the near future. While AI can certainly be used as a tool to create illustrations, it is currently not capable of replacing the creativity, skill, human touch, and most importantly, *accuracy* that editorial artists bring to their work. AI-generated illustrations are often created based on algorithms and pre-existing data and lack the originality and nuance that comes from a human artist's understanding of the subject matter and their individual style. Images created by artificial intelligence can often look generic because they are based on algorithms and pre-existing data. AI-generated images are created using machine learning models that have been trained on large datasets of existing images. Because these models are based on pre-existing data, they may not have the capability to generate truly original or unique images. They can also be limited by the quality and diversity of the data they have been trained on. These models identify patterns in the data and use them to generate new images based on those patterns. Editorial artists bring a unique perspective and style to their work, and their ability to understand the story being told and to create illustrations that connect with the audience is a critical component of effective editorial illustration.

AI is likely to continue to play an increasing role in the field of editorial illustration, potentially changing the way editorial artists work and allowing them to be more efficient and productive. Additionally, AI-generated images are often created to meet specific technical requirements or criteria, rather than being driven by artistic vision or creativity. As a result, the images generated by AI can often look formulaic or stereotypical, lacking the individuality and expressiveness of images created by human artists. Ultimately, the use of AI in editorial illustration will depend on the specific needs of the project and the goals of the illustration.

Fig. 3.32 *Various creatures and organic landscapes created in AI,* with permission from Joe Samson, 2023

3.7 Conclusion

Editorial illustration will no doubt continue to grow and evolve within the field of biomedical visualization. Because of the diversity of scientific and medical topics and specialties that require illustration problem solving, there is great latitude and opportunity for personal expression by trained artists that can culminate in creative, novel, and effective scientific communication (see examples in Figs. 3.33 and 3.34). In medical illustration, we are often tasked with creatively presenting subject matter with a traditional didactic approach. In our experience, we often observe a student's first attempt at editorial work to be met with self-doubt as first ideas are being developed. This genre, much like any other creative task, while iterative, must be met with a more experimental approach. This chapter has provided many tools, tips, tricks, and ways of thinking for planning and executing purposeful creative images for publication by artists interested in the field of editorial illustration for science and medicine. For those considering delving into this area of creative imagery, keeping the following thoughts and directions in mind will set the artist up for optimal success (Phelps and Mandala 2022):

Fig. 3.33 *How the Brain Decides Whether to Hold'Em or Fold'Em,* Jennifer Fairman, © 2019, Johns Hopkins University

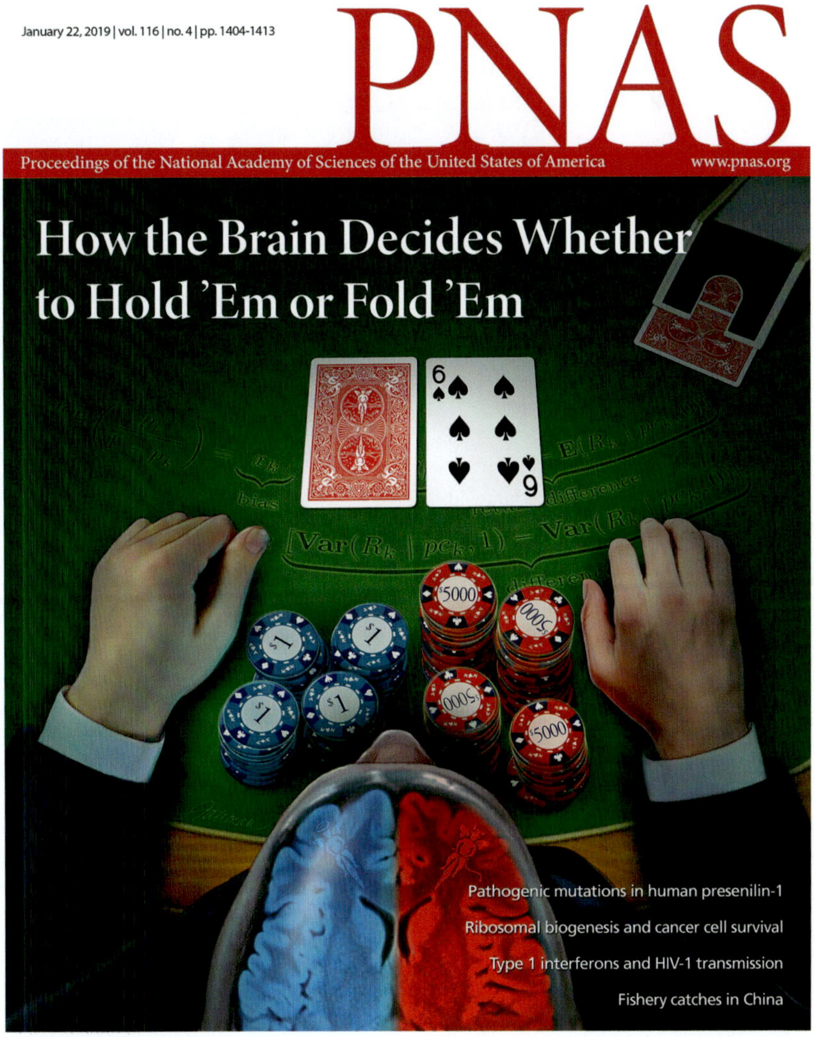

Fig. 3.34 *Cas9-induced p53 activation*, Jeff Day, MD © 2020, Johns Hopkins University

1. Search for your inner child 2. Embrace play 3. Be open to suggestions 4. Search for commonality in different disciplines 5. Allow for incubation and reengagement (step in. . .step away. . .step back in again) 6. Embrace comfort in the process of the creative act 7. Apply meaning to the images you create	14. Be inspired by others without thinking you must be just like them 15. Look for common ground in uncommon connections 16. Strive for authenticity 17. Embrace anticipation in your search for artistic expression 18. Explore your past for inspiration 19. Make art "play" and not "work"	8. Ponder what is just around the corner 9. Stay true and authentic 10. Reach down into your well of experience and inspiration 11. Dispense with your inner critic 12. Do not compare yourself to others 13. Stay true to the things that make you happy and fulfilled	20. Separate process and destination 21. Learn from mistakes 22. Seek and savor serendipity 23. Balance the Teeter-totter of work life and art life 24. Credit your mentors 25. Borrow knowledge and pass it on to others 26. Be both a teacher and a student in all that you create

References

Aparaschivei L (2022) 12 illustration styles every illustrator should know, Vectornator. https://www.vectornator.io/blog/illustration-styles

Association of Illustrators (2022) Editorial illustration basics. https://theaoi.com/resources/professional-practice/editorialillustration-basics

Association of Medical Illustrators (2022) Annual salon, ami.org

Bajema R et al (2017) The medical illustration sourcebook. The Association of Medical Illustrators, Serbin

Bossert J (1996) Editorial illustration: Step by Step Techniques, a Unique Guide from the Masters. The Society of Illustrators, Roto Vision

von Debschitz U, von Debschitz T (2017) Fritz Kahn: Infographics Pioneer, Taschen, Biblioteca Universalis, Cologne.

Department of Medieval Art and The Cloisters (2001) The metropolitan museum of art. The art of the book in the middle ages. In Heilbrunn Timeline of Art History New York: The Metropolitan Museum of Art, 2000. http://www.metmuseum.org/toah/hd/book/hd_book.htm

Ella Maru Studio (2023) Scientific design and animation, Cover Images. https://scientific-illustrations.com/portfolio/covers

Errea J (2017) Visual Journalism: infographics from the world's best newsrooms and designers hardcover. Translation edition, Gestalten

First Amendment Museum (2023) Art & Politics: 300 Years of Political Cartoons. https://firstamendmentmuseum.org/exhibits/virtual-exhibits/art-politics-300-years-of-political-cartoons/. Viewed 18 February 2023

Goldberg S (2021) Meet one of Nat Geo's amazing artists. https://www.nationalgeographic.com/magazine/article/meet-fernando-gomez-baptista-one-of-nat-geos-amazing-artists

Greenbaum Kasson Elizabeth (2016) Edward Tufte on data visualizations and art. Dice Daily. https://www.dice.com/career-advice/edward-tufte-visualizations-data-art

Heavner R (2019) Conceptual illustration: framework and methodologies. A Companion to Illustration:47–58

Ives C (2004) The print in the nineteenth century. In: Heilbrunn timeline of art history, vol 2000. The Metropolitan Museum of Art, New York. http://www.metmuseum.org/toah/hd/prnt2/hd_prnt2.htm

Levin, Jo Ann Early (1980) The golden age of illustration: popular art in american magazines, 1850–1925. ProQuest Dissertation. AAI8018571, https://repository.upenn.edu/dissertations/AAI8018571

Normal Rockwell Museum (2023) Illustration history, The middle ages. https://www.illustrationhistory.org/history/time-periods/the-middle-ages

May T (2020) 16 medical illustrators doing groundbreaking work, Creative Boom.

Rengen S (2020) Information graphics, TASCHEN, Biblioteca Universalis, Cologne.

Perlman J (2022) Conceptual illustrations: solving problems with storytelling, Blush Design. https://blush.design/blog/post/conceptual-illustration-definition

Phelps T, Mandalas P (2022) Life musing on AI. https://phelpsmandala.com/2022/09/04/life-musing-on-ai/

Current Approaches in Narrative Medical Visualization

4

Laura A. Garrison, Monique Meuschke, Bernhard Preim, and Stefan Bruckner

Abstract

In a world increasingly driven by data and technology, it is imperative to empower stakeholders across diverse backgrounds (scientists, policymakers, patients, the general public) to make informed decisions about their health with the information available. However, simply providing the information is often not enough. The key to sharing and communicating health and medicine is making the information relevant to the intended audience. This issue lies at the heart of effective science communication. Storytelling is an age-old practice, and arguably one of the most defining features of humanity. Stories enable science communicators to translate complex information from health and medicine into a narrative that is personally accessible and relevant to the individual. Many pieces come together to craft what many would call a "good story"—one of these is *narrative*. Our work seeks to tease out the role of narrative, with its various components, in telling medical stories. This chapter discusses current approaches in narrative medical visualization, with an eye toward future opportunities in developing narratives for data-driven medical stories. We explore various strategies in narrative structure and character that may be unique to telling medical stories and provide an outlook on future directions in this space.

Keywords

Narrative visualization · Narrative medical visualization · Medical visualization · Data-driven medical story · Science communication

4.1 Storytelling in Medicine

Imagine that you are in your doctor's office, trying to understand the details of the surgical procedure that you are scheduled to undergo the following week. Your doctor roughly sketches out some confusing anatomy on stray printing paper, trying their best to answer your questions. However, you feel overwhelmed and leave the appointment feeling as if you have already forgotten most of what you and your doctor discussed. If this scenario is not familiar to you, then a friend or family member has likely

L. A. Garrison (✉)
Mohn Medical Imaging and Visualization Centre, Department of Radiology, Haukeland University Hospital, Bergen, Norway

Institute for Informatics, University of Bergen, Bergen, Norway
e-mail: laura.garrison@uib.no

M. Meuschke · B. Preim
Department of Simulation and Graphics, University of Magdeburg, Magdeburg, Germany

S. Bruckner
Institute for Visual and Analytic Computing, University of Rostock, Rostock, Germany

M. Roughley (ed.), *Approaches for Science Illustration and Communication*, Biomedical Visualization 4,
https://doi.org/10.1007/978-3-031-41652-1_4

experienced feelings of confusion, lack of understanding, and feeling intimidated by visits to the doctor. This may have left you, or your loved one, inadequately prepared and unclear on the risks of the procedure or treatment they will soon receive.

The medical community is aware of this communication issue, and recent years have seen a focus on facilitating communication between health professionals and patients, along with the growth of an entire industry focused on patient and medical education. Many science communication materials developed for patient education couch the intended message in the form of a story. Storytelling is an age-old practice central to humanity and enables science communicators to translate complex information from health and medicine into a narrative that is personally accessible and relevant to the individual (Joubert et al. 2019). Software products like *EmmiEngage* (Wolters Kluwer 2023) support patient empowerment and provide reassurance through self-paced multimedia modules that can be prescribed to patients by their doctors. Medical comics, for instance, those crafted by *Artibiotics* (Kearns 2023) based in New Zealand, can engage patients and encourage them to advocate for their health with their doctor (Alkureishi et al. 2021).

Supporting visual science communication, medical illustration is a field dedicated to translating complex medical information into approachable visual stories tailored for, e.g., patient education, clinical training, and broad outreach, such as Fig. 4.1. However, while engaging, these stories are time consuming to create and are not easily tunable to the unique needs of the individual.

Narrative visualization, from the computer science domain, combines the study of traditional storytelling techniques with data visualization (Lee et al. 2015). Driven by data, narrative visualization explores the semiautomated crafting of stories for different topics that, unlike handcrafted visual stories, are more easily tunable to different goals and audiences. Narrative medical visualization may enable and facilitate more personalized patient education and outreach, with easily generated and reusable components. In the example that opened this chapter of a patient–clinician interaction, a narrative medical visualization approach could allow the clinician to draw from the patient's medical record to explain their upcoming procedure specific to the patient's own anatomy. They could discuss the predicted effects and risks of the procedure according to the patient's personal health information in a visual style that is clear and relatable for the patient. Rather than laboriously handcrafting a custom story, narrative medical visualization can provide a framework for the story crafter to prepare different versions of a story with different variants of a given condition that can be fine-tuned to fit a communication goal. Finally, while storytelling is inherently subjective, by incorporating data into the storytelling process narrative medical visualization may introduce a degree of objectivity preferable in certain situations. In the following sections of this chapter, we sketch the basic theory and elements of narrative visualization. We then discuss current approaches and challenges in adapting this study for medicine, known as narrative medical visualization. For the sake of clarity, our use of the terms *narrative visualization* or *narrative medical visualization* throughout this chapter refers to the conceptual or methodological approaches and frameworks within the field of visualization, while *data-driven medical stories* are the end products of the implementation of these frameworks.

4.2 Narrative Visualization: Theoretical and Practical Foundations

Narrative visualization combines traditional storytelling techniques with data. While not a particularly new area of study, its practical application has expanded in recent years with the advent of an increasing accessibility of the web and technology supporting Internet access from nearly anywhere. Data journalists, e.g., from the New York Times Graphics Division, and data visualization agencies, such as Fathom Information Design (Fathom 2023), often employ storytelling techniques when crafting visualizations about

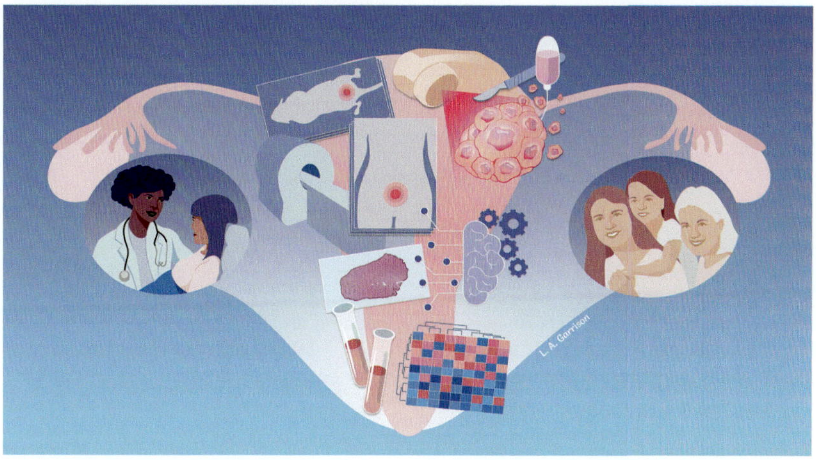

Fig. 4.1 Example of a hand-crafted medical illustration, created for broad outreach that visually describes the Mohn Medical Imaging and Visualization Centre's (MMIV) Gynecological Cancer Imaging research process and initiatives. The goal of this illustration is to visually communicate the group's research focus on gynecology, its various methods, and to capture the notion of patient care at the beginning and end of the research pipeline. Illustration © Laura Ann Garrison, used with permission

data in order to attract and appeal to their intended audience.

Spanning practitioners and researchers, narrative visualization is an interdisciplinary area of study well-suited to collaboration. Visual and interaction design, as well as some programming experience and knowledge of the particular application area, are often necessary to craft a successful visual data story. In this section, we provide a brief overview of theoretical and practical foundations of narrative visualization, including:

- Narrative intent
- Story pieces: data facts and visualization techniques
- Narrative structure and genre
- Narrator and story personalization
- Evaluation in narrative visualization

From these foundations, we discuss applications of narrative visualization to the sciences which feed into the ideas central to narrative medical visualization.

4.2.1 Narrative Intent

The story creator(s) has two main objectives when telling a story: (1) engage the audience and (2) hold their interest. Defining a *narrative intent* is arguably the most important aspect of story development, as this dictates nearly all other aspects of narrative construction. Narrative intent essentially is the take-home point or key objective of the story. As a story creator, what do you want your audience to understand, or remember, from your story, and how can you achieve that? Narrative intent can more formally be framed in the context of *narrative patterns*, which are low-level narrative devices that, alone or in combination, serve a specific intent in a story. Defined by Bach et al. (2018), these narrative patterns can be categorized into five high-level groups: argumentation (e.g., *compare*), flow (e.g., *repetition*), framing (e.g., *address the audience*), emotion (e.g., *humans behind the dots*), and engagement (e.g., *call to action*).

Having set the narrative intent, the story creator can then determine the necessary material to include in the story and the visualization techniques to use for this material. They can also consider how to group and sequence the story material, i.e., how to construct the story chronology), and determine their narrative strategy, i.e., how to tell the story through narrative genres, structure, and narrator perspective. These aspects of a story are illustrated in Fig. 4.2. Other issues, such as story personalization,

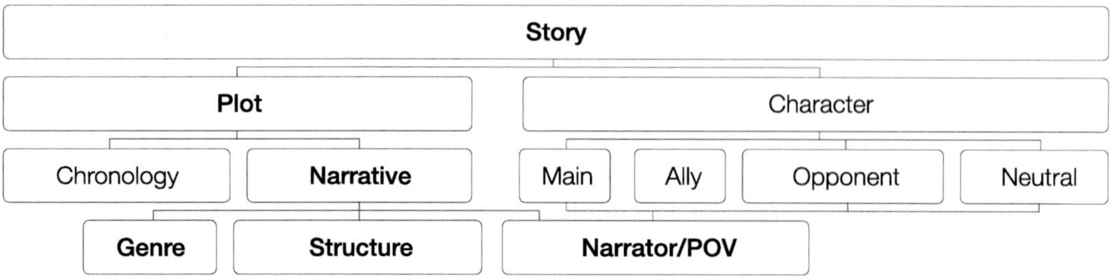

Fig. 4.2 Elements of a story, where *narrative* is the element that drives how the story is told. Narrative in turn is influenced by many elements. The three that we focus on are: structure, genre, and narrator/point of view (POV). Image adapted from Beemgee GimbH (https://www.beemgee.com/blog/story-vs-narrative/) 2017

accessibility, and credibility, can also be considered. Depending on the domain, these issues have various degrees of importance (Böttinger et al. 2020; Ynnerman et al. 2020). For instance, credibility is vital for medicine, as medical information from a trusted source often means that the content and advice are vetted and safe to follow without detriment to one's health.

4.2.2 Story Pieces: Data Facts and Visualization Techniques

The raw material used to construct a visual data story includes concepts, images, data facts, and textual descriptions (captions, labels), and visualizations derived from the data. Story pieces can be static or animated in cases where demonstrating temporality is key to the narrative intent. Interactivity is another important dimension to consider in narrative visualization. User interactions with various story pieces may include highlighting and filtering behaviors, or transitions between different events (Segel and Heer 2010). These interactions can serve as integral narrative elements that guide the user through the story.

The visualization technique chosen in narrative visualization should always be appropriate to the data it represents and align with the narrative intent or user task(s). This corresponds with standard practice in visualization design. Classic visualization design often tends toward widely used, unembellished techniques that are known to be safe and effective, such as bar charts. However, a risk of using such simple or generic charts is that these are easily forgotten (Kosara 2016). Narrative visualization seeks to engage the audience and create memorable data stories. This means that there is greater use of charts incorporating specific visual aspects of the data for better recall. For example, disease incidence can be depicted with human isotypes to show disease occurrence as a fraction or percentage, rather than simply text. Annotation of visual elements is another important aspect of narrative visualization, tying in with important concepts such as data provenance and acknowledgment of uncertainty (Hullman and Diakopoulos 2011). Various storytelling techniques can be realized by means of annotation, e.g., labels, whereby annotations are revealed incrementally for different aspects of the visualization to guide the user through the story without being inundated with information.

Scatterplots appear often in data stories to compare variables against each other and to highlight possible correlations and trends. In narrative visualization, the story creator often adds a trend line to the data to send a clear message to the audience. Scatterplots can also incorporate a time dimension through animation. One of the most successful and well-known examples of an animated scatterplot telling a clear visual data story is Hans Rosling's World Income Distribution plot from the Gapminder Foundation shown in Fig. 4.3. Rosling plots the health vs. wealth of all countries of the world, and then guides the audience through the evolving story of this relationship over the last few centuries. It is a

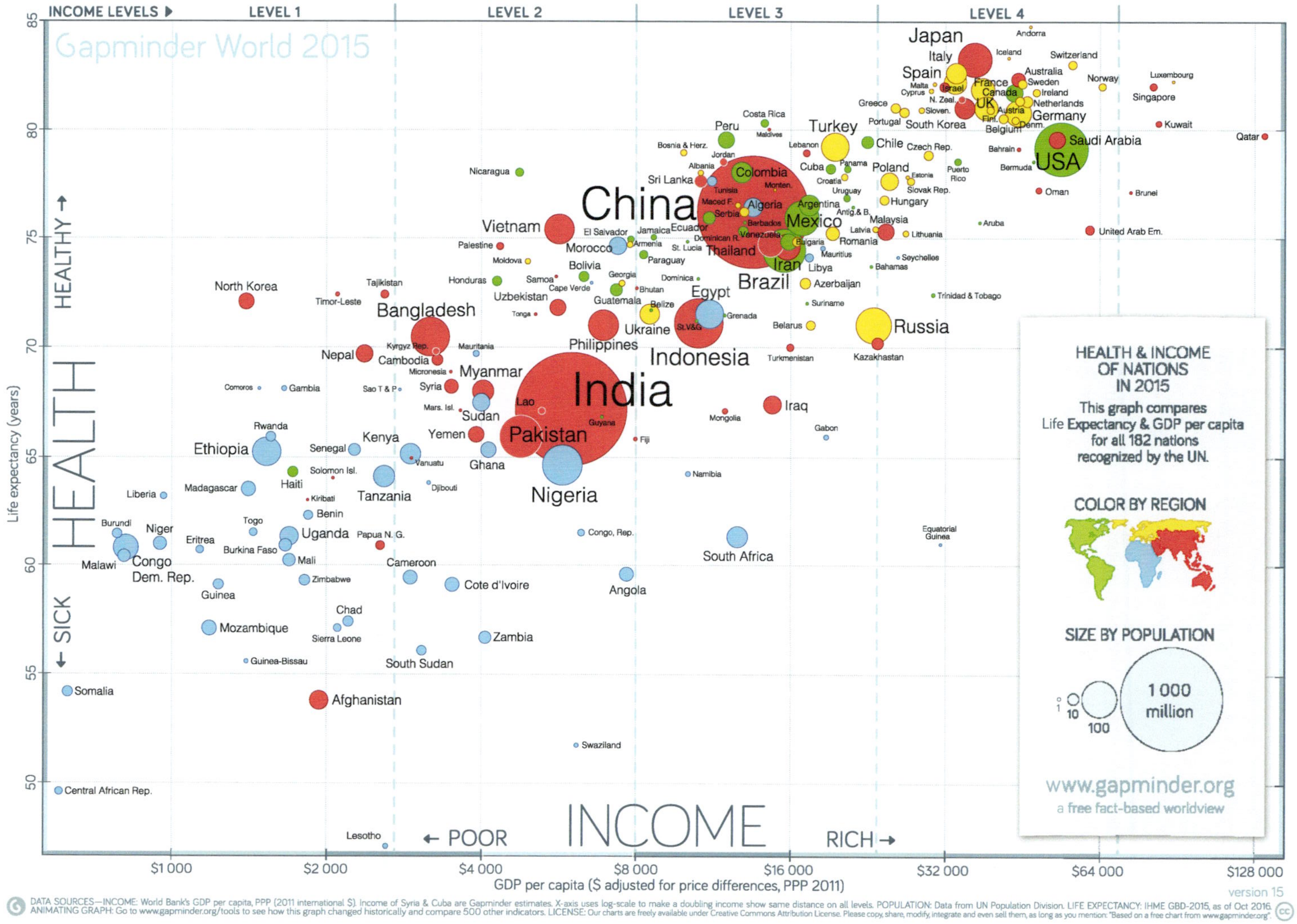

Fig. 4.3 Gapminder visualization of the countries of the world, plotted in terms of life expectancy (*y*-axis) and median income (*x*-axis). The colors represent the regions of the world, while the size of the circles indicates the population of the country in question. Image from Gapminder Foundation, reused under Creative Commons Attribution 4.0 International License (https://en.wikipedia.org/wiki/File:Gapminder-World-2015.pdf, https://creativecommons.org/licenses/by/4.0/deed.en)

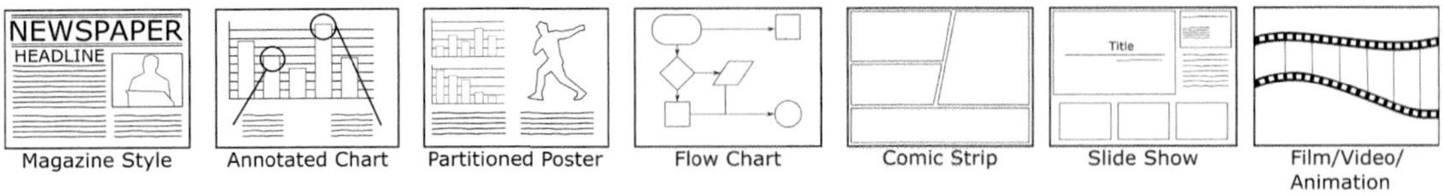

Fig. 4.4 Seven narrative genres for storytelling with data graphics, defined by Segel and Heer (2010). Image adapted from Segel and Heer (2010), reproduced with permission from Meuschke et al. 2022

fascinating and insightful story, where his narrative intent is to demonstrate that world development is not simply a division between rich and poor or developing versus developed countries. This visualization is complex. Rosling's narrative guidance, paired with the animation, is helpful for the audience to follow the narrative thread. Absent animation or Rosling's guidance, a story creator could consider incorporating filters into the visualization, e.g., by region, to reduce the amount of visible information at one time.

4.2.3 Narrative Structure and Genre

Following decisions on narrative intent and assembly of the story pieces, narrative visualization considers the structure and presentation of the story. The chronology of the story, and how it is presented, depends on the narrative intent and the target audience. We can borrow from ancient storytelling techniques like Aristotle's suspense arc (Aristotle 2008) to arrange our story in a logical and engaging way. Aristotle's suspense arc consists of three acts: after an introduction in which characters and a conflict are introduced, the story moves toward a climax in which the conflict intensifies. In the third act, the conflict is resolved; the characters succeed or fail. These three acts easily expand to five acts in Freytag's pyramid, which includes an exposition, rising action, climax, falling action, and denouement (Madej 2008). Shakespearean plays, such as Macbeth and Hamlet, follow this structure. Another popular storytelling structure is Campbell's Hero's Journey (Campbell 2008). This is a character-driven structure that describes typical stages a hero reaches throughout a story. The Hercules story, for example, follows this narrative structure. In Sect. 4.5, we include a detailed case study of the Hero's journey in a data-driven medical story.

Whatever storytelling structure is followed, presentation of the story's narrative is another consideration. Segel and Heer (2010) characterize the design space of storytelling with data graphics through an analysis of 58 visual data stories including content produced by news media outlets, blogs, instructional videos, and visualization and communication research. Their resulting classification of these visual data stories introduced seven visual narrative genres, which are illustrated in Fig. 4.4. These genres include (1) magazine style, (2) annotated chart, (3) partitioned poster, (4) flow chart, (5) comic strip, (6) slideshow, and (7) film/video/animation. These genres are distinguishable mainly by their ordering and number of frames. Much of our research in narrative medical visualization has explored the slideshow genre and the scrollytelling genre, a newer addition to the set of visual narrative genres. Scrollytelling is generally agreed to have been introduced by the New York Times in 2012 with their story, "Snowfall" (New York Times 2012), and has more recently been formalized in visualization research (Seyser and Zeiller 2018).

Genre selection can in part decide the linearity, or non-linearity, of the story structure. Stories may have branching events that return back to the main storyline, or that diverge in a "choose your own adventure" type of scenario. Our case study in Sect. 4.4 discusses a simple incorporation of storyline branches in the context of narrative medical visualization. Stories with branching structures naturally require a degree of interactivity with the audience, while more

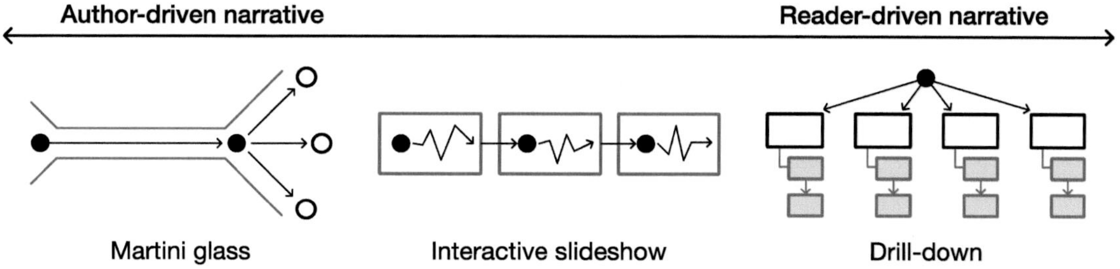

Fig. 4.5 Narrative approaches range along a continuum from author- to reader-driven structures. The martini glass (left) begins as an author-driven narrative with limited user choice, but later expands to a reader-driven narrative, allowing the user many options to explore. The interactive slideshow (middle) is mainly author-driven with restricted user freedom between each slide. However, within each slide the user may have many options to explore the story. This narrative structure is well-suited to a slideshow genre (Segel and Heer 2010). A drill-down structure (right), falls furthest along the continuum toward a reader-driven narrative, with many possible options for the user to explore before drilling down into a particular narrative branch. Image adapted from Segel and Heer (2010)

linear, non-branching structures allow for a more passive audience experience.

The story creator's decisions about user interactivity, intended messaging, and narrative genre choice lead to data-driven stories that lie along a continuum ranging from author- to reader-driven narratives (Segel and Heer 2010), illustrated in Fig. 4.5. In a completely author-driven narrative, the ordering of events is linear and prescribed, with strong messaging and no interactions, as in a static, magazine-style genre narrative. Stories may also blend author- and reader-driven approaches. For instance, a *martini-glass* structure (Fig. 4.5, *left)* begins as an author-driven narrative with little choice or interactivity allowed to the audience, but once the author has delivered their intended message the story structure opens up for exploration on the part of the user. Another structure that lies closer to the author-driven side of the continuum includes *interactive slideshow*, shown in Fig. 4.5, *middle.* This typically adopts a slideshow narrative genre and allows for limited interactivity within a given slide. The *drill-down* structure, shown in Fig. 4.5, *right*, enables even more user exploration with a free selection of a theme or storyline that they can then drill into for further details. On the furthest end of the continuum, completely reader-driven narratives can support multiple event orderings, which results in less strong messaging from the story creator and open user interactions.

An important cautionary note is that the ways in which the story creator chooses to group and sequence data and visualizations have an impact on the amount of information that the audience can directly access. The omission of, or different approaches to categorizations or aggregations of information, are effectively *framing effects* on the information that can lead to vastly different outcomes and opinions from the audience (Hullman and Diakopoulos 2011). While connected to narrative intent, this question of what to omit and how to aggregate information can also raise questions of ethics and transparency.

4.2.4 Narrator and Story Personalization

The point of view from which a story is told can dramatically effect how an audience perceives the story. A narrator can help deliver a story's message in the tone that the story creator intended. Choice of narrator can also personalize a story, and is a device used often in journalism to emotionally engage an audience. A narrator can be a character within the story itself, perhaps the main character, an ally, an antagonist, or a neutral party (Campbell 2008; Giovannelli 2009). The narrator may also be an objective, omniscient presence residing outside of the story's cast of characters.

One of the open questions in narrative visualization is the degree of influence that the identity of the narrator exerts on audience perception, trust, and subsequent behaviors after completing a data-driven story. Section 4.5 explores this question and considerations in a narrative medical visualization case study with three different narrators in a story about diagnosis and treatment of a neurodegenerative condition.

4.2.5 Evaluation in Narrative Visualization

Evaluation is critical to determine the success of a story, e.g., its narrative intent, story engagement, and story coherency. These aspects can be evaluated through baseline interviews or surveys prior to the story experience, and with follow-up interviews after the audience has experienced the story to assess story memorability, knowledge gain, and suitability of the narrative intent. Evaluation criteria often used in classical visualization design, such as user task speed and precision (Preim et al. 2018), are less relevant in narrative visualization.

4.2.6 Narrative Visualization in the Sciences

Research into applying narrative visualization techniques and strategies in climate research (Böttinger et al. 2020), astronomy (Ynnerman et al. 2020), and the biological and medical sciences are gaining traction. For instance, the World Health Organization incorporates narrative visualization techniques into their web content for public health education and outreach, such as the W.H.O. COVID-19 Dashboard (W.H.O. 2023). Semiautomatic authoring tools to tell visual stories are also in development. A notable recent example of applying narrative visualization to the life sciences is Molecumentary (Kouřil et al. 2023), a tool for producing documentary-style content that renders structural models from molecular biology in real time. Artistic expertise is not necessary, and this tool is able to prepare a story structure and construct a narrative from this structure. Text can be incorporated as narration, and this textual input may also guide semiautomated camera moves to guide the narrative.

Investigations into audience preferences when crafting biomedical stories are another growing branch of study from narrative visualization that can shed light on appropriate visualization techniques to use within a life sciences-oriented story. An early study in this space examined the practice and preferences of different audience groups for data-driven visualizations and medical illustrations of five well-known biomedical processes: signal transduction between molecules, constitutive activation of molecules, normal blood flow, aneurysm formation, and tumor movement in metastasis (Garrison et al. 2021). This study found that aesthetics, while important, is not the most critical criterion for topic experts to choose a visualization, and that, unsurprisingly, a degree of information abstraction is more often preferred by broad audiences relative to expert audiences. These audience-specific findings are useful to keep in mind when developing visual stories in the life sciences and other complex topics.

4.3 Narrative Medical Visualization

Narrative medical visualization takes and adapts the basic ideas and framework of narrative visualization and applies these concepts to the specific needs and challenges of the medical domain. Different flavors of these adaptations are likely true for any specific domain, e.g., adapting narrative visualization to stories for climate sciences. Adapting the concepts of narrative visualization to medicine is important because this framework can, for instance, provide an efficient and engaging means to educate patients and the general public about disease occurrence and, perhaps more importantly, disease treatment and prevention. Despite the wealth of medical data collected for every patient throughout their lifetime, understanding how to communicate real medical data encoding complex anatomical relationships and diseases to a lay audience remains an immense and open challenge. Data-driven stories showing real medical data are promising in their possibility to allow nonartists, such as researchers, to create medical stories that visually communicate their science. Our discussion in this section of approaches for narrative medical visualization blends theory and concepts from narrative visualization with practical advice for crafting data-driven medical stories.

4.3.1 Expertise and Audience Involvement

The key to any successful story is ensuring that the right voices are heard and included in the story creation process. In narrative medical visualization, this means that the storytelling team should include a physician or medical expert with competence in the topic of interest. Involving members of the target audience or those in a position to sympathize with the needs and interests of the target audience is invaluable, but this is often where corners are cut in the story creation process. Continued engagement with these stakeholders, e.g., through interviewing and testing concepts, and sharing early sketches and prototypes of a data-driven medical story, will ensure a more effective and targeted end result.

4.3.2 Narrative Intent

Narrative intents for medical stories can focus on communicating preventable risk factors for a given disease that can leave the audience feeling empowered to make actionable, positive health choices. For example, smoking is a risk factor for numerous diseases, with multiple studies reporting probabilities of disease occurrence with smoking. A story creator can combine the *call to action* and *human behind the dots* narrative patterns (Bach et al. 2018) to achieve a narrative intent inspiring the story viewer to quit smoking. This narrative intent could be realized by introducing a relatable story character who initially is a smoker, and reporting the higher risks of a given disease amongst smokers in comparison to non-smokers. The story can then wrap up with an outlook on reduced disease risk and other health benefits the story character has experienced when they quit smoking. Rhetorical questions to hook and engage an audience with a story can be useful narrative patterns to incorporate, e.g., *Why is cancer often curable today?* or *How can I strengthen my immune system?*

Human aspects are particularly important to consider for medical stories. Many data-driven medical stories focus on patients or loved ones of a patient. These people are often in a stressful, emotional situation where they are scared or overwhelmed by a disease or condition that they do not fully understand. Empathy at the outset can go a long way in ensuring that the message lands as intended, without painful repercussions.

4.3.3 Story Pieces: Data

Medical data are common story pieces in narrative medical visualization that can deliver a high personal impact. Clinicians use medical imaging data, e.g., X-ray, magnetic resonance imaging, ultrasound, or computer tomography data in diagnostics and treatment planning. If processed appropriately, these data can also form the basis of a medical story for broader audiences. Appropriate processing requires that any diseased or abnormal structures are delineated, highlighted in color, and provided with suitable textual information. Time-series data can be animated to help illustrate the progression or treatment of a particular disease. For example, if the data are available, a medical story can be crafted that shows the shrinkage of a tumor with successful therapy over time. Beyond medical imaging data, other medical data such as statistics on disease prevalence, disease burden, and commonly affected age groups are useful story pieces adopted for narrative medical visualization.

4.3.4 Story Pieces: Visualization Techniques

The visualization techniques available within narrative medical visualization are numerous and their selection depends on the data, narrative intent, and user tasks. Scatterplots, similar to Hans Rosling's world health vs. wealth example that we discussed earlier, are useful for narrative medical visualization to show relationships between clinical variables, e.g., smoking and disease occurrence. This type of visualization can support a narrative intent of comparing and revealing patient-specific disease information.

Imaging data can play a decisive role in patient diagnostics and treatment and is an important facet to consider when crafting a medical story. Imaging data can not only provide a more tailored story to a patient by using their personal data but also means that artistic training or skills may not be necessary to craft a visualization. These data can be rendered and presented in different styles, from static abstracted 2D to interactive realistic 3D structures. The decision comes back again to the narrative intent and the target audience. While interactivity has been shown to engage users and contribute to memorability (Richter Lipford et al. 2010), this may not be the right call for a patient audience with low computer literacy. Understanding when and where to leverage different visual representations of imaging data is an ongoing research question. We encourage incorporation of text labels and descriptions to accompany imaging visuals (Oeltze-Jafra and Preim 2014), as this information can help with identification and clarification of the structures, especially for audiences less familiar with anatomy and physiology (Meuschke et al. 2022). Icon-based visualizations, such as isotypes as shown in the far-right panel of Fig. 4.6, can also be meaningfully connected with imaging data visualizations.

Increasing data complexity necessitates considered approaches to story guidance and reduced information complexity. This can be achieved through thoughtful interaction and navigation techniques that are part of the visualization. In a 3D environment, e.g., a liver model with embedded tumor tissue, preset views, limited rotation capabilities, and/or preset parameters (Fig. 4.6, bottom left panel) can be helpful to restrict the exploration space and ensure that the audience

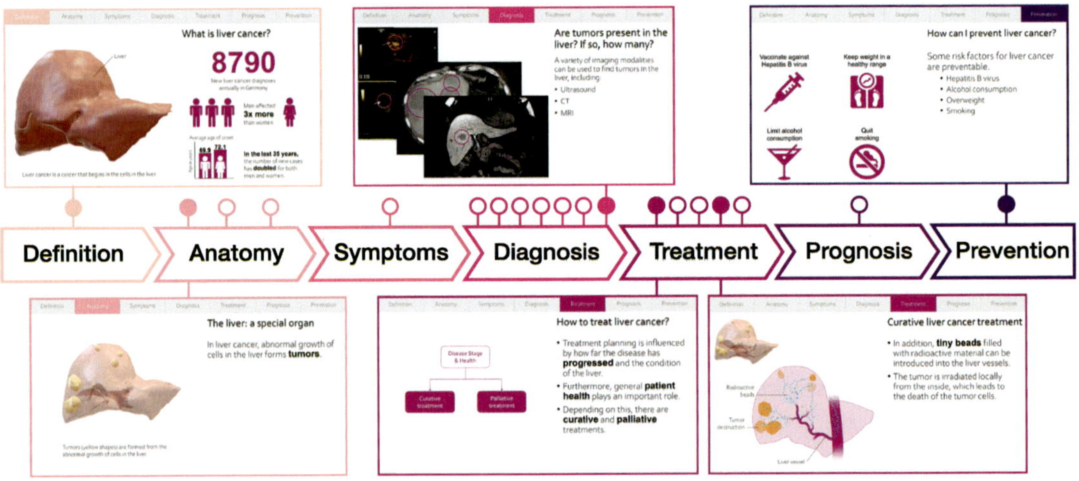

Fig. 4.6 Liver cancer story overview with scene excerpts using Meuschke et al.'s (2022) proposed seven-stage narrative structure for data-driven medical stories. Using the slideshow narrative genre, this story about liver cancer uses diverse story pieces including imaging data, 3D models, medical illustrations, diagrams, free text, annotations, and isotype graphics. Image modified with permission by Meuschke et al. (2022). Illustrations created by Laura Ann Garrison

does not become disorientated (Ynnerman et al. 2020). "Hotspot" annotation mechanisms are another option to filter information or to facilitate transitions. On selection, these may trigger, e.g., movement to a new camera angle or reveal further contextual information about the region of interest.

Another consideration for 3D environments is managing occlusion. Elmqvist and Tsigas 2008 identify five approaches to manage object occlusion, including multiple viewports, virtual X-ray tools, tour planners, volumetric probes, and projection distorters. Virtual X-ray tools are especially exciting for narrative medical visualization, as the audience can be empowered by bringing them closer to the clinician's perspective. They can investigate and reveal hidden information, like turning a liver to "x-ray mode" to localize the tumors buried within. Other smart visibility techniques, like ghosted views and cutaway techniques, are good options for such "invisible" structures (Lawonn et al. 2018).

4.3.5 Medical Narrative Structure and Genre

To date, narrative medical visualization has explored possibilities for a base narrative structure for telling data-driven medical stories. These efforts have centered on stories about diseases for broad audiences (Meuschke et al. 2022). Similar to the data story design space analysis conducted by Segel and Heer 2010, Meuschke et al. (2022) devised their medical narrative structure by analyzing health and medical online content developed by hospitals and other health and wellness websites. Their narrative structure for disease stories, shown in Fig. 4.7, outlines seven steps that follow the path of Aristotle's tension arc.

The narrative opens with (1) a definition of the disease and brief statistics on its incidence in the population. This serves as a hook to draw the audience into the story. An (2) anatomical overview follows, which orients the audience to the anatomical structures that are involved or affected by the disease. Tension is introduced to the narrative by (3) describing common symptoms of the disease, and tension reaches a climax with (4) disease diagnosis. The story progresses toward resolution with (5) a discussion of treatment options and their associated benefits and risks. Tension continues to drop with (6) disease prognosis, often presented as a 5-year prognosis, and the story finally concludes with an overview of (7) preventative measures for the disease that leave the audience with a sense of agency. This structure provides a logical framework for the grouping of data and other story pieces. For example, Fig. 4.6 shows excerpts from a liver cancer story that Meuschke et al. (2022) created following this template. Imaging data story pieces are used in (4) disease diagnosis, while 3D models and abstracted illustrations show (5) possible treatment options. While this template is linear by design, it can be easily extended with side branches to provide, e.g., more in-depth information on a certain surgical procedure in the treatment section before rejoining the main story line. From this basic structure, more complex branching structures may also emerge, e.g.,

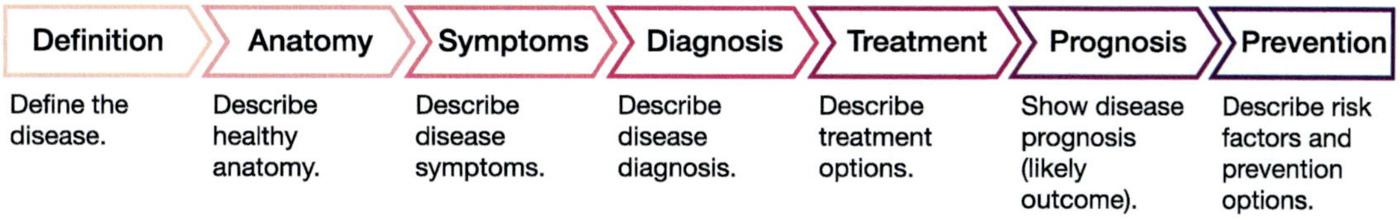

Fig. 4.7 Meuschke et al. (2022) proposed a narrative structure for visual disease stories told to broad audiences. The steps include (1) disease definition, (2) an anatomical overview, (3) symptoms, (4) diagnosis procedure, (5) treatment options, (6) prognosis, and (7) prevention. Image was modified with permission from Meuschke et al. (2022)

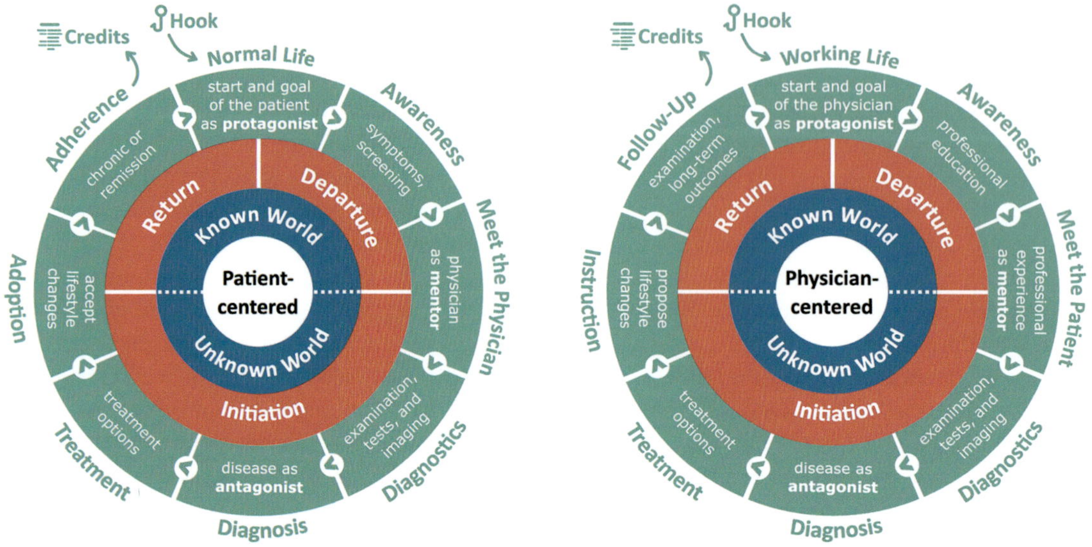

Fig. 4.8 The disease journey adapted from Campbell's Hero's Journey shows a structure for the patient-centered (left) and physician-centered (right) disease story. The structures depict the different stages (green) the protagonist must experience. These stages divide into three acts (red): departure, initiation, and return, which are split between the protagonist's known world and the unknown. The story begins with a narrative hook, before showing the first stage of the disease journey. After the last stage, the story ends with a call for prevention and credits showing the story authors' affiliations for information transparency and credibility. Image reproduced with permission from Mittenentzwei et al. (2023)

the audience chooses a specific treatment, which leads to a diverging branching structure for each possible treatment option. Other structural variants may skip one or more steps altogether if not relevant to the narrative intent or the content of the story.

Campbell's Hero's Journey (Campbell 2008) is another story structure which can be applied to disease journeys that can be experienced by, e.g., a patient or clinician protagonist, as conceptualized by Mittenentzwei et al. (2023). Illustrated in Fig. 4.8, the disease journey crosses between the protagonist's known world of prior experiences and the unknown in three main acts: (1) departure, (2) initiation, and (3) return. In the first act, the protagonist, e.g., a patient, is introduced to the audience living their normal life, before an event, e.g., symptoms or diagnostic screening, alters their awareness and begins their journey. Meeting a mentor, e.g., a healthcare professional, builds to the second act: initiation. During the second act, the protagonist encounters the antagonist, e.g., the disease, and experiences the main conflict of the story, e.g., diagnosis of the disease. In the final act, the protagonist navigates from the unknown back to the known world, e.g., through treatment and follow-up.

4.3.6 Personalizing Data-Driven Medical Stories

Narrative medical visualization can introduce personalization through the choice of narrator, or by including additional details about story characters to encourage relatability and empathy between the audience and the story. To facilitate broader relatability between story characters and a diverse audience, the story creator may represent characters as abstract and androgynous figures as shown in Fig. 4.9, such that characteristics like gender, body shape, hair color, or age are left vague and open to the audience's imagination (Kleinau & Stupak et al. 2022).

Many medical stories that are targeted to patient audiences adopt a patient character as the main narrator and protagonist of the story. The

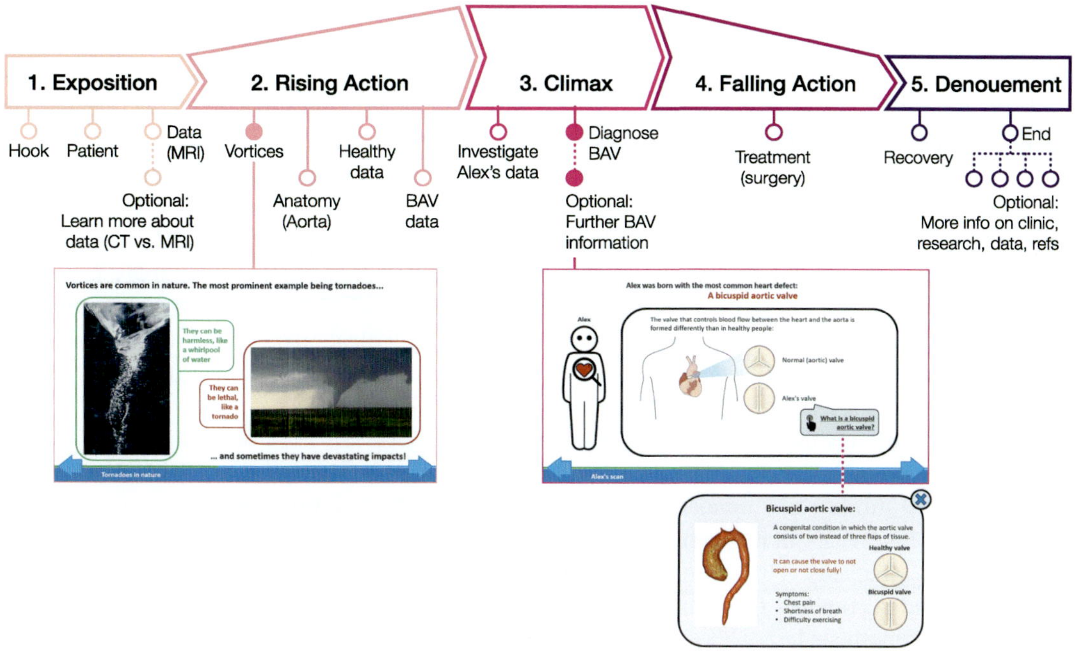

Fig. 4.9 Freytag's Pyramid story structure (Madej 2008) in five acts as applied to a bicuspid aortic valve disease story. Image modified with permission by Kleinau & Stupak et al. (2022) under Creative Commons Attribution 4.0 International License (https://creativecommons.org/licenses/by/4.0/deed.en). Heart and valve illustrations created by Laura Ann Garrison

main conflict of the story is often the presence of a disease or other medical condition affecting the patient's quality of life. Following the structures that Meuschke et al. 2022 and Mittenentzwei et al. 2023 describe, the audience joins the patient through their experience of disease diagnosis and discussion of treatment options, all while the disease continues to progress, building to a climax with disease diagnosis. The story tension begins to recede with our patient receiving treatment, their subsequent recovery, and their progressive return to normal life. Throughout the narrative, sharing our patient's inner struggles and emotions can help the audience empathize with our patient in this journey.

Data-driven medical stories may, instead of patients, focus on the clinician as the main character and story protagonist. The case study we discuss in Sect. 4.5 explores a story variant where the narrative follows a clinician as the story protagonist (Mittenentzwei et al. 2023). The narrative intent of such stories may be to garner sympathy and to promote cooperation with an overworked and overburdened health system in the face of public health emergencies, as experienced at the height of COVID-19. Personalization for this type of story can focus on a doctor or nurse who works long hours under intense time pressure, all the while wearing masks that leave deep and painful imprints on their faces. This can help the audience see that clinicians are not, as we may sometimes want to believe, superheroes—they are human, and are generally trying their best to help.

These personalization examples are in the spirit of positive emotional empathy, or to encourage choices for the advancement of public health, e.g., getting vaccinated. However, a critical perspective on stories triggering feelings of empathy from an audience may find that these approaches are deliberately manipulative and impact story credibility. Introducing counterfactuals or alternative viewpoints in a narrative can mitigate this perception of bias or manipulation, and ultimately produce a more holistic and informative story.

4.3.7 Credibility in Medical Stories

Credibility is vital in narrative medical visualization. Regardless of their level of medical education, people can and should be critical of their information source and demand transparency when related to their personal health and medical records. The fact that the pharmaceutical industry, the medical technology sector, and the medical profession have not only the welfare of the patient in mind but also tangible economic interests is well-known in times of rampant hospital privatization. This means that a narrative intent to encourage patients to participate in vaccinations and screenings could be perceived as economically motivated.

To establish and retain credibility, it is important to avoid oversimplifying or providing only one-sided information on the benefits of medical interventions or any other aspect of medical care. The story creator should be careful to use data from reliable, high-quality sources, and to avoid bias in their interpretation. To allay concerns about data source or quality, data provenance should be retained and displayed where and whenever possible. It should also be clear whether the data has been altered for the preparation of the story and, if so, by whom and with what objectives. Does a renowned medical professional, e.g., a hospital director, vouch for this story? If so, this can help establish trust. This vetting by a formal health entity should be conveyed early in the story, and repeated at the end to remind the audience of this validation.

4.3.8 Evaluating Medical Stories

Evaluation in narrative medical visualization is challenging, as there are numerous ways to tell a story and evaluation can, on some level, become a subjective exercise. Audience engagement and memorability are meaningful objective measures of evaluation, with an early between-subject study ($n = 90$) finding that a carefully designed narrative structure, comprehensible visualizations, easy-to-use interactions, and credible references contribute to more memorable data-driven medical stories (Meuschke et al. 2022). The following two case studies provide concrete examples of story development and evaluation in narrative medical visualization research, beginning with a case study exploring the process and various challenges of adapting complex medical data into a coherent and engaging data-driven medical story.

4.4 Case Study: Bicuspid Aortic Valve Disease

This case study illustrates an application of a narrative medical visualization structure presented in two different narrative genres. We also explore some of the complexities and challenges of translating medical data into a broadly-accessible visual narrative. The story describes a common type of congenital heart disease and its treatment (Kleinau & Stupak et al. 2022). This story was developed in close cooperation with clinicians and researchers affiliated with the Heart Center Leipzig in Germany, which often encounters and treats this condition.

The aortic valve is one of four valves of the heart, providing the connection between the left side of the heart and the aorta, which is the largest blood vessel in the body, and which carries oxygen-rich blood from the heart to the rest of the body. In healthy individuals, the aortic valve consists of three leaflets. With bicuspid aortic valve disease, two of the three aortic valve leaflets are fused at birth. Bicuspid aortic valve disease is the most common congenital heart defect, affecting up to 2% of the population (Masri et al. 2017). The fused valve leaflets may mean that the valve cannot seal completely, leading to a so-called "leaky valve," or the valve may be narrowed. In either case, blood flowing through this valve becomes more turbulent. Over time, this turbulence can stress and weaken the wall of the aorta, causing it to bulge into what is called an aneurysm. Aneurysms are potentially life-threatening and need careful monitoring. If an aortic aneurysm ruptures, survivability is low. Early detection and observation of this disease are critical to ensure patient health and mitigate risks of such a rupture.

The *narrative intent* of this story is to inform and engage a broad, non-expert audience on the complexities and risks associated with bicuspid aortic valve disease, and more specifically to help the audience understand the connection between turbulent blood flow and aneurysm formation from this valvular disease. Metaphor can be a powerful tool to give meaning and context to unfamiliar phenomena. For this disease, clinicians often use a tornado metaphor to describe turbulent blood flow in the aorta to their patients. Tornados are well-known extreme weather occurrences with varying degrees of strength and severity, similar to the varying strength and severity of flow vortices in the aorta. The story creators adopt this metaphor into the narrative of this story, shown in Fig. 4.9, to achieve familiarity and bridge gaps in medical knowledge.

The *story pieces* include clinical imaging data known as 4D PC-MRI (phase-contrast magnetic resonance imaging) that capture blood flow dynamics in the human body. These data are processed through software that detects and classifies the different types of blood flow turbulence (Köhler et al. 2019). Blood flow data are complex. At each point in space, what is known as a vector describes the direction and strength of the blood flow at that point. These vectors change over the course of a cardiac cycle. Common convention in medical illustration and in flow visualization is to place blood flow in an anatomical context, in this case, the aorta, to make the information less abstract. To reduce visual complexity, another common convention in flow visualization is to show flow vectors only in anatomical regions of interest, e.g., low-velocity regions where vortices may be more likely to occur (van Pelt and Vilanova 2013). These are areas where flow is especially turbulent, which is what the story creators want to highlight. To make the flow visualization more readable and more illustration-like, another convention is to integrate the flow vectors visually into pathlines that show the path of blood flow, rather than a single point in space. While the fluid mechanics community often maps the velocity of the flow to color, this mapping choice is not intuitive to an audience outside of this domain. To facilitate communication, the story authors instead represent the blood flow pathlines as one color, as shown in Fig. 4.10.

This story uses Freytag's Pyramid and follows the basic *narrative structure* for data-driven medical stories about diseases from Meuschke et al. (2022). Figure 4.9 shows a high-level overview of this structure, which consists of five acts: (1) exposition, (2) rising action (3) climax, (4) falling action, and (5) denouement. The story creators adopt a martini-glass story structure as in Fig. 4.5 (Segel and Heer 2010), with an author-driven opening that has limited interactions and a strong message in the form of a narrative hook before transitioning to a reader-driven experience with more interactive elements and limited optional story paths.

In the first act, the story opens with a narrative hook phrased as a rhetorical question, with different hook options and accompanying visuals that the story creators explored shown in Fig. 4.11. To establish credibility at the outset, the story creators place the logos of the involved institutions, which include universities and hospitals, on the opening screen of the story. The audience is also introduced to the main character, Alex, who is the patient experiencing this disease. The last segment of the exposition introduces the user to the blood flow data in the context of Alex's upcoming examination. At this point, the story pivots from a fully author-driven experience to a partly reader-driven experience, where the audience may choose an optional story path to learn about the difference between CT and MRI medical data, before returning to the main story path.

The second act builds story tension by reintroducing the tornado metaphor, shown in Fig. 4.9, which is referenced in the title of the story. The narrative continues, describing blood flow vortices as "tornadoes in the blood." While at first illustrated as a harmless case in a healthy person, the narrative escalates tension by showing

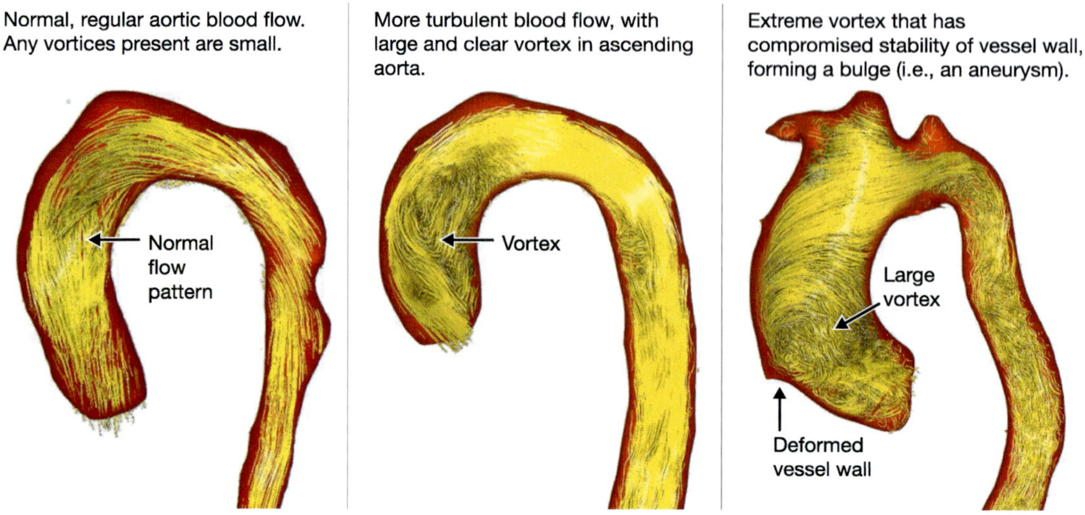

Fig. 4.10 Data-driven representation of blood flow through the aorta in three different levels of severity. The blood flow data are represented as pathlines. While fluid dynamics convention maps the velocity of these data to color, for the purposes of this story, the data are mapped to a single color to minimize information complexity and audience confusion. Image reproduced and modified with permission by Kleinau & Stupak et al. (2022) under Creative Commons Attribution 4.0 International License (https://creativecommons.org/licenses/by/4.0/deed.en)

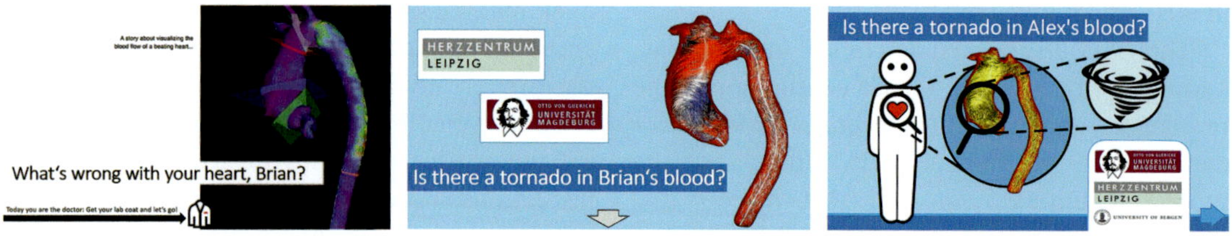

Fig. 4.11 Early variations of a narrative "hook" with different visuals to begin the data-driven medical story about bicuspid aortic valve, a common congenital heart disease, including What is wrong with your heart, Brian? (left), Is there a tornado in Brian's blood? (middle), and Is there a tornado in Alex's blood? (right). Image reused with permission by Kleinau & Stupak et al. (2022) under Creative Commons Attribution 4.0 International License (https://creativecommons.org/licenses/by/4.0/deed.en)

a blood flow vortex causing serious health risks, with the possibility of aneurysm formation and subsequent rupture. The climax of the story is the point where the user investigates if Alex has blood vortices by comparing a 3D visualization of Alex's blood flow against the normal flow of a healthy patient and the flow of a patient with an aortic aneurysm. On determining that Alex does indeed have vortices in their blood, tension begins to resolve when the narrative identifies bicuspid aortic valve disease as the cause of the vortices. Often in medicine, having an answer to the cause of the symptoms that a patient is experiencing can already help relieve their anxiety. Another optional story path is available at this point for the interested audience to learn more about the signs and symptoms of bicuspid aortic valve disease, shown in Fig. 4.9.

The fourth act follows the narrative's falling action by describing Alex's surgical procedure, which uses medical illustrations of the relevant heart and aortic anatomy in a clear and simple style intended to inform without intimidating the patient or their loved ones who may be following the story. In the fifth and final act, story tension is resolved by allowing a side-by-side visual comparison of Alex's aortic blood flow before and after the surgery, where the audience discovers that Alex's blood flow patterns are no longer life threatening. The story closes with Alex returning to normal life, with links to further reading for the audience who would like to learn more about bicuspid aortic valve disease.

This narrative is presented in two genres: (1) a PowerPoint-style *slideshow* and (2) a long, vertically structured "*scrollytelling*" experience. These genres produce different interactive and navigational experiences that require thoughtful user interface and experience design. While scrolling naturally initiates transitions between story acts and is generally intuitive on touch screens, slideshows still have a strong association with clicking to advance to the next slide, which can be unintuitive on tablets or other touchscreen solutions. Speed and smoothness of transitions between story pieces are another important consideration. Scrollytelling experiences set an expectation for a smooth and rapid transition between pieces. If the transition feels slow or awkward, the audience may disengage from the story. Some audiences may not understand that navigation forward and backward in the story is possible, or that alternative story paths are available. For both genres, graphical user interface elements that make these navigational options clear are integral to a complete user experience. In a preliminary evaluation with a broad audience ranging from older children to middle-aged adults ($n = 24$), Kleinau & Stupak et al. (2022) found that many users prefer the click interactions of a slideshow to navigate the story and that some users were unsure of how to move forward from the story opening with the scrollytelling genre. While this is just one case study, these findings demonstrate the importance of interaction and navigation experience planning in narrative medical visualization.

4.5 Case Study: Cerebral Small Vessel Disease

Our second case study explores the challenges and considerations of narrator and story protagonist choice in a data-driven medical story about cerebral small vessel disease (Mittenentzwei et al. 2023). Cerebral small vessel disease (CSVD) is a broad term for a variety of changes in small blood vessels in the brain that can cause further damage to surrounding brain tissue (Wardlaw et al. 2013). These vascular changes can have effects ranging from mild to severe, e.g., slight dizziness, dementia, or a life-threatening stroke. CSVD affects about 5% of people in their 50's, but this number increases to 100% of the population older than 90 years (Cleveland Clinic 2022). In spite of such widespread occurrence, most people have never heard of this disease. Awareness of CSVD, and an understanding of protective measures to mitigate severe effects later in life, can have a profound positive effect on public health.

The *narrative intent* of this story is to familiarize an early- to middle-aged adult audience with the possible causes and symptoms of CSVD, focusing in particular on the causal relationship between high blood pressure and CSVD. Through empathy and engagement with the narrative, a second intent is to motivate personal lifestyle changes and regular blood pressure checks to mitigate the more serious risks of CSVD later in life.

The story follows Campbell's Hero's Journey *narrative structure*, depicted in Fig. 4.8, which we discussed as a narrative structure for disease stories earlier in Sect. 4.3 of this chapter. In this structure, a disease story protagonist is the story character who is affected by a disease or its risk factors. Mittenentzwei et al. (2023) designed two human protagonist variations of the CSVD disease journey story, one with a patient protagonist and the second with a physician protagonist.

In the patient-centered variation of the story, we meet Emma, a 59-year-old patient with chronic high blood pressure. Emma's disease journey begins with her awareness of high blood pressure as a possible symptom of an underlying disease. She departs her known world and enters the unknown, meeting her mentor in the form of an unnamed physician who helps and guides her through the next stages of the story. She undergoes diagnostic imaging, examinations, and tests, before arriving at a diagnosis where she finally meets her disease antagonist: cerebral small vessel disease. She is then exposed to and chooses from a set of possible treatment options, and re-enters her known world and adopts lifestyle changes, e.g., quitting smoking, to manage the disease. She visits her mentor, the physician, for follow-up to assess whether the disease is under control or in remission, or if further treatment steps must be taken. The story concludes with a call for prevention by Emma advising her family to take steps to manage high blood pressure before serious symptoms of CSVD could appear later in life.

The physician-centered story follows the same basic structure, but from the physician's perspective: the first act begins with meeting Dr. Schreiber in her position as senior physician in her hospital's neurology department. Her medical education leads to her awareness of CSVD and its signs and symptoms in her patients. Her professional experience on the hospital floor serves as an abstract mentor in this narrative, which could instead be another more senior physician. She is initiated into the unknown world when she begins examinations and tests on her patient whom she suspects may have developed CSVD, but she is unsure until she amasses enough diagnostic evidence to meet the main story antagonist, CSVD, and make her diagnosis. She researches different treatment options that may work and discusses these with her patient, and re-enters the known world by proposing lifestyle changes to help her patient lower their blood pressure to slow or stop progression of the disease. She schedules follow-up visits with her patient and assesses possible long-term health outcomes for her patient. Her narrative ends with a call for prevention, where she outlines preventative measures that all people can take to mitigate risks of CSVD later in life.

This case study introduces several important design questions to consider when developing a character-centered narrative. Centering a story's narrative around a human character whom the audience can more easily identify with, rather than a more impersonal recitation of statistics, can help an audience empathize and engage with the story (Mittenentzwei et al. 2023). Story creators may consider spending more time developing the character at the outset of the story to help the audience form a bond with this character before their journey begins.

Chen et al. (2016) suggest guidelines for audience identification and self-referencing with the story and its protagonist. Allowing the audience to identify with and see themselves in the story is key, as people tend to underestimate their personal risk. In health, this underestimation leads to less personal motivation to make big lifestyle changes (Wachinger et al. 2013). The story creator can also consider how credibility can be differently perceived if a disease story is presented through the eyes of an experienced physician, or through a patient who the audience may more closely identify with and relate to. Physicians are often upheld as authority figures, but pressing too hard on this point could sacrifice the narrative's relatability to the audience.

To assess credibility, the Gaziano-McGrath scales of accuracy, fairness, trustworthiness, bias, and completeness are useful guides (Gaziano and McGrath 1986). Lastly, the story creator can consider audience engagement by measuring the users' perceived versus actual length of time spent with a story. Tying back into other questions connected to protagonist relatability and empathy, an optimal design choice would see a story that the audience wants to spend time in, where the story feels shorter than it is.

4.6 Outlook and Challenges for Narrative Medical Visualization

By blending traditional storytelling structures and devices with best practices for data visualization, narrative medical visualization can help to tell better data-driven medical stories. In this chapter, we outlined the core ideas behind narrative visualization, and discussed how these ideas transfer and must be adjusted to address specific challenges for telling stories in medicine. By adopting narrative medical visualization practices, we can tell medical stories that can transform patient education and public health outreach. Narrative medical visualization is a growing area of study, with several challenges as well as opportunities to be aware of in this developing research space.

4.6.1 Community-Driven Stories

Stories are told by people, about people, for people. Without involving those for whom the stories are to be told by, or to, in the development and evaluation process, data-driven medical stories can fail to reach their target audience with their intended message. As story creators, our approaches must involve members from diverse communities if our goal is to communicate medicine to the general public. This involves thinking beyond Western structures of storytelling that have been investigated in narrative medical visualization to date. Furthermore, mounting evidence shows that the traditional "knowledge deficit model" of science communication is not particularly effective in inciting behavioral changes (Bucchi and Trench 2021; Simis et al. 2016). We need to adopt different, situated approaches to communicating data-driven medical stories. People need to feel personally motivated to make such changes, and understanding how to motivate requires a dialogue with the target audience communities.

4.6.2 Data Challenges

Data is a core component in narrative medical visualization, with an array of associated challenges ranging from data access and ethics to handling the myriad data types and quality. Most medical data is simply inaccessible for use in narrative medical visualization due to patient privacy. While some open-source data sources exist, these are limited. Initiatives for producing data compliant with government regulations, e.g., data derived from data donors, could help increase medical data access.

Data quality is another challenge, especially for stories built on data from longitudinal clinical studies. The story creator must consider if the dataset is mature enough to draw and present valid conclusions, which involves assessing the completeness and comprehensiveness of the data. A dataset may be intrinsically biased in some way, leading to different conclusions than related datasets. If this happens, the story creator must carefully evaluate how to reconcile these differences, and if there is a more trustworthy source.

Clinical data can vary over broad scales of space and time. For instance, organs measure in centimeters but consist of tissues and cells measuring several orders of magnitude smaller. The molecular events triggering cellular mechanisms that drive organ functionality can occur in the space of nanoseconds, while treatment of a disease that results from a repeating series of misconfigured molecular events can take years. Given such variation, it is not surprising that the data describing these broad scale ranges are also numerous (Garrison et al. 2022). Medical data can include radiological and histological image data, numerical values from lab studies, and statistical information. These data can be acquired across myriad different devices, often with different calibration settings that tie into our earlier questions on data quality. Simulation data for biomedical processes may also be useful in a narrative, which further expands the data space.

4.6.3 Story Experience Challenges

With so many possible data sources, types, and story scenarios, authoring systems in narrative medical visualization should be flexible enough to support a range of interactions and rendering styles that the story creator can select for the story that they wish to tell. The system should also be robust enough to be made available to and usable by a broad audience. However, while many such options should be available for the story creator, it is often not a good idea to expose so many options to the audience. Curated and thoughtful interactions and stylistic choices can lead to a more positive and impactful user experience.

While we discussed earlier the issue of data accessibility, user accessibility is a separate issue. Around 15% of the global population lives with some form of disability, which can include visual, motor, or cognitive disabilities. Narrative medical visualization seeks to reach a broad audience that encompasses people with such disabilities, but understanding the mechanisms that we as story creators can use to tell more inclusive stories is by no means a solved problem. Revisiting the first challenge regarding community involvement, this problem can only be solved by including those with disabilities in the story development and evaluation process.

Credibility is important in any science communication but can come with life-or-death consequences in the medical domain, e.g., if a person chooses to follow bad medical advice from an untrustworthy or unreliable source. An active research question in narrative medical visualization is how to best establish the credibility of a story, and how credibility can be differently perceived with different story narrators. Data provenance is another aspect of credibility. Apart from including the data source and providing annotations in the story that reveal relevant data source information, what other methods can we incorporate to better link our stories to the underlying data that drive these stories?

Closely connected to credibility is the question of ethics in storytelling. Although narrative medical visualization espouses a data-driven approach to storytelling, it is undeniable that humans are heavily involved in the process, and may, wittingly or not, insert their personal subjectivity or biases into the story. Considering how to send the desired story message without unduly influencing or manipulating the audience is a nuanced and challenging question. Ensuring that a data-driven medical story addresses both sides of an argument, or discusses both positive and negative aspects of, e.g., a particular treatment, can produce a more ethical data-driven medical story.

Finally, the often highly specific needs and requirements of different topics in medicine mean that it is difficult to develop meaningful evaluation guidelines in narrative medical visualization that are both comprehensive and general enough for use across a range of use cases. We can look to other disciplines, e.g., in the learning theory and education domains, to develop more robust and general evaluation guides for the stories that we create, and to help prioritize the most important story characteristics for evaluation.

4.7 Conclusion

Much work remains to bridge narrative medical visualization research into practice, and to understand drivers for engagement and driving positive lifestyle-changing behaviors for audiences of such visualizations. Building on broader research and practice in narrative visualization and science communication, this chapter has provided a foundation for current approaches in narrative medical visualization, offering readers a basis for effective communication and directions for future research.

References

Alkureishi MA, Johnson T, Nichols J et al (2021) Impact of an educational comic to enhance patient-physician-electronic health record engagement: prospective observational study. JMIR Hum Factors 8:e25054. https://doi.org/10.2196/25054

Aristotle (2008) The poetics of Aristotle, by Aristotle. Translated by S.H. Butcher. Available at: https://www.gutenberg.org/files/1974/1974-h/1974-h.htm. Accessed: March 26, 2023

Bach B, Stefaner M, Boy J et al (2018) Narrative design patterns for data-driven storytelling. In: Riche NH, Hurter C, Diakopoulos N, Carpendale S (eds) Data-driven storytelling, 1st edn. A K Peters/CRC Press, Boca Raton, Florida, pp 107–133. https://doi.org/10.1201/9781315281575-5

Böttinger M, Kostis H-N, Ynnermann A (2020) Challenges and open issues in visualization for broad audiences. In: Chen M, Hauser H, Rheingans P, Scheuermann G (eds) Foundations of data visualization. Springer International Publishing, Cham, pp

381–389. https://doi.org/10.1007/978-3-030-34444-3_21
Bucchi M, Trench B (2021) Rethinking science communication as the social conversation around science. J Sci Commun 20(03):Y01. https://doi.org/10.22323/2.20030401
Campbell J (2008) The hero with a thousand faces. New World Library
Chen M, Bell R, Taylor L (2016) Narrator point of view and persuasion in health narratives: the role of protagonist–reader similarity, identification, and self-referencing. J Health Commun 21:1–11. https://doi.org/10.1080/10810730.2016.1177147
Elmqvist N, Tsigas P (2008) A taxonomy of 3D occlusion management for visualization. IEEE Trans Vis Comput Graph 14:1095–1109. https://doi.org/10.1109/TVCG.2008.59
Fathom (2023) Fathom information design https://fathom.info/. Accessed 20 Mar 2023
Garrison L, Meuschke M, Fairman J, et al (2021) An exploration of practice and preferences for the visual communication of biomedical processes. Proc of Eurographics Workshop on Visual Computing for Biology and Medicine. https://doi.org/10.2312/vcbm.20211339
Garrison LA, Kolesar I, Viola I et al (2022) Trends & opportunities in visualization for physiology: a multiscale overview. Computer Graphics Forum 41: 609–643. https://doi.org/10.1111/cgf.14575
Gaziano C, McGrath K (1986) Measuring the concept of credibility. Journal Q 63(3):451–462. https://doi.org/10.1177/107769908606300301
Giovannelli A (2009) In sympathy with narrative characters. J Aesthet Art Critic 67:83–95. https://doi.org/10.1111/j.1540-6245.2008.01337.x
Hullman J, Diakopoulos N (2011) Visualization rhetoric: framing effects in narrative visualization. IEEE Trans Vis Comput Graph 17:2231–2240. https://doi.org/10.1109/TVCG.2011.255
Joubert M, Davis L, Metcalfe J (2019) Storytelling: the soul of science communication. JCOM 18. https://doi.org/10.22323/2.18050501
Kearns C (2023) Artibiotics comics portfolio. https://artibiotics.com/comics. Accessed 19 Mar 2023
Köhler B, Grothoff M, Gutberlet M, Preim B (2019) Bloodline: a system for the guided analysis of cardiac 4D PC-MRI data. Comput Graph 82:32–43. https://doi.org/10.1016/j.cag.2019.05.004
Kosara R (2016) Presentation-oriented visualization techniques. IEEE Comput Grap Appl 36:80–85. https://doi.org/10.1109/MCG.2016.2
Kouřil D, Strnad O, Mindek P et al (2023) Molecumentary: adaptable narrated documentaries using molecular visualization. IEEE Trans Vis Comput Graph 29:1733–1747. https://doi.org/10.1109/TVCG.2021.3130670
Lawonn K, Smit NN, Bühler K, Preim B (2018) A survey on multimodal medical data visualization. Computer Graphics Forum 37:413–438. https://doi.org/10.1111/cgf.13306
Lee B, Riche NH, Isenberg P, Carpendale S (2015) More than telling a story: transforming data into visually shared stories. IEEE Comput Grap Appl 35:84–90. https://doi.org/10.1109/MCG.2015.99
Madej K (2008) "Traditional narrative structure"–not traditional so why the norm?
Masri A, Svensson LG, Griffin BP, Desai MY (2017) Contemporary natural history of bicuspid aortic valve disease: a systematic review. Heart 103:1323–1330. https://doi.org/10.1136/heartjnl-2016-309916
Meuschke M, Garrison LA, Smit NN et al (2022) Narrative medical visualization to communicate disease data. Comput Graph 107:144–157. https://doi.org/10.1016/j.cag.2022.07.017
Mittenentzwei S, Weiß V, Schreiber S, Garrison L, Bruckner S, Pfister M, Preim B, Meuschke M (2023) Do disease stories need a hero? Effects of human protagonists on a narrative visualization about cerebral small vessel disease. Computer Graphics Forum 42: 123
New York Times (2012) Snowfall. https://www.nytimes.com/projects/2012/snow-fall/index.html#/?part=tunnel-creek. Accessed 20 Mar 2023
Oeltze-Jafra S, Preim B (2014) Survey of labeling techniques in medical visualizations. In: Proc. of Eurographics Workshop on Visual Computing for Biology and Medicine, pp 199–208. https://doi.org/10.2312/vcbm.20141192
van Pelt R, Vilanova A (2013) Understanding blood-flow dynamics: new challenges for visualization. Computer 46:60–67. https://doi.org/10.1109/MC.2013.121
Preim B, Ropinski T, Isenberg P (2018) A critical analysis of the evaluation practice in medical visualization. Proc. of Eurographics workshop on visual computing for biology and Medicine, pp. 45–56. https://doi.org/10.2312/VCBM.20181228
Richter Lipford H, Stukes F, Dou W, et al (2010) Helping users recall their reasoning process. pp 187–194. https://doi.org/10.1109/VAST.2010.5653598
Segel E, Heer J (2010) Narrative visualization: telling stories with data. IEEE Trans Vis Comput Graph 16: 1139–1148. https://doi.org/10.1109/TVCG.2010.179
Seyser D, Zeiller M (2018) Scrollytelling–an analysis of visual storytelling in online journalism. In: 2018 22nd International Conference Information Visualisation (IV)., pp 401–406. https://doi.org/10.1109/iV.2018.00075
Simis MJ, Madden H, Cacciatore MA, Yeo SK (2016) The lure of rationality: why does the deficit model persist in science communication? Public Underst Sci 25:400–414. https://doi.org/10.1177/0963662516629749
Stupak E, Kleinau A, Mörth E, et al (2022) Is there a tornado in Alex's blood flow? A case study for narrative medical visualization. Proc. of Eurographics Workshop on Visual Computing for Biomedicine. https://doi.org/10.2312/vcbm.20221183

Wachinger G, Renn O, Begg C, Kuhlicke C (2013) The risk perception paradox–implications for governance and communication of natural hazards. Risk Anal 33: 1049–1065. https://doi.org/10.1111/j.1539-6924.2012.01942.x

Wardlaw JM, Smith EE, Biessels GJ et al (2013) Neuroimaging standards for research into small vessel disease and its contribution to ageing and neurodegeneration. Lancet Neurol 12:822–838. https://doi.org/10.1016/S1474-4422(13)70124-8

World Health Organization (W.H.O.) (2023) WHO health emergency dashboard. https://covid19.who.int/. Accessed 20 Mar 2023

Wolters Kluwer (2023) Guide patient engagement journeys with emmiengage. https://www.wolterskluwer.com/en/solutions/emmi/emmi-engage. Accessed 19 Mar 2023

Ynnerman A, Ljung P, Bock A (2020) Reaching broad audiences from a science center or museum setting. In: Chen M, Hauser H, Rheingans P, Scheuermann G (eds) Foundations of data visualization. Springer International Publishing, Cham, pp 341–364. https://doi.org/10.1007/978-3-030-34444-3_19

Cleveland Clinic (2022) Microvascular ischemic disease: symptoms & treatment. https://my.clevelandclinic.org/health/diseases/22927-microvascular-ischemic-disease. Accessed 19 Mar 2023

Animation for Science Communication: A Description of the Production Process and Contexts for Using Science Animation

5

Mhairi Claire Towler

Abstract

Animation as a visualization tool for explaining scientific concepts to specialist and non-specialist audiences is widely used. At Vivomotion we have developed a process over the last 10 years, when working with scientific collaborators to develop a visual narrative for their research, as follows: (1) Write a script, (2) Create a voice-over, (3) Storyboard the script, (4) Illustrate the storyboard, (5) Animate the illustrations, (6) Receive comments on the draft, and (7) Produce and deliver the final animation. These steps encompass the pre-production, production, and post-production stages of animation production and simplify the process so we can explain it with ease when embarking on new collaborations.

This chapter discusses, using case studies, how this process is executed, where the pitfalls lie, and how to select a 2D animation versus a 3D animation. This process is proposed as a methodology for those collaborating on scientific animation productions and those involved in animation production of technical subject areas. It highlights the different contexts, with case studies, where animation can be used for scientific communication.

Keywords

2D animation · 3D animation · Animation production process · Animation methodology · Scientific animation · Scientific visualization · Scientific communication

5.1 Introduction

Since the first scientific animations were published, Harvard Universities' "*The Inner Life of the Cell*" animation being one of the first in 2006 (XVIVO 2006) (https://xvivo.com/examples/the-inner-life-of-the-cell/), the number of scientific animations available online has increased dramatically, and their use for various aspects of scientific communication has grown rapidly. Whether this is to illustrate the mechanism of drug action (Aqdot 2023) (https://aqdot.com/products/advanced-academic-research/), how the immune system works (Rebhahn et al. 2014), or how gene synthesis occurs inside a cell (Stromme and Mork 2021), the applications are endless. Animation is particularly useful when the dynamic spatiotemporal relationships of molecules functioning at the sub-microscopic level cannot be seen with the naked eye, and where static imagery cannot do these relationships justice (Jantzen 2015).

The audiences also vary; ranging from teaching science to students, communicating science to a general audience through outreach projects,

M. C. Towler (✉)
Vivomotion, Dundee, UK
e-mail: mhairi@vivomotion.co.uk

M. Roughley (ed.), *Approaches for Science Illustration and Communication*, Biomedical Visualization 4,
https://doi.org/10.1007/978-3-031-41652-1_5

helping investors understand innovative technologies, or for peer-to-peer communication at scientific conferences. Audience consideration informs what style of animation to choose, whether 3D computer-generated animation, to show protein/protein interactions at a level understood by experts, or 2D, for storytelling of a more generic nature.

Vivomotion (Vivomotion 2023) (https//www.vivomotion.co.uk) is a science animation company that was launched by Dr. Mhairi Towler in 2012 after she spotted a gap in the market to help scientists communicate what they do using visuals. Towler has a background in life sciences and carried out a degree in Biochemistry followed by a PhD in Cell and Molecular Biology. Following this she carried out postdoctoral research at the University of California San Francisco and at The University of Dundee. As part of her training, she carried out a Postgraduate Research Certificate in Teaching & Learning in Higher Education and had to address how to make large group lectures more engaging. After some research, she decided that animation would be one way to do this. Fortuitously, there was an animator, Janice Aitken, from Duncan of Jordanstone College of Art & Design, also studying for a Postgraduate Certificate, so they teamed up for their action research project to create an animation of a cell for Towler's lectures to second-year undergraduate students studying a module in Cellular and Molecular Biology. The project won "The Principals Prize for Innovative Teaching" from the University of Dundee in 2009. Towler enjoyed the project so much that she decided to develop the idea for a business competition run through the University of Dundee called "Venture."

Part of the Venture competition involved market analysis. Having attended many scientific conferences Towler was aware that sometimes the visual presentation of ideas and concepts could be improved. She appreciated that animation would be a solution to illustrate dynamic processes that occur within the cell. In addition, many protein/protein interactions would benefit from being seen in a 3D animated format, thus enhancing scientific communication. Towler also discovered that scientific animation could be done as a profession, and scientific animation companies already existed in the USA, Australia, and Canada. After being placed runner-up in the Venture competition, she decided to put the idea into practice but realized she would have to undertake some training in animation, and so applied for a Masters in Animation and Visualisation at Duncan of Jordanstone College of Art and Design. On completion of the Masters, Towler launched Vivomotion and has since engaged several freelance animators to work with her on animation commissions, dependent on the animation style requested. 3D animator Fraser Murdoch (Murdoch 2023) (https://www.frasermurdoch.co.uk), 2D animator Dan Thompson (Thompson 2023) (https://danthompson.tv) and illustrator Andrew Torrens (Torrens 2023) (andrewtorrens.com) have worked frequently with Vivomotion. Voiceover artists Alison McKay (McKay 2023) (https://www.alisonmckay.com), Hannah Visocchi, and Charlotte Thompson, have recorded scripts for Vivomotion productions.

A second branch of Vivomotion comprises training for postgraduate research students and academic staff on the visual presentation of research, including academic poster design, data visualization, animation, visuals for public engagement, figure preparation, and graphical abstracts (Vivomotion 2023) (https://www.vivomotion.co.uk/training). These topics are taught by Towler online, on-demand and face-to-face and have been developed based on her uniquely placed perspective, as someone who moved from science into the creative sector. She delivers the training at Universities across the UK and Europe.

This chapter describes, from previous experiences at Vivomotion, the process involved in animation production when collaborating specifically with academic scientists. This pipeline has evolved over time and is a simplified version of the traditional animation pipeline used by animation studios such as DreamWorks Animation, as outlined by Lepito 2018. Additionally, it is discussed why consideration of the context for the animation delivery, informs whether to produce a 2D or a 3D piece.

5.2 Why Use Animation as a Communication Tool?

It is said that a picture is worth a thousand words. If this is true, then a moving picture or animation must be worth so much more. A visual narrative is powerful—creating emotion, tension, and empathy with the viewer. Animation allows complex information to be made accessible to everyone. As a medium, it immediately connects with people, as they think back with nostalgia to animations and cartoons, they watched when they were children, associating it with entertainment, which immediately creates engagement. Within a research context, it allows for technical information to be conveyed to the audience, whether they are specialists or not. Using a visual language surpasses language barriers in international settings or where much jargon is involved and allows inclusivity for students who are visual learners. In a medical context, multi-language animation was shown to support explanations of mechanical ventilation, to families needing reassurance who had relatives in intensive care during the COVID-19 pandemic (Wachtl 2022). All these different contexts for the use of animation demonstrate what a dynamic medium it is and why it is such a powerful method for communication.

Vivomotion has collaborated with academics at various stages of their careers and produced animations for a variety of reasons described below. First, let us consider the use of animations by postgraduate research students. For PhD students, outreach is an important part of their development as researchers (Vitae 2023) (https://www.vitae.ac.uk/vitae.ac.uk/vitae-publications/rdf-related/public-engagement-lens-on-the-vitae-researcher-development-framework-rdf-apr-2013.pdf). The ability to communicate their research to non-specialist audiences is a transferable skill they can use throughout their career. To this end, Towler developed a training workshop for postgraduate researchers to help them make a 2 min, 2D animation, explaining what their research is about and importantly, why they are doing it, for a lay audience. This training provides an opportunity for students to foster and apply their digital literacy skills. Topics covered include, the importance of visual storytelling, script writing, voiceover production, storyboarding, and software options. The outcome can be seen as a "visual elevator pitch" to convey the essence of their research to a wide audience. It is also a chance to develop crucial employability skills including effective communication, digital literacy, and visual storytelling. It builds an awareness of open access to information and the laws around copyright. It is an approach to outreach/research dissemination that can utilize social media as a channel to reach more people while allowing students to establish themselves as experts in their own fields.

This workshop was adapted to provide training for PhD students at The University of Glasgow and The University of Newcastle, who wanted their students to enter the "Visualise your Thesis," competition, initiated by The University of Melbourne (The University of Melbourne 2022) (https://sites.research.unimelb.edu.au/visualise-your-thesis). This competition is open to those who are currently enrolled as PhD, MPhil, and Professional Doctorate (Research) candidates at universities who have registered for the competition. Researchers can be from any discipline. The competition encourages entrants to use visual storytelling, either live film or animation, to communicate their research for a lay audience in 1 min. It was rewarding to see the outputs from the "Animation Production for Social Media" training workshop, being entered into this competition after several students were mentored through the animation production process. An example animation produced by PhD student Deborah Molloy from The University of Glasgow can be seen here (Molloy 2021) https://www.youtube.com/watch?v=BYe5iSVmthw. For PhD students, learning to communicate their research in this way early on in their careers, is a good skill

Fig. 5.1 Stills from animation about how quantum communications work. Created for Dr. Ross Donaldson at Heriot-Watt University for public engagement purposes

to take forward in their career, whether academic or non-academic.

For academic researchers, using animation, is a way to communicate what their research is about to a wider audience using online platforms. Grant applications require a section on impact and public engagement activities which can encompass animation as a method of research dissemination or pathway to impact. A recently produced animation for Dr. Ross Donaldson at Heriot-Watt University in Edinburgh is a good example of this (Donaldson 2022) (https://www.youtube.com/watch?v=yocfFT1XeAo). Dr. Donaldson wanted to use this animation to educate the public and tell other academics about his research on Quantum Communications and wanted to use social media platforms such as Twitter, YouTube, and LinkedIn to reach more people. He also wanted to use it as a tool when visiting schools to talk about his research with young people. Thus, animation was used as a medium to make his work more accessible to more people. Stills from this animation, illustrating aspects of quantum communication, can be seen in Fig. 5.1.

Other individuals Vivomotion have worked with have used animation in education for teaching purposes. Mr. Rodney Mountain, a retired ear, nose, and throat surgeon who operated and taught at Ninewells Hospital in Dundee, wanted an animation of a new surgical instrument he had developed to teach medical students how to use it during surgery. He also used it to promote this innovative instrument to companies that wanted to sell the instrument to surgeons across the world (Fig. 5.2). The benefit of using animation in this context, over live film, is that some viewers get squeamish when observing surgery, therefore using animation can convey how the surgical instrument works while being removed from the real thing.

Animations have also been produced to accompany the publication of a scientific journal. Two examples are given below. First, Prof. Nicola Stanley Wall, University of Dundee, and Prof. Cait MacPhee, University of Edinburgh, who work on biofilms wanted an animation to accompany their publication (Stanley-Wall and MacPhee 2015) (Fig. 5.3). In this case, a mixture of real data with representative data was used. The protein structures were collected from the protein data bank (RCSP PDB 2022) (https://www.rcsb.org), a repository of experimentally determined 3D structures of proteins and computed structure models and placed into an imagined environment to demonstrate how biofilms formed over time.

Secondly, Vivomotion worked with Dr. Tam Mignot from The University of Marseille to create an animation to support his publication in Nature Microbiology (Mignot 2016) (https://microbiologycommunity.nature.com/videos/video-43) (Fig. 5.4). In this case, having previously thought about bacterial movement with 2D illustrations, visualizing proteins in a 3-dimensional way helped Dr. Mignot and his collaborators shape their hypothesis about how the proteins were working together to create movement of bacteria across a substrate.

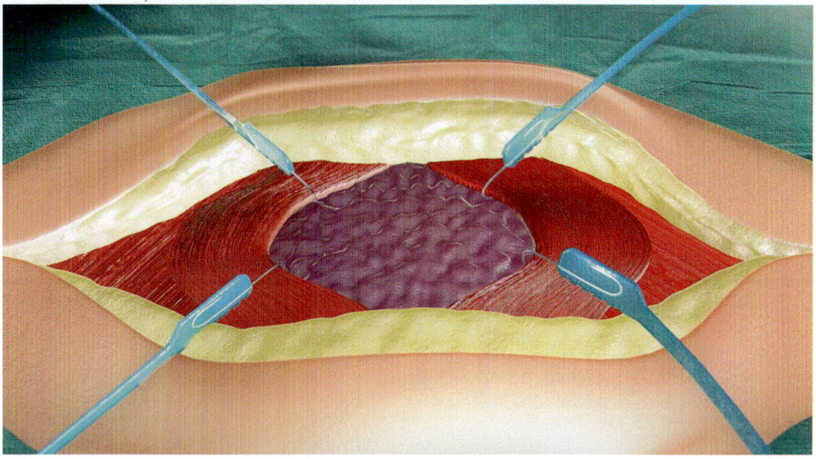

Fig. 5.2 Still from animation illustrating a surgical elasticated retraction instrument pioneered by Dundee surgeon Mr. Rodney Mountain. Mr. Mountain designed this instrument to make surgery less invasive for patients and thus safer. Animation helps medical students to learn the technique

Fig. 5.3 Still from animation produced for Prof. Nicola Stanley-Wall and Prof. Cait MacPhee illustrating how protein Bs1A creates a hydrophobic biofilm

Finally, another context that animations can be used to generate significant impact is in public engagement with research through art exhibitions. An example to illustrate this comes from work completed by Dr. Towler during her Master in Animation and Visualisation at Duncan of Jordanstone College of Art and Design at The University of Dundee in 2012. She was conducting a placement with Dr. Paul Harrison, a printmaker, and Lecturer in Contemporary Art Practice. Dr. Harrison was Artist in Residence for a European consortium of scientists working on epigenetics and systems biology called EpiGeneSys. He was looking retrospectively at the work of scientist Conrad Waddington, a leading geneticist from the 1930s onward at The University of Edinburgh. The aim was to produce a range of visual outcomes of the "Epigenetic Landscape" concept conceived by Waddington (Waddington 1957). The basic premise for this concept, described by Prof. Murdo Macdonald in EpiGeneSys: (re) visions (Harrison and Nickel 2016), is that, "*any developmental system can be visualised as a sloping landscape, and any developmental state within that system can be visualised as a ball rolling through that landscape.*" A still from the animation created to represent the Epigenetic Landscape is shown in Fig. 5.5.

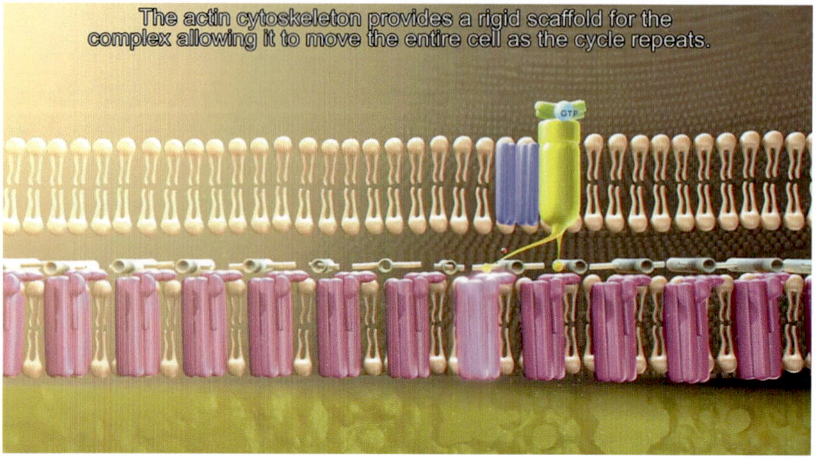

Fig. 5.4 Still from animation produced for Dr. Tam Mignot illustrating bacterial focal adhesion complexes (Reproduced from Faure 2016)

Fig. 5.5 Still from Epigenetic Landscape animation produced by Dr. Mhairi Towler, Link Li, and Dr. Paul Harrison 2012. In this case, the visual metaphor of the "Epigenetic Landscape" represents how cells differentiate from stem cells into a specific cell type, for example, a neuronal cell or a muscle cell, dependent on the pathway the cell takes through a particular environment

The animation created for this, as well as prints from the project, have been exhibited in art exhibitions globally, one example being a multimedia exhibition in the Cite Internationale des Arts in Paris in May 2015, as the culmination of the EpiGeneSys project (Fig. 5.6). A book about the outcomes of the project called "(re) visions" was also published (Harrison and Nickel 2016). To provide easy-to-understand but scientifically based information, the EpiGeneSys project held public lectures, commissioned production of scientific films, animations, and artworks, to promote discussion on the topic of epigenetics and systems biology with the wider public.

Frequently, once an animation has been completed for its primary function, it is used in a secondary context, whether that is teaching, gaining investment for business, grant funding, public engagement, art exhibitions, or for academic presentations at scientific conferences/symposiums. The examples highlighted above exemplify the plethora of contexts that animation can be utilized in.

5.3 The Value of Visual Storytelling

Academics must tell their story throughout their career. Whether that is a personal career story or

Fig. 5.6 Work created by Dr. Harrison, as well as digital prints from the Epigenetic Landscape Animation exhibited in Cite Internationale des Arts Paris, May 2015. Photo credit Andy Rice

the story of their research, they must tell it and sell it. When applying for grants they must craft their research into a story to achieve funding to continue the story. Collaborators are encouraged to ask, "Can your story be retold?" "Would people want to retell it?" and "What action does someone take as a result of hearing your story?" In animation studios, "Story is king," is a mantra that is often used and this philosophy is supported by all those involved in the production process (Lepito 2018). Scientists should remember this mantra when composing scripts, and try to become more creative with their stories, than merely reporting facts and figures. Often, the question of "Why are you doing this research?" serves as a good prompt when writing a script to help audiences care about what is being said.

In addition, the modern academic must be able to communicate their research to non-experts to increase societal impact. Non-experts include the public, policy makers, journalists, and other academics out with one's own discipline. This brings benefits to both universities and individuals. Funders and policy makers expect universities to do it and it helps universities adapt to a changing world. Storytelling is thus a skill that should be practiced by scientists for a variety of audiences.

5.3.1 Context

Is a script needed or not? Different versions of an animation can be provided for varied reasons. It comes down to the purpose of the animation and in what context it will be shown. If it is to be used for teaching purposes, then an academic may be talking over the animation themselves, therefore, a sequence of events might be all that is needed. This can be decided through preliminary discussions. Animation may also be required for a website, in which case a version with a script and closed captions may be required. The first decision is therefore to ascertain the purpose/ function of the animation.

When collaborating with academics to construct a narrative for their animation it is usual to begin with the conversation that we had on the first meeting to discuss their ideas. This provides an overview of the information they would like to convey to their given audience. When the purpose is clarified, the second thing to consider is audience. Who is going to consume the information? Who is the end viewer? Purpose and audience typically go hand in hand. Is it for a group of students, the public, for policy makers, or for other academics? This also informs a decision on the style of the animation that suits the purpose. If it is for public engagement, and specifically for school children, a 2D motion graphic style animation works well. If the context is to wow colleagues at an academic conference, then 3D animation is more sophisticated. With the audience and style of animation considered, writing the script is the next step.

5.3.2 Writing the Script

When writing the script, it is paramount to always keep the audience at the forefront of one's thoughts. Forget what your interests are and focus on what you want your audience to know. What facts do you want to convey? How do you want them to feel? What do you want them to think? Do you want them to take action as a result of viewing your animation?

Consider the length of the script when constructing this part. A question often asked is, "How long should the script be?" This of course depends on how long the piece of animation is to be. A guide for 2 min of animation is 300–340 words. This also depends on how quickly the voiceover artist speaks. Research has shown that on social media, anywhere between 7 s and 2 min, will keep the attention of the viewer, with the first 10 s being the most important (Socially Sorted 2023) (https://www.sociallysorted.com.au/ideal-video-length-social-video). This depends entirely on the platform being used, the interest the viewer has for your subject, and what they are doing at that point in time. Keeping it short, sharp, and to the point is preferable. Although researchers may enjoy talking at length about their research, again, the main take-home messages and key points that we want our audience to leave with need to be considered, and focus placed on those, bearing in mind that 2 min of animation takes a lot longer than 2 min to create! In general, beyond 30 s, the audience may begin to switch off.

To communicate effectively, it is advisable to grab the attention of the audience at the beginning. This is what is referred to as "the hook." It could be a shocking statistic, an interesting fact, or an element of surprise that the audiences were not expecting. Whatever strategy is chosen, it should grip the audience and make them keen to find out more about a given topic.

Connecting emotionally is one of the more complex nuances of script writing, especially from a scientific angle, when research topics can be abstract. This can be achieved subconsciously through the use of color, music as a backing track, or through a personal story that the audience can relate to (not always possible). Even selecting the right voice-over artist can be effective to connect with an audience. It can also be achieved by giving real-world applications of the research. Again, this is not always achievable depending on discipline/research topic. We are telling stories, so any anecdotes that can be included will bring a realism and human element to a piece. Case studies also work well.

Collaborators are encouraged to think of the beginning, middle, and end of a story, just as though they were writing a children's story. Begin by setting the scene and putting the piece into context for the audience. The middle is where the action occurs or the key points are highlighted, and finish with a memorable message—a metaphor, a thought-provoking question, or proposing a call to action. Finalizing, a piece can include acknowledging funding bodies and providing contact information. Keeping in mind that most often, animations will be online, it is important to include contact information for people to reach out as a researcher if they would like to open a dialogue.

Getting started can be the most difficult hurdle to overcome. In this instance "Free Writing" is suggested as a useful aid. This technique involves setting a timer for a set amount of time, for example, 3 min, thinking about a specific topic, and scribbling down anything that comes to mind. This information can then be reordered, edited, and put into meaningful sentences retrospectively. It overcomes writers' block and helps the writer to "begin" their narrative. It does not have to be grammatically correct, or make sense, but merely a mind dump onto the page to get the script started.

Another recommended tip, to get into a creative mood, is to take a walk in nature, or change environment. This can stimulate creativity by allowing the brain to relax and drift sideways, promoting lateral thinking and allowing innovative ideas to seed. A first draft of the script should be edited, removing anything that is not necessary to get a point across, eliminating any jargon that would not be understood by a general audience and making sure everything is as concise as possible. It is also a good idea at this point to have a practise of reading the script aloud. Conversational language should be used, bearing in mind that the spoken word is vastly different from the written word. When we read things aloud, we realize where changes need to be made. The script can also be highlighted where the voice-over artist should adjust their tone and emphasis to communicate a particular emotion. Once the script is written, polished, and ready to be recorded, voice-over production can begin.

5.3.3 Creating a Voice-Over

When producing an animation, it is best to listen to a few voice-over artists to cast the correct voice-over artist for your film. Again, it is worth keeping the audience in mind and the tone of the animation should also be considered. Are you pitching to young kids? In which case, a younger voice-over artist would be more relatable. It could be that you have decided to record your own voice. If this is the case, you will need to have a few rehearsals to familiarize yourself with the script and if there are any sticking points for you to get your teeth and tongue around. It is also important to warm up your voice. "The tip of the teeth, the tongue, and the lips," repeated several times is a good warm-up exercise. Remember to breathe, and mark pauses in the script, if you are using pauses for emphasis. Practise the delivery for pacing. Everyone has a natural rate of word delivery and when recording, adrenaline may speed up delivery. It is good to stand when reading the script, which helps project the voice, and makes your voice sound more animated. In addition, it is advisable to make a few recordings. On the first take, you might be more nervous, but on the second take you will relax into the process. A third take might sound even more confident. Talking with a smile on your face is a trick to make yourself sound more animated. Call center employees are taught this in training.

For the recording, choose a high-quality microphone and assess your microphone by reading a few lines first. Find a quiet room where you will not be disturbed and one that has no echo. If you want to dampen down sound, you can hang a duvet up behind you. Once you have recorded at least three versions, choose the one you are happy with.

Timing is important in animation. Images must fit the voice-over so must appear on screen at the same time as the voice is narrating the image. This sounds obvious, but on many occasions, animations have been observed where the timing of the animation does not fit the narration. Voiceovers should be created in one take if possible. If this is not possible then audio editing software should be used, such as Audacity. This allows editing of the audio file to remove unwanted pauses and mistakes, and to adjust the volume of the recording.

Adding music to an animation is an option. The main point of a backing track is to lift the film to another dimension and add personality, flare, and mood to the production. It is vital to choose a piece of music that is in keeping with the tone/mood of your animation. If you are working on a disease, for example, an upbeat jazzy piece of music would not be appropriate. Conversely, it does not have to be all doom and gloom, and a nice piece of classical music would suit well. Sites like Jamendo (Jamendo 2022) (https://www.jamendo.com) and Bensound (Bensound 2022) (https://www.bensound.com) are great for selecting royalty-free music to accompany short animations. If budget allows, then commissioning a musician to write a piece of music for a film would be beneficial, as it would be created in a bespoke manner to suit the idiom of the film. Whatever the approach, the music adds an extra dynamism to a piece of film and brings the animation to life. Sound design is a further option, depending on what aspects need to be highlighted in the piece. Professional sound designers can be commissioned to create special sound effects.

5.3.4 Storyboarding the Script

When collaborating with clients, it is preferable to finalize the script and finish voice-over production before moving on to the storyboard. Drawing can take time, and it is good practice, to be able to hand on a storyboard to my illustrator, which does not have panels crossed out or changes to be made later. It makes it easier for them and saves time drawing the storyboard and illustrating. However, there are times when creating the storyboard can result in changes made to the script, so having an open mind here is recommended.

What is a storyboard? A storyboard is a visual translation of the written script. Often, these hand-drawn panels will be stitched together to create an animatic, a pre-visualization of the final animation. The animatic can give a sense of timing and clarify a sequence of events, to make sure everything is in the correct order. The storyboard is a series of illustrated panels strung together in a sequence and in scientific animation, sometimes a simple word/picture association can work to create the visuals, as imagery can be a literal translation of the script. The storyboard gives us a visual template of the film and allows our thoughts to be turned into something tangible.

Storyboarding was first developed at Walt Disney Productions during the early 1930s. Simplifying concepts at the storyboard stage can be extremely effective, and often stick figures can be used to represent people, and blobs or shapes can be used to represent genes or proteins. Color can be added for effect, and this often highlights how someone may or may not have wanted something to be represented. Through personal experience, it has been observed that scientists like to use bright primary colors, whereas artists would tone these shades down. This is a generalization of course, and perhaps it is a legacy of the use of PowerPoint in the scientific world for delivering talks, where the default is to use brighter colors when drawing shapes to represent genes and proteins. It is an observation made when collaborating with artists and scientists.

There are many distinct styles of storyboard, dependent on the storyboard artist. In Fig. 5.6, the scientist, Prof. Leonie Ringrose from Humboldt University in Berlin, took charge of the storytelling and she knew that a 3D animation would best suit to illustrate the way that proteins were binding and releasing from the DNA. Prof. Ringrose, as a visual thinker, had a clear idea in her head of how she wanted the animation to play out. Prof. Ringrose had been invited to give a special public engagement lecture at The University of Bergen in Norway entitled, "Epigenetics: myths, mysteries, and molecules" (IMBA - Institute of Molecular Biotechnology 2014) (https://www.youtube.com/watch?v= MmKHVSl5Wlo). The rough script was written out down the right-hand side of the page, with illustrations to match on the left-hand side (Fig. 5.7). Above each of these panels, directions were given for the action of the proteins as well as camera actions detailed. It is not usual to receive such accurate descriptions from a scientist, however, in this case, Prof. Ringrose knew exactly what she wanted, and it made it very easy for interpretation. Camera angles and shots are usually planned at the storyboard stage, and clients are encouraged to think of themselves as the camera operator. If they were behind the camera, what action would they like to see happening?

In the next storyboard example (Fig. 5.8), a sequence of events was narrated by scientist, Prof. Miratul Muqit, when he gave the Francis Crick Lecture for The Royal Society in 2018 (The Royal Society 2019) (https://www.youtube.com/watch?v = 5vMZMNls_po). A still from the final animation is illustrated (Fig. 5.9). Prof. Muqit is based at the Medical Research Council Protein Phosphorylation Unit at The University of Dundee. No voiceover was requested, to allow Prof. Muqit to narrate the animation himself. In some respects, having no recording of the voice-over, made creating the timing of the animation more complex, but guidance on this was requested from Prof. Muqit, so that the action fitted what he wanted to say. His audience was made up of scientists and members of the public, including Parkinson's patients, so had to be understood by all. The lecture was awarded to Miratul "in recognition of his research on cell signalling linked to neurodegeneration in

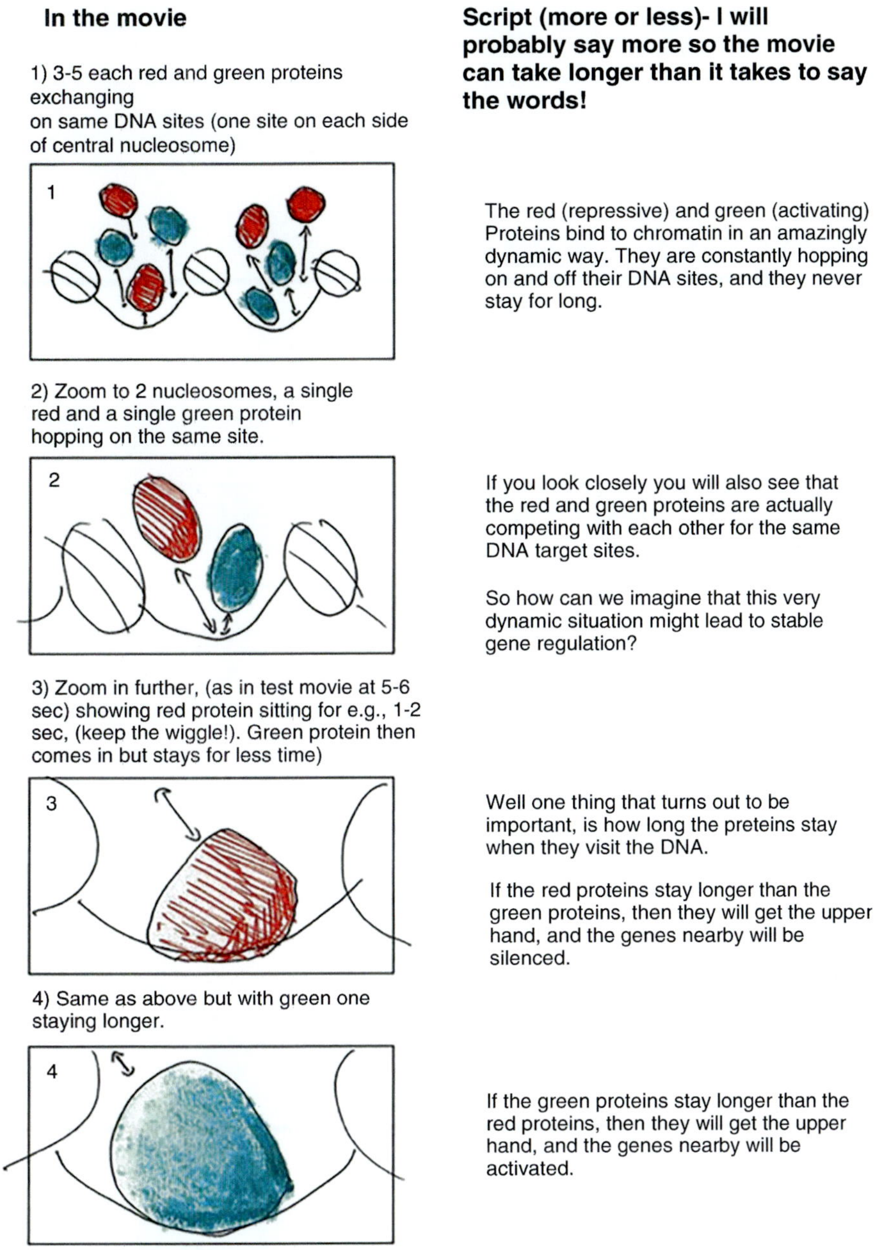

Fig. 5.7 Storyboard created by Prof. Ringrose for a public lecture on Epigenetics. The storyboard in this case includes sketches of the proteins involved, actions for the proteins, and a rough narrative down the right-hand side

Parkinson's disease." Prof. Muqit's lab studies how mutations in the PINK1 kinase and Parkin ubiquitin ligase enzymes lead to Parkinson's disease.

In this third example, a storyboard created for clinicians working at The University of Glasgow, Dr. Barbara Nicholl, and Professor Sara Macdonald is shown (Fig. 5.10). They wanted a way to communicate their research findings with

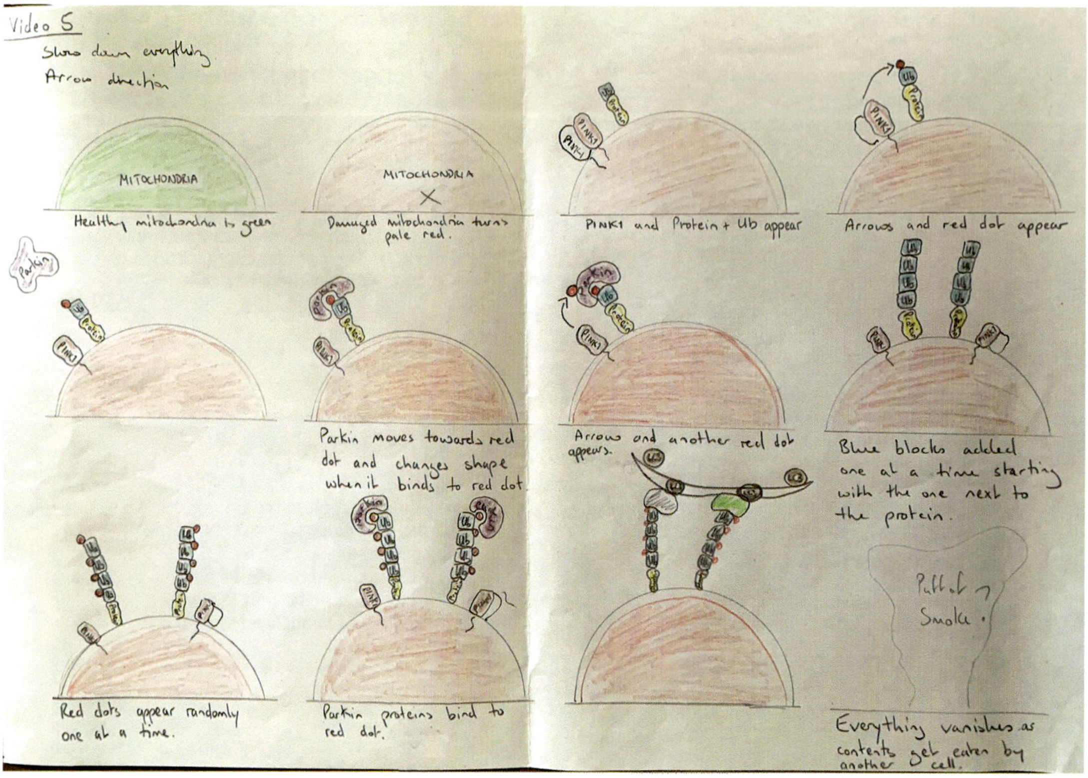

Fig. 5.8 Storyboard developed with Prof. Muqit about his research on Parkinson's disease for a public lecture. It shows a sequence of events that Prof. Muqit narrated during his talk

Fig. 5.9 Still from animation used by Prof. Muqit for The Francis Crick Lecture 2018. The title of his talk was, "Parkinson's disease: decoding the mysteries of neurodegeneration." The still shows how the panels in Fig. 5.8 were translated for the animation

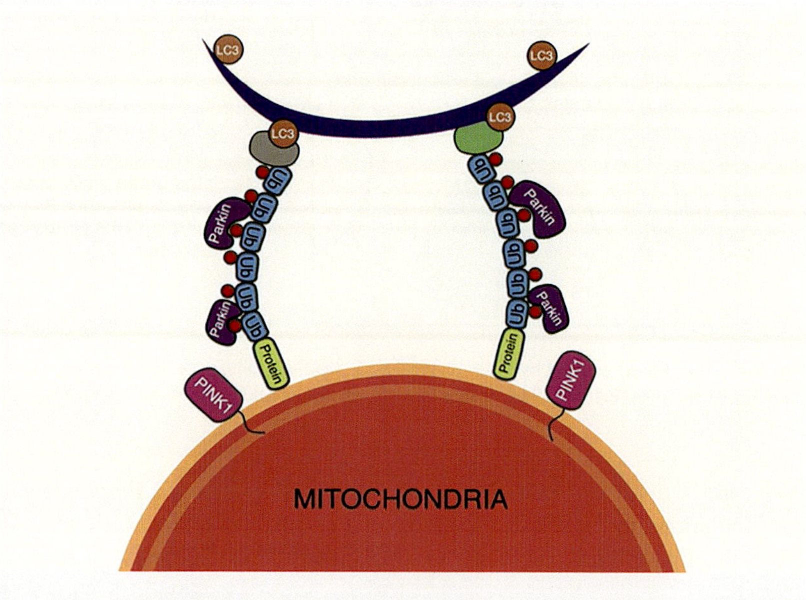

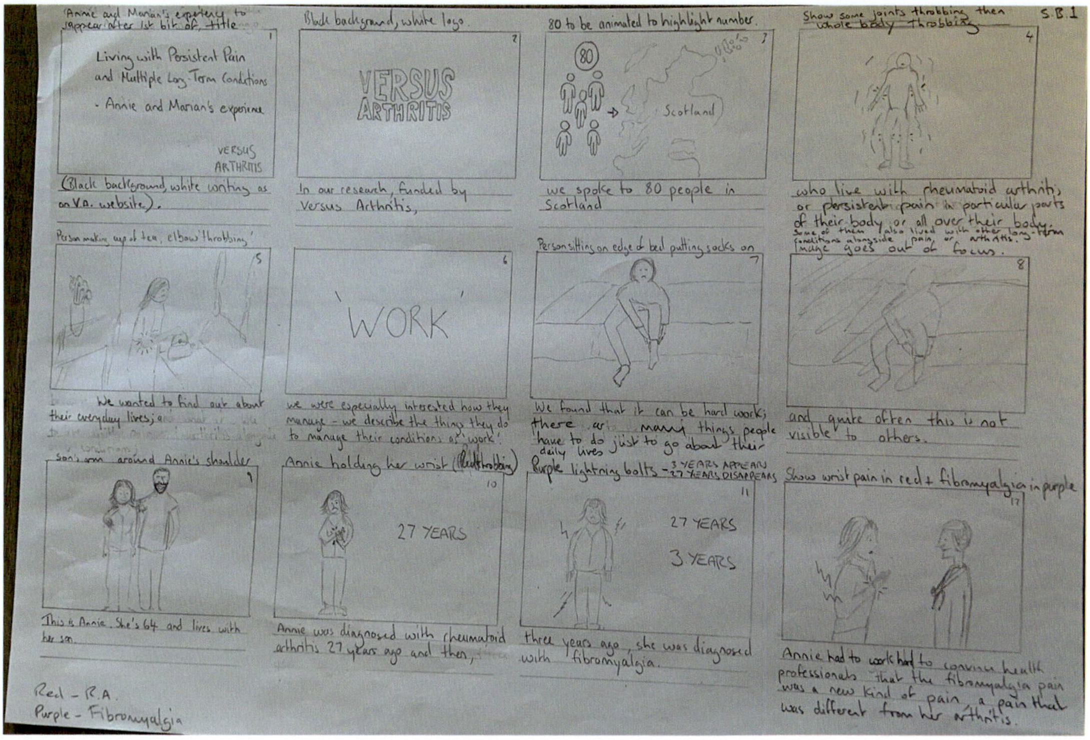

Fig. 5.10 Storyboard developed with Dr. Nicholl and Prof. Macdonald for an animation about arthritis to aid discussions with their patient group in Glasgow

their patient group. The script was developed and clarified before moving on to the storyboard. In this case, the narration is written out in full beneath the illustrated panels. If creating flat 2D animations, as in this case, the storyboard is sent on to illustrator, Andrew Torrens (Torrens 2023) (http://andrewtorrens.com), who brings the concepts to life with digital drawing.

5.3.5 Illustrating the Storyboard

The talent of a skilled digital illustrator is required to make the leap from stick figure sketches above (Fig. 5.10) to real characters. Color, personality, and environments from the storyboard are created (Fig. 5.11). This requires someone who is especially talented at interpreting the "feel" of the storyboard, again being sensitive to the subject matter. The illustrator interprets color choices and imagery, and this is a job only trusted to someone who has a wealth of experience in doing so. This process also adds an element of depth to the narrative, as often, additional elements are added by the illustrator, not drawn in the original storyboard, that provides extra context. Once the client approves the illustrations, it is over to the animator to bring the still imagery to life.

5.3.6 Animating the Illustrations

At this point, information is transferred to the animator to bring everything to life. For 2D animations, Adobe After Effects is used to bring the characters in the stories into existence. Small movements are subtle but effective and add to the storytelling. Timing is also crucial, to make sure that movements match what the voiceover is narrating. Music is selected to be sympathetic to the tone of the piece. For 3D animations, a mixture of Autodesk Maya®, Maxon Cinema 4D, and Nuke

Fig. 5.11 Panels from the storyboard in Fig. 5.10, further illustrated by Andrew Torrens (Torrens 2023) (http://andrewtorrens.com). Torrens adds color, personality, develops characters, and adds extra visual elements to the digital illustrations, not present on the initial storyboard, to add depth of meaning to the storytelling

compositing software are used. Models are prepared and environments are created before wireframe animatics are sent to clients for approval. Concept art with lighting and color are rendered out and sent for approval before final renders are created. Drafts of 2D and 3D animations are produced with sound, and subtitles, if requested.

5.3.7 Production and Delivery of the Final Animation

Once a draft has been produced it is sent for review using Vimeo (Vimeo 2023) (https://vimeo.com) as a platform for sharing. Client's comments are taken on board and amendments are made. Two rounds of comments are permitted before a final draft is completed. Subtitles are checked for accuracy. Once all amendments are applied, the final animation is ready for delivery. Vimeo is also used for a downloadable version to be sent. Alternatively, if the client prefers, we use Dropbox (Dropbox 2023) (https://www.dropbox.com), or WeTransfer (WeTransfer 2023) (https://wetransfer.com). Very rarely have files been copied onto an external storage device for mailing, but it has happened once as a special request. After completion of this process, if further amendments are requested, the budget must be reviewed, and a fresh quote sent out for any further work.

5.4 Conclusions

Scientists create visualizations using animation for a variety of reasons; to validate experiments, explore data sets, and communicate hypotheses and research findings to specialist and non-specialist audiences. In a digital age, the reach of these visualizations is ever increasing, but with this increased access, we need to ensure that the visualizations are well explained, so as not to misrepresent or misinform an audience.

3D scientific animation needs to come with a disclaimer, or at least a rigorous explanation of where information to create the animation has come from and whether it is based on fact or fiction. There is a danger that an audience may be misinformed, and credibility of the work is put into question if this information is not forthcoming (Jantzen 2015). Has real data been used to show protein interactions or is it a representation of events? Further, with the advent of artificial intelligence predicting protein structures (Callaway 2022), how can the information we are using for 3D animation be trusted by a given audience? Transparency from the creator as to where they have sourced their material, and rigorous citation is important if scientific animations are to be held in high regard in the academic world.

Having worked with several scientists over the last 10 years at Vivomotion, we have discovered that it is best at the outset of a new project to gather as much information as possible in terms of the style of animation requested and, if possible, to get the scientists to sketch out their ideas. If the scientist can provide reference material and source examples of animations they like, it is useful for the design process. This avoids playing the role of a mind reader and often saves time second guessing what they are looking to achieve with their visuals. Each project is unique, and every scientist is different to work with, some having very clear ideas of what they want, while others require more probing and guidance. Often during the act of describing what they are looking to achieve, new ideas or concepts emerge, or hypotheses become crystalized, so the initial discussions can inform part of the scientific discovery. The practice of vocalizing the science for a lay person, in this case, myself as an animator, can help to clarify the thoughts of the scientist. Knowing what questions to ask in the initial stages of the project, to get "inside the head" of the scientist, is something that has been learned over time. Other important questions to ask at the outset include: What is the timeline? What budget is there? Who is the audience? How long should the animation be? All these questions will inform the course each individual project will take.

In this chapter, we have highlighted the multitude of contexts "animation as a communication tool" can be used by scientists, as well as what aspects to consider when choosing the style of the animation, for example, who is your audience and what will appeal to them and importantly, what do they understand? The step-by-step process/pipeline described here is one that animation production teams can follow in the future as an approach when collaborating with academic scientists.

References

Aqdot (2023) [online] Available at: https://aqdot.com/products/advanced-academic-research/. Accessed 26 Apr. 2023

Bensound (2022) [online] Available at: https://www.bensound.com. Accessed 25 Mar. 2023

Callaway E (2022) The entire protein universe: AI predicts shape of nearly every known protein. Nature 608:15–16

Donaldson R (2022) [online] Available at: https://www.youtube.com/watch?v=yocfFT1XeAo. Accessed 25 Mar. 2023

Dropbox (2023) [online] Available at: (https://www.dropbox.com). Accessed 25 Mar. 2023

Faure LM (2016) The mechanism of force transmission at bacterial focal adhesion complexes. Nature 539(7630):530–535

IMBA–Institute of Molecular Biotechnology (2014) [online] Available at: https://www.youtube.com/watch?v=MmKHVSl5Wlo. Accessed 25 Mar. 2023

Jamendo (2022) [online] Available at: https://www.jamendo.com. Accessed 25 Mar. 2022

Harrison PL, Nickel D (2016) (re) visions, 1st edn. Discovery Press, Dundee

Jantzen SG (2015) Transparency in film: increasing credibility of scientific animation using citation. Nature Methods 12(4):293–297

Lepito A (2018) Where animation and science meet. Integr Comp Biol 58(6):1279–1282

McKay A (2023) [online] Available at: https://www.alisonmckay.com. Accessed 25 Mar. 2023

Mignot T (2016) [online] Available at: (https://microbiologycommunity.nature.com/videos/video-43). Accessed 25 Mar. 2023

Molloy D (2021) [online] Available at: https://www.youtube.com/watch?v=BYe5iSVmthw. Accessed 25 Mar. 2023

Murdoch F (2023) [online] Available at: https://www.frasermurdoch.co.uk. Accessed 25 Mar. 2023

Rebhahn JA, Deng N, Sharma G, Livingstone AM, Huang S, Mosmann TR (2014) An animated landscape representation of CD4+ T-cell differentiation, variability, and plasticity: insights into the behavior of populations versus cells. Eur J Immunol 44(8):2216–2229

RCSP PDB (2022) [online] Available at: https://www.rcsb.org. Accessed 25 Mar. 2023

Socially Sorted (2023) [online] Available at: https://www.sociallysorted.com.au/ideal-video-length-social-video. Accessed 1st May, 2023

Stanley-Wall NR, MacPhee CE (2015) Connecting the dots between bacterial biofilms and ice cream. Phys Biol 12(6):063001

Stromme TA, Mork SM (2021) Students' conceptual sense-making of animations and static visualizations of protein synthesis: a sociocultural hypothesis explaining why animations may be beneficial for student learning. Res Sci Educ 51:1013–1038

The Royal Society (2019) [online] Available at: https://www.youtube.com/watch?v=5vMZMNls_po. Accessed 25 Mar. 2023

The University of Melbourne (2022) [online] Available at: https://sites.research.unimelb.edu.au/visualise-your-thesis. Accessed 25 Mar. 2023

Thompson D (2023) [online] Available at: https://danthompson.tv. Accessed 25 Mar. 2023
Torrens A (2023) [online] Available at: http://andrewtorrens.com. Accessed 25 Mar. 2023
Vimeo (2023) [online] Available at: (https://vimeo.com). Accessed 25 Mar. 2023
Vitae (2023) [online] Available at: (https://www.vitae.ac.uk/vitae.ac.uk/vitae-publications/rdf-related/public-engagement-lens-on-the-vitae-researcher-development-framework-rdf-apr-2013.pdf). Accessed 26 Apr. 2023
Vivomotion (2023) [online] Available at: https://www.vivomotion.co.uk/training. Accessed 25 Mar. 2023
Wachtl M (2022) Animation supported communication on intensive care; a service improvement initiative. J Inten Care Soc 23(4):433–438
Waddington CH (1957) The strategy of the genes, 1st edn. Allen and Unwin Book Publishers
WeTransfer (2023) [online] Available at: https://wetransfer.com. Accessed 25 Mar. 2023
XVIVO Scientific Animation (2006) [online] Available at: https://xvivo.com/examples/the-inner-life-of-the-cell/. Accessed 25 Mar. 2023

Effect of the Rings: A Visual Story Design Comparing Three Chemical Characters

6

Hana Pokojná, Farhan Rasheed, and Konrad J. Schönborn

Abstract

For millennia humans have interacted with a range of chemical substances that are an integral part of human history and culture. The chemical characteristics of the three similar-looking chemical species psilocybin, nicotine, and caffeine have remarkably different biological effects. The narratives of these chemical characters are compelling—their role traverses the unobservable submicroscopic world, influences our biological essence, and impacts societal values and scientific judgement. Visual storytelling through modern display technologies offers an educational method to communicate the nature and effect of these chemical actors. Biological information that transcends multi-directionally and constantly from submicroscopic through to macroscopic processes can be made accessible and meaningful to students and the public. In doing so, links are forged between scientific knowledge and the manner we view these chemical forms from a human and societal context. In attempting to capture these complexities, the aim of this chapter is to conceptualise and design a visual story to communicate the effect of psilocybin, nicotine, and caffeine on humans for public engagement. Whilst remarkably similar in chemical structure, these chemical species have dramatically different psychological and physiological effects on the human body. At the same time, human interaction with these substances is intertwined with historical associations, cultural norms, as well as perceived and enacted taboos.

Keywords

Visual storytelling · Public engagement · Visual communication · Scientific communication · Macroscopic to microscopic · Widescreen display technology

6.1 Background to the Story

6.1.1 Historical and Societal Aspects of Human Interaction with Chemical Substances

In a detailed and eloquent historical and cultural perspective of human use of substances, Crocq (2007) demonstrates how interaction with psychoactive chemical forms has always been part of human history. Their use features in religious ceremonies, for medicinal intervention, as well as

H. Pokojná (✉)
Faculty of Informatics, Masaryk University, Brno, Czech Republic
e-mail: hpokojna@mail.muni.cz

F. Rasheed · K. J. Schönborn
Department of Science and Technology (ITN), Linköping University, Norrköping, Sweden
e-mail: farhan.rasheed@liu.se; konrad.schonborn@liu.se

M. Roughley (ed.), *Approaches for Science Illustration and Communication*, Biomedical Visualization 4,
https://doi.org/10.1007/978-3-031-41652-1_6

for socially oriented recreational purposes. As is in the present day, the notion of addiction was already a point of debate in the seventeenth century that is intertwined with dynamic flux in psychological, societal, attitudinal, medical, and moral factors associated with human use and interaction with substances. Crocq (2007) points out that forging social and moral opinions around human relationships with chemical substances is complex, and it is necessary to be aware that addiction is born out of the intersection of both the substance properties as well as the consumer's psychological make-up, which makes social and moral opinion and assertion about addiction a delicate one.

Chemical substances, such as caffeine and nicotine, have psychoactive and psychotropic properties and are a central part of consumption in society. Indeed, stimulant substances are catalysts for social interaction and have been a documented part of recorded human history throughout the world since at least 7000 BC (Crocq 2007). Following the common means of use through inhalation, nicotine is swiftly absorbed by the human circulatory system and reaches the brain in seconds. It is also a substance correlated with high physical dependence, at levels comparative to morphine and heroin (Gable 2006). Despite its previous promotion and large-scale media advertisement as a socially imperative adjunct to enjoyment and satisfaction, in recent decades, nicotine has been classified as a highly addictive substance and its use considered somewhat of a behavioural taboo. Caffeine enhances alertness and cognitive performance, whereupon it takes about 1–2 h for maximum blood concentration to be reached following drinking a caffeine-containing beverage. Notwithstanding caffeine being a regulated substance in the 1900s, including various bans across the world in the 1700s and 1800s, it has emerged as the most socially accepted and used psychoactive substance on the planet (Weinberg and Bealer 2001).

Human interaction with psilocybin has been shown to have occurred prior to recorded human history, and is associated with spiritual ceremony and self-awareness experiences. Whilst ingesting psilocybin induces mental changes in perception, in contrast with nicotine (and caffeine), it has a very low physical dependence potential (Gable 2006). In this regard, and in a case of exemplifying a significant shift in the societal assertion of taboo to promise, despite previous forbidden and strict restrictions on researching the substance, recent scientific avenues have opened to exploring potential beneficial psychological properties of the substance for treating depression symptoms in particular (Więckiewicz et al. 2021).

At the submicroscopic level, all three chemical substances contain double rings, and each exhibits a six and five-membered atomic ring structure. Caffeine is a stimulant, as is nicotine; both chemical species exist as naturally produced by plants, and both substances feature prominently in human recreation and social interaction. Psilocybin is a psychedelic compound produced naturally by fungi, and used to influence mental perception. It is a compelling reflection that whilst these three substances share such molecular similarities in the microcosmos, they yield completely different effects on human physiology and behaviour as one traverses their story across biological, psychological, and societal realms. At the same time, each of these molecular characters is a causal actor in a multitude of cultural, societal, and moral paradigms, choices and opinions that resonate in a dynamically changing continuum of acceptance and taboo.

6.1.2 Visual Storytelling Techniques for Communicating Scientific Phenomena

Visual storytelling is used as a means to effectively communicate information and knowledge (Akiba et al. 2010; Tong et al. 2018), and serves as a compelling means to embed information into memory—it is often the case that great scientists are great storytellers (see Cropper et al. 2015; Hadzigeorgiou 2021). It combines multiple media such as visual representations, text, and audio to construct meaningful narratives for a target audience (Botsis et al. 2020).

Visual storytelling techniques aim to dynamically relate multiple combinations of these in the form of communicating coherent and meaningful messages to entertain, disseminate information or provide opportunities for persuasion (Williams 2019). In turn, visual storytelling can deliver opportunities for accessing, developing, and sharing new and enhanced understanding (Tong et al. 2018). At the same time, given the significant increase in the complexity of information transmitted to the public (one only has to think of the recent plethora of visual narrative associated with the SARS-CoV-2 pandemic), the challenge is to forge careful connections between the information and intended impact on the audience (Cropper et al. 2015).

It follows that storytelling can be a powerful tool for engaging an audience with scientific topics (Hadzigeorgiou 2021). On this note, Ma et al. (2012) refer to the idea of scientific storytelling, the construction of narrative that deploys scientific data, wherein visualisation and visual images serve as an important component of the narrative. When it comes to applying storytelling for communicating bioscientific phenomena, Botsis et al. (2020) suggest that communicating 'solid data' (scientifically obtained datasets) and providing 'comprehensive analysis' (rather than mere presentation of facts) are paramount. This is especially important when the desire is for the audience to ascertain potential risk and benefit scenarios.

An extensive review by Tong et al. (2018) revealed that emerging visualisation storytelling techniques have concentrated more on information visualisation than scientific visualisation, with emphasis placed on exploration (exploring scientific data) rather than on presentation (presenting scientific data in a story). When it comes to presenting complex stories, Ma et al. (2012) assert that a well-designed scientific story is paced in the sense of where the meaningful 'plot points' occur. These authors designed a scientific storytelling visualisation where users could interact with climate or astronomical data via story nodes. When delivering scientific visual stories, Ma et al. (2012) affirm that the story designers should know their intended audience and also take ownership of the story's essence.

In contrast with interactive visualisation as a storytelling technique alone, in work that demonstrated the power of animation for storytelling, Akiba et al. (2010) developed an interface where a user could transform their own unique exploration of complex visualised temporal and multivariate volume data (e.g. MRI data, hurricane data, and combustion) into animated content. Animations are often used by scientists to effectively express and explain chemical and biological processes. Ma et al. (2012) have asserted that although it remains far from a trivial endeavour, scientific visual storytelling successfully integrates the components of perspective, characters, imagery, and language.

Work by Botsis et al. (2020) showed how visual storytelling techniques can be effective for communicating information to target audiences, whilst Cropper et al. (2015) have shown how science communication skills can be improved through scientific storytelling. In the example of disseminating information about biomedical science, Botsis et al. (2020) demonstrated how visual storytelling could be used to overcome obstacles in engaging with and perceiving information around health.

When communicating bioscientific information, successful application of visual storytelling requires designers to 'put themselves in the shoes of their audience and think as non-experts' (Botsis et al. 2020, p. 2). Doing so also involves communicating visual information in a way that acknowledges the interplay between the audience's conceptual knowledge of the bioscientific phenomena in focus, the reasoning demands associated with interpreting the displayed representations, as well as the nature and characteristics of the visual language used to represent the phenomena itself (Schönborn and Anderson 2010). This means that the storyline should take the 'visual literacy' of the audience into account and provide narrative points that scaffold their understanding of the visual narrative (Williams 2019). For example, The National Literacy Trust (UK) estimates 7.1 million people in England have poor reading ability, meaning that they can understand short simple texts on familiar topics but reading information on unfamiliar topics can be challenging (Teravainen-Goff 2019). Delp and Jones (1996) suggest that adults with lower literacy benefit from the addition of pictures to text when reading health information, and that cartoons improve comprehension of instructions. It also requires the storyline to entice the audience to think about the bioscientific data presented in the forging of opinions about what is communicated. Herein, it is important that the storytellers refrain from offering personal conclusions; emphasis must be on inducing the audience to reflect and craft their own viewpoint on what is communicated.

6.2 Conceptualising a Visual Story Comparing Three Chemical Characters

Our visual storytelling design process follows a typical communication-oriented visual design pipeline (Fig. 6.1) comprising five building blocks, namely, determining the brief, building the concept, detailing, and evaluating the final prototype (Botsis et al. 2020). In addition, prior to these typical blocks, an explorative analysis was carried out on different visualisation environments available at Visualization Center C, Norrköping, Sweden, in order to select a viable environment effective for our story. Following each step through the design process to the completion of the final prototype took roughly 6 months from August 2021 to January 2022. The individual steps of the design pipeline are summarised in the following section.

6.2.1 Visualisation Platforms

An explorative and experiential analysis was carried out on seven different visualisation environments available at Visualization Center C, Norrköping, Sweden (see Ynnerman et al. 2020). These included *head-mounted VR* (affording a virtual reality experience for the wearer), *handheld AR* (affording an overlayed digital and real-world experience), *SIMlab*–simulation environment (a semi-immersive visualisation environment built for task simulation), *Dome* (affords an immersive experience of high-resolution 3D projection in a 15-metre diameter space with 99 seats), *Multitouch-tables* (affords the of use finger-based gestures to interact with visualised information), *Decision Arena* (a 360-degree projection system, allowing for immersive and collaborative knowledge work, presentations and meetings), and *VR Arena* (a headset-less widescreen projection display technology that immerses viewers and also provides options to interact with displayed content through handheld consoles). The focus was on experiencing the environments through the lens of the audience and reflecting upon their immersive and experiential qualities, as well as

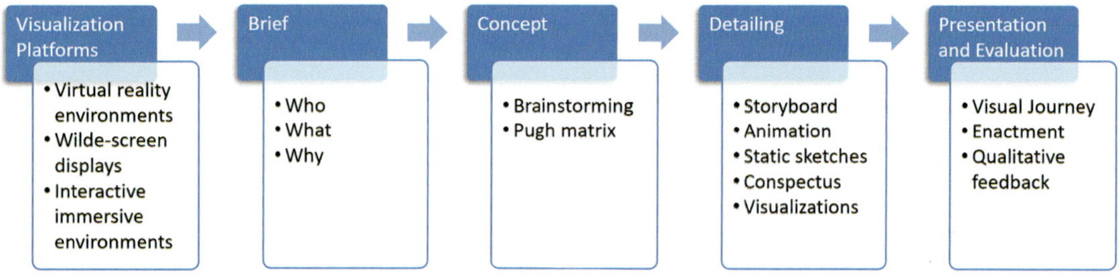

Fig. 6.1 Illustration of the visual story design process highlighting individual steps with accompanying details related to each building block

their suitable use scenarios in visual communication.

6.2.2 Brief

The design process commenced with reflecting on three key questions; *who* your audience is, *what* message do you want to deliver, and *why* are you addressing the issue (intended effect). These factors are essential for any storytelling design and define the scope of the project. Usually, these questions are answered by persons who want to convey a message to a specific audience (consumers, students, the general public etc.) to make an impact.

Answers to these questions can emerge from immediate need, or they might require explorative fieldwork with possible interviews to clarify the problem that shapes the core message.

In the application of the brief in the current story design, the *Who* component targets the general audience with the focus on adolescent teens (14–18) who may be exposed to these substances. In the *What* component, the core message focuses on comparing three chemical compounds at several levels, specifically nicotine, psilocybin, and caffeine. We will focus on their effects on the body and behaviour, their potential use in healthcare, their social connotations, and their cultural role through history. The explanations range from the sub-micro level of biomechanical binding and modulation, through the neural micro-level, to the macro-level and its effects on bodily organs. The *Why* components in the current project brief aim to educate the public domain about the effects of these substances. We would also like to point out that the 'public' consists of both laypersons and experts in fields that are encompassed in the topic we address (health, drugs, law making, and historical contributions of chemicals). Hence, it would be difficult to create a format that does not exclude one of the levels of expertise. The target group was chosen at the said age because they can make an informed decision about the future of the mentioned substances and potentially contribute to their legal status in the future. This refers to their personal use of drugs and their uses in research, attempting to give an objective overview of the subject rather than persuade the viewers to take specific action for or against drugs.

6.2.3 Concept

With the current story's brief in place, synthesising a concept that effectively conveys the message to the target audience with the intended effect is an essential step in any storytelling design. Ideation commenced with collecting many radical ideas through brainstorming sessions keeping critical qualities in mind, followed by generating an affinity diagram that organises these ideas based on their natural relationship. The systematic selection from the pool of ideas was driven by the application of a Pugh matrix (Pugh 1981, 2009), a criteria-based decision matrix comparing several alternatives in a structured way and helping to determine the optimal one. It assigns a score to each potential candidate, (+) for better, (−) for worse, and (0) for same, against a baseline, based on specific evaluation criteria and reports the one with the highest aggregation score. Any candidate can be treated as a baseline in the evaluation process. The key task in generating a Pugh matrix is identifying the relevant critical qualities that define the evaluation criteria.

These tools informed sophisticated choices about the visual story including structure (temporal and spatial), visual effects, colors, interactivity, and visualisation display technologies. Implementing these evolving choices play a key role in ushering perceivable and meaningful communication that enhances audience engagement.

In the utilisation of these ideation tools in the current visual story design, the brainstorming session results in four potential concepts:

1. A lecture-oriented presentation moderated by a teacher.
2. A video containing rendering with complementing audio.

Table 6.1 Depiction of comparing the visual presentation formats through a Pugh matrix. The animation and interactive simulation concepts emerged as the most favourable

Critical quality	Concept 1: Lecture	Concept 2: Animation	Concept 3: Interactive simulation	Concept 4: Interactive game
Interactivity	0	0	1	2
Law-making, social-value	0	1	−1	−1
Inspire to do research	0	1	2	−1
Make decisions about health	0	1	1	−1
Sound/text complementarity	0	0	−1	−2
Immersion	0	1	2	1
Fun / entertaining / engaging	0	0	1	1
Cost / resources / feasibility / implementation / sustainability	0	−1	−2	−2
Intended outreach	0	1	1	1
Score	0	**4**	**4**	−2

3. An interactive simulation indicating respective effects triggered by external input.
4. A game with a penalty reward-based gamification element.

These viable concepts are then arranged in the Pugh matrix for systematic comparison based on critical qualities covering important aspects essential to yielding the intended effect. The final Pugh matrix listing all concepts along with critical qualities is shown in Table 6.1.

In the next step, choosing concept 1 as a baseline and evaluating other concepts against it results in two leading candidates with the highest but equal rank. This concludes a possible hybrid of animation and interactive simulation, with animation providing a context for the audience whilst interactive simulation providing a confined space/environment that allows users to explore. The animation/interactive hybrid provides a semi-linear narration.

Since the current story aims to compare three chemical substances, the VR arena is a natural choice among other visualisation environments we experienced early in the design process because it facilitates interactive support on a large wide display allowing for simultaneous direct comparisons. At this stage of the process, the brief and choices made during ideation must be clear and rigid, because changing them in later steps would be costly and might affect the success of the whole story pipeline.

6.2.4 Detailing

Following ideation, this step outlines the key elements of the story with sufficient details to craft a simple experimental prototype of the design, which can be a semi-functional mock-up of the final production. The level of detail in the prototype can vary from a quick and rough paper sketch (low-fidelity) to a fully formed and detailed version (high-fidelity) depending on need. The low-fidelity prototypes are helpful for early-stage testing, whilst the high-fidelity prototypes are closer to the final production, allowing for better feedback. Prototyping involves converting the ideation outcomes and transforming scenarios into tangible representations such as visualisations, wireframes, sketches, enactments, storyboards, and mood boards. These representations can be used to test and validate the idea and design assumptions, which help to explore problems and adapt appropriate changes early in the process (McCurdy et al. 2006).

Prototypes are useful tools for exploring, experimenting, and communicating design, can save time and resources, help to create a clear picture of benefits, can provide an opportunity to showcase, can provide feedback, can estimate risks, and form the basis for production (Erickson 1995). The realisation of the concept in the current story through prototyping emerged as a storyboard with detailing expressed in the form of sketches, biomolecular renderings, graphs, textual adjuncts, and animations for production and visualisation on a widescreen display environment. It also combined both 2D and 3D content.

6.2.5 Evaluation

The final stage of the design process is to predict how likely a final product will provide the desired impact on the intended audience given an existing prototype. Evaluating a prototype is an important step in any design process since it helps to indicate potential shortcomings and areas for improvement. This becomes imperative when confirming in addition, whether the design assumption made in previous steps can be executed or implemented in accordance with hardware implications. The overall idea is to test or validate design assumptions and refine the prototype to maximise the impact in an iterative manner. In turn, this provides an opportunity to receive constructive feedback with possible improvement suggestions from experts or/and potential audiences. Typical tools for the evaluation of design prototypes consist of usability testing, collaborative exploration, think-aloud, Wizard of Oz methods, and impact versus effort matrix descriptions (Bertzen and Basjuka 2022).

To evaluate the current visual design, we create an effective interactive audio-visual journey of the overall concept, the selected detailed parts, and the reasoning behind the design decisions. This was then presented to a mixed audience of student peers and design experts in January 2022, during a 'Visual storytelling and immersive media day', which was organised by Linköping University's Media and Information Technology division, where an expert review assessed the usability of the final visual story prototype (e.g. Demian and Fruchter 2009).

Previous empirical research into effective visual storytelling has a set of rules discussed in the previous section. This theory was then applied in the creation of the prototype whose visually specific imagery, voiceovers, camera angles, scientific explanations, and other design choices are explained in the following section.

6.3 Designed Visual Story Prototype

6.3.1 Overall Narrative Structure

The resulting prototype is a combination of visual and auditory stimuli that provides an immersive experience and utilises storytelling to transfer historical, biological, and societal information to the target audience. The developed storyline compares three chemical species on a VR Arena display (Fig. 6.2a) that can also independently project a triple split view on a wide screen (Fig. 6.2b). The display screen is equipped with three overlapping projectors (placed behind the screen not visible to the audience) resulting in a total screen resolution of 5592x1600 pixels with both mono and stereo support. Our prototype offers historical context, a description of metabolic processes, molecular function, effects on the body, and perception of content through different temporal and scale dimensions. Overall, the prototype story raises questions and reflections about the future of the three chemical forms and their place in current and future society, all rooted in biology and chemistry.

6.3.2 Setting and Technical Components of the Visual Prototype

The visual prototype consists of a detailed visual story comprising 44 sequences, each of which describes a combination of imagery, animations, and simulations. These include camera

Fig. 6.2 Depicting the prototype in the VR Arena, Visualization Center C, Norrköping (**a - top image**), and depiction of the technical viewer parameters of the VR Arena on the screen (**b - bottom image**)

movements, voiceovers, visuals, and potential interactions between the viewer and the visual elements on the screen. Such interactions are possible through various potential scenarios. For example, the pilot of the story delivery could interact directly with the visual playback whilst the audience observes. In another scenario, an individual viewer in the audience could interact with the interactive features of the visual display through an external device such as a game controller or digital tablet. A further possibility is for multiple members of the audience to interact with different visual elements simultaneously, offered by the triple-view split-screen capability (Fig. 6.2a and b).

6.3.3 Design Choices for Communication

Technological advancements provide helpful means of teaching different concepts. One of the most used technological aids in education is animation. A recent review has shown that animations are very useful tools in education due to their attractive design, visuals, and audio (Mansor et al. 2020).

Animation in education has several different trends, among which belong expository animations and interactive animations (Xiao 2013). 3D animations have been shown to be helpful when they aid the viewer with cognitive processes that would be difficult to perform without the 3D depiction (Solomon 1994). An example of this would be displaying molecular bonding that cannot be seen with the naked eye and therefore visualising would make it more accessible. Animation was also appraised as an effective tool in the learning of STEM subjects because of helping to maintain focus and interest simultaneously (Bedrina 2016). Moreover, research has found that self-generated animations aid in learning chemistry and experiments (Berg et al. 2019).

To make animation an effective educational tool, one must consider its design prior to creating it so as to not induce high levels of cognitive load. Multimedia Learning Theory proposed by Mayer (2014) details how the information from multimedia is processed and stored from short-term memory into long-term memory. Mayer and Moreno (2003) have identified ways of reducing cognitive load when using multimedia for learning, which were also considered in this project. For example, by combining audio and visual stimuli, which allows for referential connection of different perceptual channels. Another strategy is to use segments, which allows the audience to process information in smaller and more digestible 'chunks' rather than being overwhelmed with much information at once. A further strategy concerning visual and auditory stimuli is to synchronise the visual and auditory cues and to reduce visual clutter. Examples of such considerations were applied in our design of *Effect of the Rings.*

6.3.4 Results of Selective Detailing

The comparative story is divided into three parts: introducing the three substances, their effects on the body, and their societal implications. The story uses a comparison of the attributes of each of the chemicals giving a tri-parallel visual experience storyline to provide a novel direct comparison (see Fig. 6.3a *comparing chemical absorption time*) as well as overlapping attributes of the chemicals (e.g. see Fig. 6.3b *comparing activated brain regions*).

The story commences with an introduction to the three chemical forms and their importance in society throughout history. This is followed by a focused comparison of the biophysiological

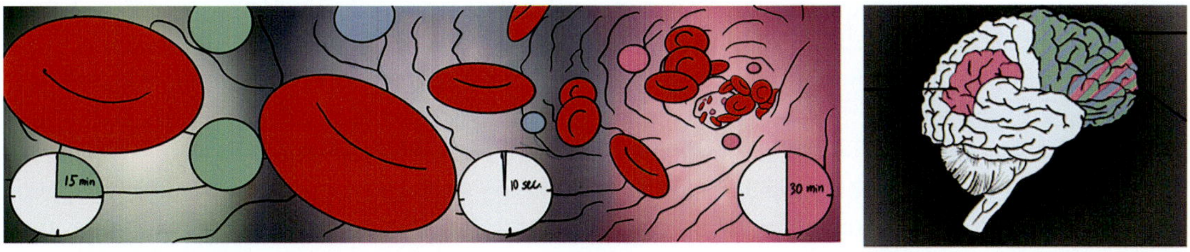

Fig. 6.3 Depicting the chemical absorption times (**a - left image**) and the comparison of the overlapping brain regions affected by the three chemicals (**b - right image**)

Fig. 6.4 Ligand–receptor interaction at submicroscopic level

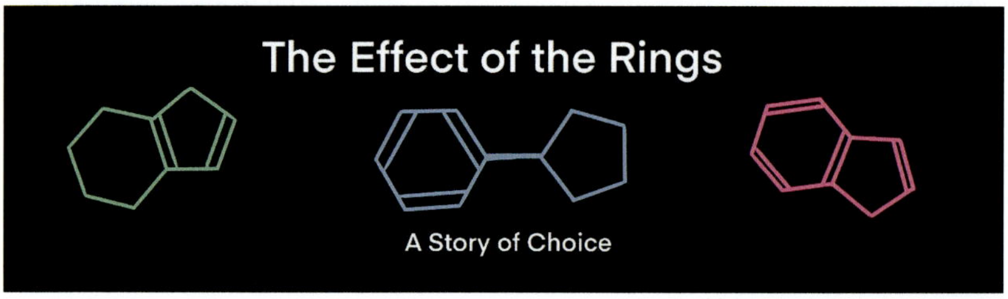

Fig. 6.5 The opening scene depicting the title and the three chemical actors

aspects and health effects of the three chemical actors on the body by transcending the submicroscopic dynamics (e.g. submicroscopic ligand–receptor interactions, *see* Fig. 6.4), and microscopic events (e.g. blood flow, *see* Fig. 6.3a), to macroscopic effects (e.g. brain activation, *see* Fig. 6.3b) scale. The concluding part summarises the shared ring-like structure, the societal comparisons of the substances, their different yielded effects, and their current and past legal status. In closing, the visual story maintains neutrality in challenging the viewer to consider the future of the three chemical forms.

This section of the story describes how the biological, historical, and cultural concepts are communicated in the final product through detailed description sequences. Selected sequences are shown to illustrate what the final product looked like. The complete story consists of three parts: the overall introduction of the substances with historical context and their importance and their brief biological summary. This section provides an introduction to the entire product and starts with an opening scene (*see* Fig. 6.5) showing simplified artistic ring structures of the molecules and the title 'Effect of the Rings: A Story of Choice'. The story then continues to introduce the origin of the substances and how they were obtained through gathering plants.

The next sequence is the first example of forking used in this project and divides a single screen into three split parallel stories, showing the similarity between gathering plants, and the differences in collecting different types of plants. It specifically shows how caffeine comes from shrubs, tobacco from plants, and mushrooms are found growing from the ground. From left to right, coffee is depicted with green colours, tobacco in blue, and mushrooms in pink. The next transition depicts the form in which these substances are extracted from coffee beans, rolled-up tobacco, and raw mushrooms. The fifth sequence (Fig. 6.6) shows people consuming the substances and the voiceover talks about Ethiopian coffee ceremonies, Native Americans smoking a Peace Pipe, and indigenous people of Central America consuming mushrooms they thought to be 'god-given'.

Fig. 6.6 Scene showing historical social context in relation to consumption of the three substances; Ethiopian coffee ceremonies, Native Americans smoking pipes, and the consumption of what were thought to be god-given mushrooms in Central America

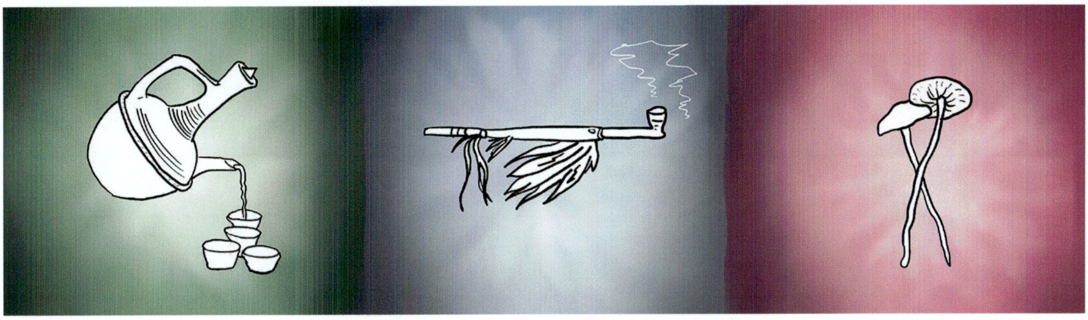

Fig. 6.7 The different vessels used to consume the substances. Beverages for drinking caffeine, pipes for smoking tobacco, and ingestion of mushrooms by eating

In the next sequence (*see* Fig. 6.7), the camera zooms in and focuses on the processed substances/vessels for the substances: specifically a coffee container, a pipe, and raw mushrooms. From here the chemical structures of the molecules appear on the screen. This is a starting point of the micro-level that will continue to traverse through the levels of scaling.

The next sequence highlights the two-ring structure of these molecules, showing the important relation they all share at the molecular level (*see* Fig. 6.8). The rings then transform into spheres representing the simplified molecular structure of each substance, each with its respective colour. Next, the forking of the three molecules joins into a single display and the molecules travel through the bloodstream, followed by crossing the one-cell thick blood–brain barrier into organ tissue. The camera follows the molecules zooming out slightly and showing the molecules crossing the synaptic cleft between neurons. The camera zooms out even more, revealing neurons interacting, then continuing to show the extensive neural network conducting electrical current (*see* Fig. 6.9).

The camera continues to zoom out, traversing into the meso-depiction, the organ-brain level, which the molecules have influenced. A 3D model of a human brain rotates, and as it does so it becomes more transparent whilst highlighting and naming specific areas that are affected by the substances (*see* Fig. 6.10). The voice-over states the different functions and responsibilities of the brain areas. The subsequent sequences zoom out even further onto a full human body whilst the brain is visible in the head, reaching the full macro level. The series of sequences serves as an outro from the first part of the introduction and introduction to the second interaction part. The brain, now in proportionate size to the full human body depicted on screen, disappears.

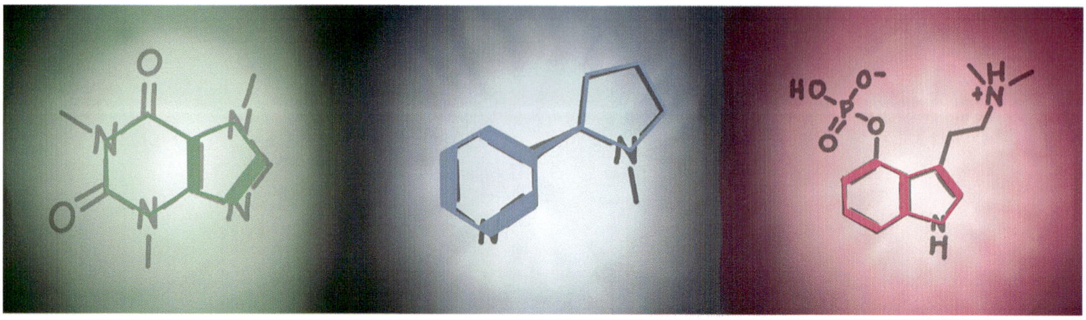

Fig. 6.8 The two-ring chemical structure in all three of the chemical compounds

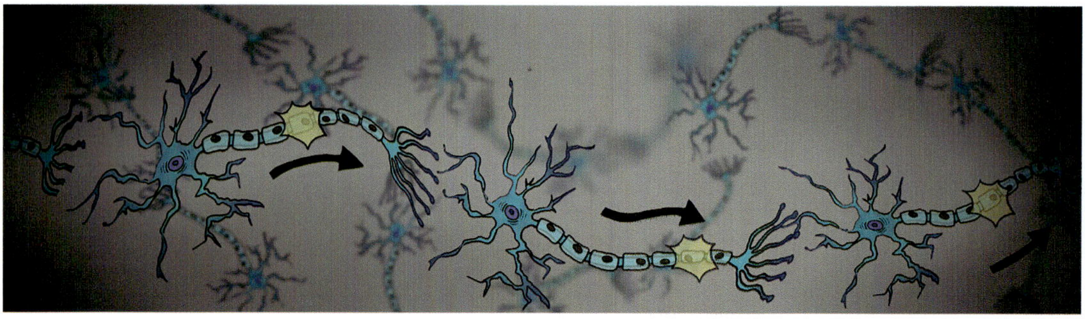

Fig. 6.9 Neural connections at the microscopic level depicting electrical current passing through the neurons in various brain regions

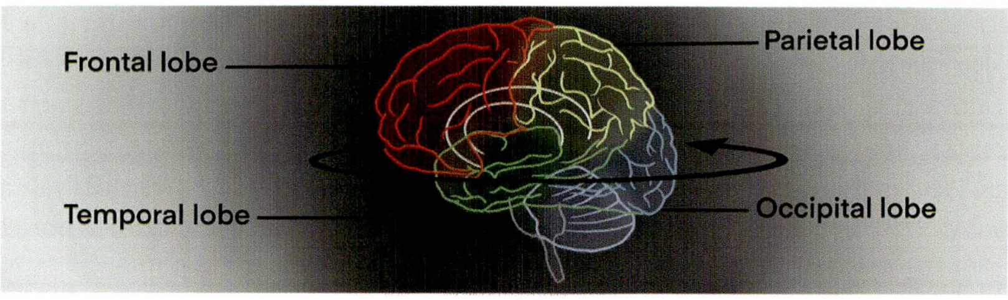

Fig. 6.10 The brain model showing the different brain regions affected by the different substances

The focus from the body moves to a table that appears and has a cup of coffee, a 'vape', and a bowl containing mushrooms (*see* Fig. 6.11). These three objects are the three choices that can be selected in the forthcoming interactive segment. This frame is the end of the introduction animation. The next frames are the interactive component where the audience can interact with the animation. The interaction commences with the audience dragging and dropping the substances into a designated box. It can contain between one and three substances, and then 'play' is selected to enter the comparison of substances stage.

Depending on which substances were chosen, the first part of the side-by-side comparison begins. This example (*see* Fig. 6.12) is a comparison between all three substances. The intake is as follows: drinking coffee, vaping, and eating mushrooms. The sequences are divided into 3 parallel narratives communicating the same story with different chemical actors. This part of the story takes a detailed look and comparison of the

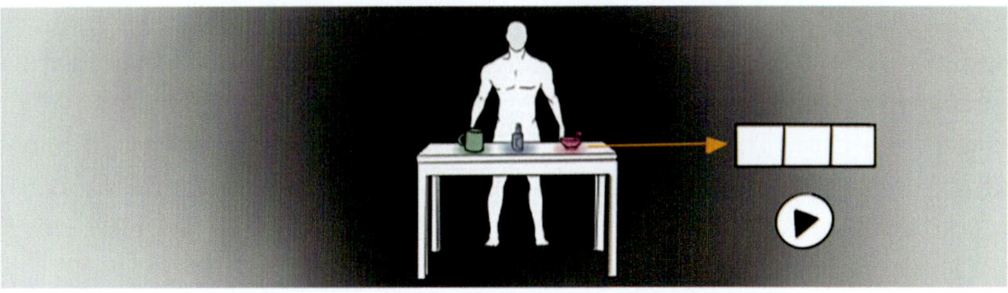

Fig. 6.11 Screenshots give the viewer an option to choose which substance they would like to explore first. This is represented by coffee, a vape, and a bowl of mushrooms

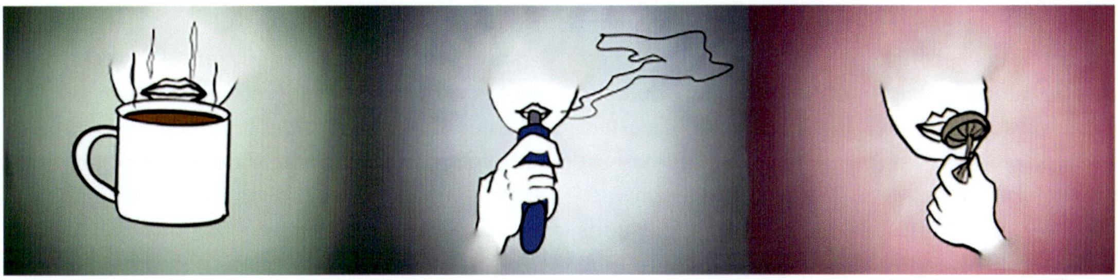

Fig. 6.12 Screenshots showing the consumption of the chemical and comparing it in a tri-parallel fashion

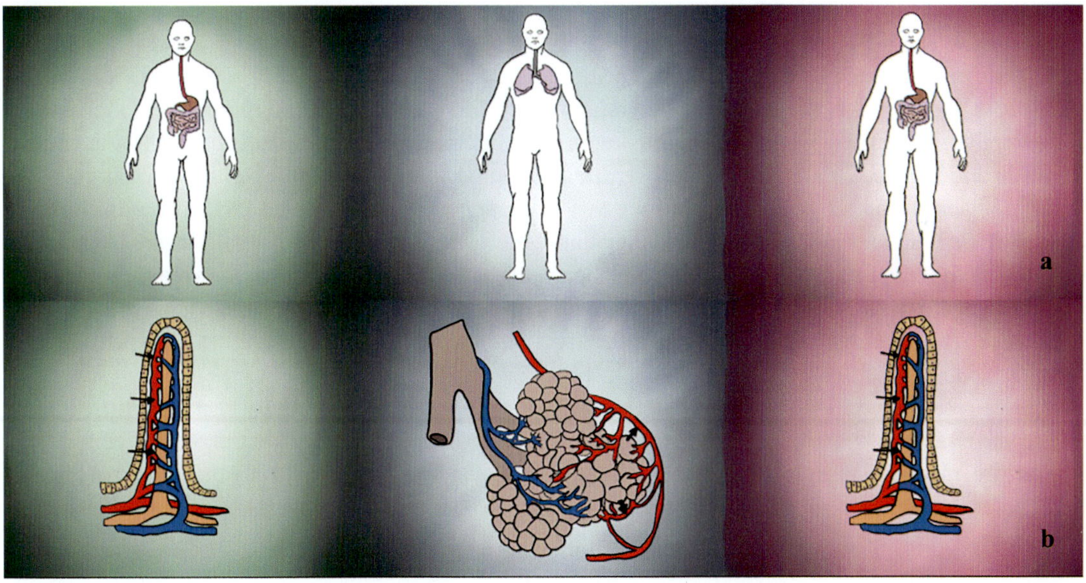

Fig. 6.13 Tri-parallel fashion of comparing the different modes of ingestion. The first focuses on consumption itself and which organ system it involves (**a**) and then zooms in into the microscopic level within the organs involved (**b**)

segments that were communicated during the first part.

The tri-camera follows the ingestion of the substances. It flows into the biological explanation of ingestion starting with the systems responsible for it (*see* Fig. 6.13a). The coffee and mushrooms are depicted in relation to the digestive system whilst the vaping shows the lungs within the outlines of human models. The camera zooms in on the alveoli in the lungs and the microvilli in the intestines in the other substances (*see* Fig. 6.13b). The voiceover talks about how after ingestion, the substances enter the bloodstream.

The coffee and the mushrooms get digested and absorbed in the gut through the microvilli, which are one cell thick. The vapour inhaled from vaping gets into the bloodstream through the one-cell thick alveoli. However, the vapour has a much quicker effect because the substance gets into the blood system much more rapidly than the other substances which first need to be digested. The coffee takes less time to be absorbed than the mushrooms because as a liquid it is easier to process than the mushrooms. The sequence joins into one large, semi-divided screen to show the three different coloured molecules traveling through the bloodstream (Fig. 6.14).

Due to how they are absorbed, the different substances take different durations to take effect on the brain. The different times are indicated by a clock on each respective screen. The molecules then cross the blood–brain barrier. The camera follows the molecules from the bloodstream. There is zooming out to show the synapse, then zooming back in, but now directly onto the synapse surface and the enzymes where molecular docking onto receptors occurs. The camera then zooms in on the molecules. The molecules bind to the different receptors. The divided screen shows how each substance affects the brain at the molecular level (*see* Fig. 6.15). The caffeine traverses adenosine receptors by mimicking them. Nicotine blocks the nicotinic receptors. Psilocybin binds and passes through the serotonin receptors.

The camera once again zooms out to show the molecular effects on the level of the synaptic cleft. Whether molecules passed through, the release of other chemicals is possible by mimicking other ligands or blocking the receptors and causing a lack of chemicals in the synapse. The camera continues to zoom out and shows how an electric impulse travels through the neutrons (Fig. 6.16).

Depending on the effects of each substance, the speed and frequency of electrical impulses are shown. Zooming out continues into a tri-parallel comparison of mesolevel detail across three brains. Parts of the brain on 3D models are shown on each screen. The viewers can see that all three substances affect brain areas responsible for attention. To illustrate the comparison of brain areas affected by each of the substances, the 3D models of the brain merge into one 3D rotating transparent model of a brain with affected brain regions overlapping (Fig. 6.17).

The sequence is then divided into three separate screens depicting the effects of each substance on a person's psychology. For example, the coffee renders a person alert but may cause tremors. The nicotine provides a feeling of stress release and the psychedelic induces hallucinations. The three screens show a respective person and a short repetitive animation of the person that has ingested coffee moving rapidly now and then, a person with a cigarette relaxing, and a person that has consumed mushrooms lying down while perceiving abstract mental images. The sequence then becomes interactive (see Fig. 6.18).

The user can point to each location on the person's body to read about the effects of the substances. At the bottom, the user can choose to go back to selecting substances to revisit their mechanics again (frames 18–29). The user can press the 'Health Pros' button which will list

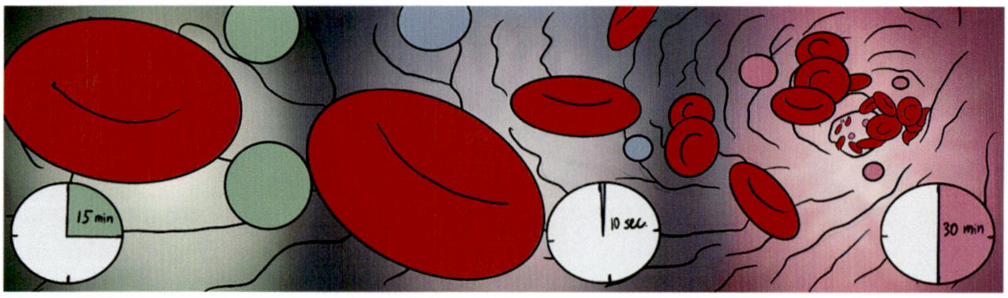

Fig. 6.14 Sequence comparing the absorption time in the bloodstream

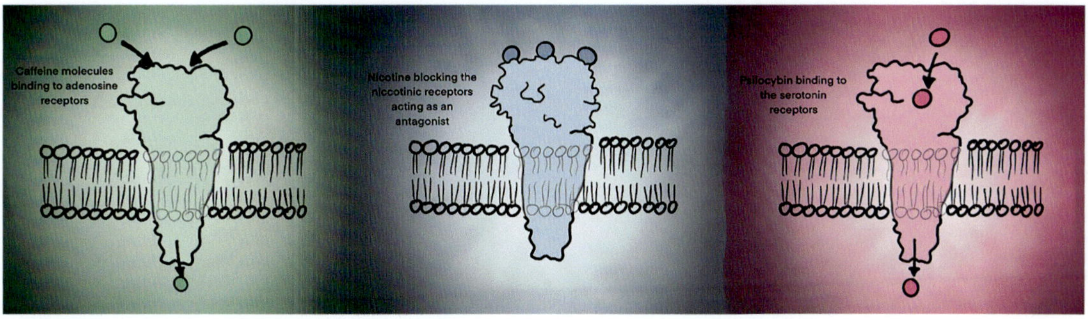

Fig. 6.15 Comparison of the different substances affecting the brain at a molecular level through the receptor–ligand interaction

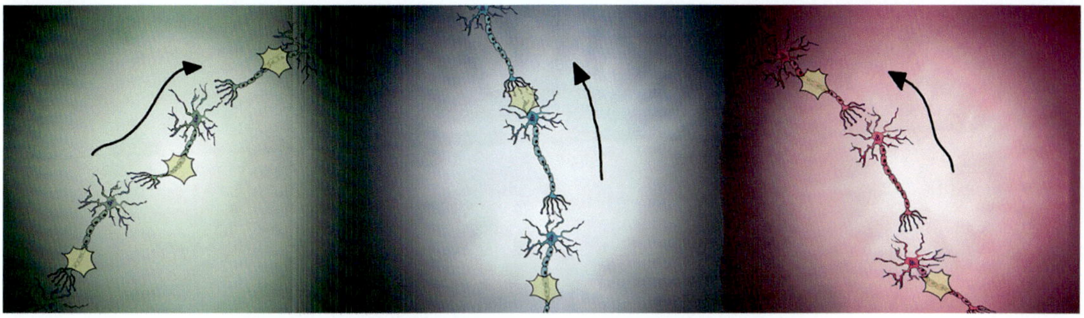

Fig. 6.16 Zooming out to show the electrical impulse travelling through the neurons

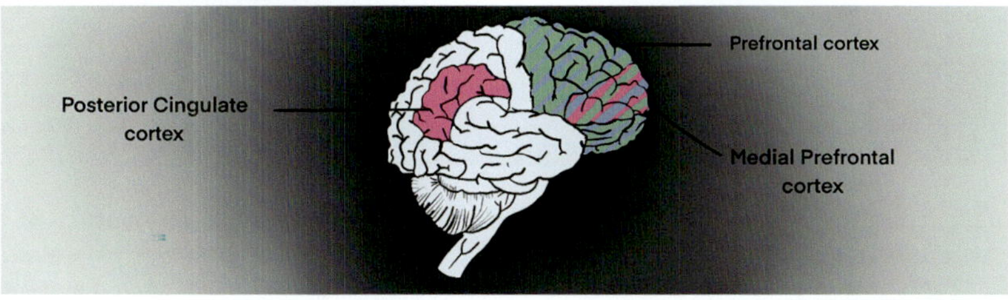

Fig. 6.17 Single screen showing a 3D model of a brain with different active brain regions highlighted. This shows that some of the chemicals overlap in brain activation regions whilst others are completely different

health benefits and current research into each substance. The 'Health Cons' button will show the negatives of each substance on health. The 'Dosing' button will display an interactive screen for increasing the dosage of each substance and its effects on the body. The button on the far right will show the windows with 'Health Pros', 'Health Cons', and 'Dosing', respectively. The health pro button (see Fig. 6.19a) conveys health benefits and research into positive health effects of the substances divided into three screens. The 'Health Cons' buttons (see Fig. 6.19b) communicate damage to health and withdrawal symptoms as a result of using these substances.

The 'Addiction Map' at the top left of the view takes the user to a chart communicating how addictive the three substances are and how lethal they are in comparison to other addictive

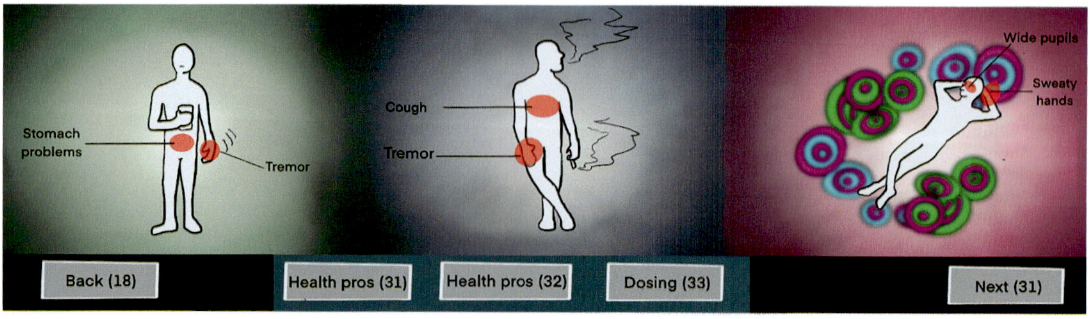

Fig. 6.18 Interactive screen comparing the physical effects of the drugs on the body

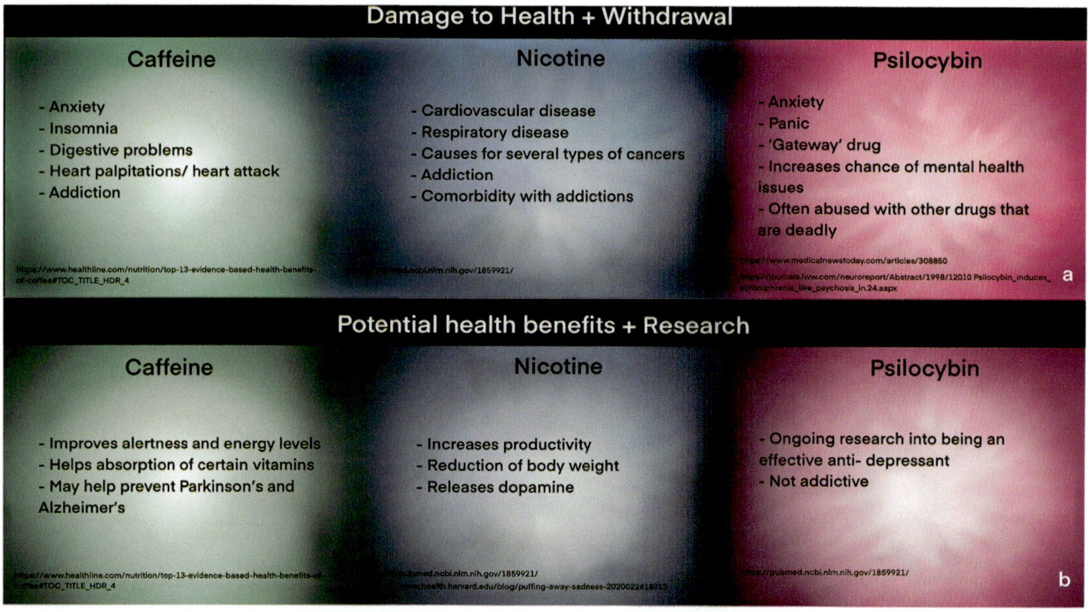

Fig. 6.19 The three screens allow the viewer to compare the potential health benefits (**b - bottom of the image**) and health risks (**a - top of the image**) caused by the ingestion of each of the chemicals

substances. In the same fashion, the 'Withdrawal Symptoms' button shows the comparison between listed withdrawal symptoms. The other features show the Addiction Map (see Fig. 6.20a) comparing the lethal dose and dependency potential of several drugs, pointing out the three substances in focus. The other feature, 'Dosing' (see Fig. 6.20b), allows the user to move sliders to see how the person is behaving depending on the dosage. The maximum dose would result in death with respect to caffeine and nicotine overdose. The maximum dose for psilocybin indicates strong hallucinations lasting several days, which are documented. There are currently no known lethal doses of psilocybin related to overdose (other than deaths related to activity whilst experiencing hallucination, such as falls and accidents). The minimum dose at the bottom of the metre indicates sobriety. This interactive slider dosing feature concludes the second part of the story.

The third and last part focuses on the historical and political views associated with each substance and a summary of the biological benefits and harm in relation to the laws surrounding their use. The sequence shows the screen with figures

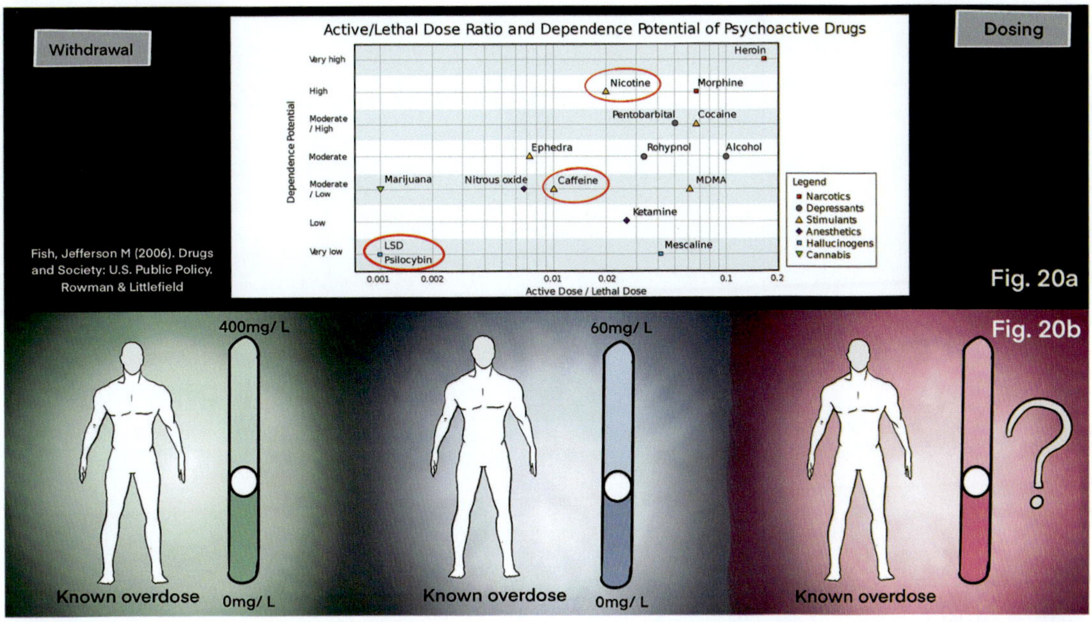

Fig. 6.20 Screens allowing the viewers to compare the danger of overdose by (**a**) the overdose incidence map in comparison to other drugs beyond the three chemical actors and (**b**) interactive comparison of drugs to each other controlled by the viewer's dosing through the slider

Fig. 6.21 One screen showing all three substances in chemical ring form, the three communities that consume each substance, serves as a summary whilst the narrator highlights the important aspects of their historical use and their similarities

consuming the substances, namely the Ethiopians, Native Americans, and Mayans whilst the chemical structures appear (see Fig. 6.21).

The narrator summarises the cultural importance and the chemicals' similarities on the micro-level, then continues to summarise the different effects of the similarly structured substances while traversing to the macro-level. This ties into the topic of their different uses, differences in legality, and their diverging societal acceptance. The sequences then show a comparison for each substance in societal perception (see Fig. 6.22 *a-c*). The narrator states that history shows that the same chemical forms were viewed in different ways by society. Caffeine in coffee

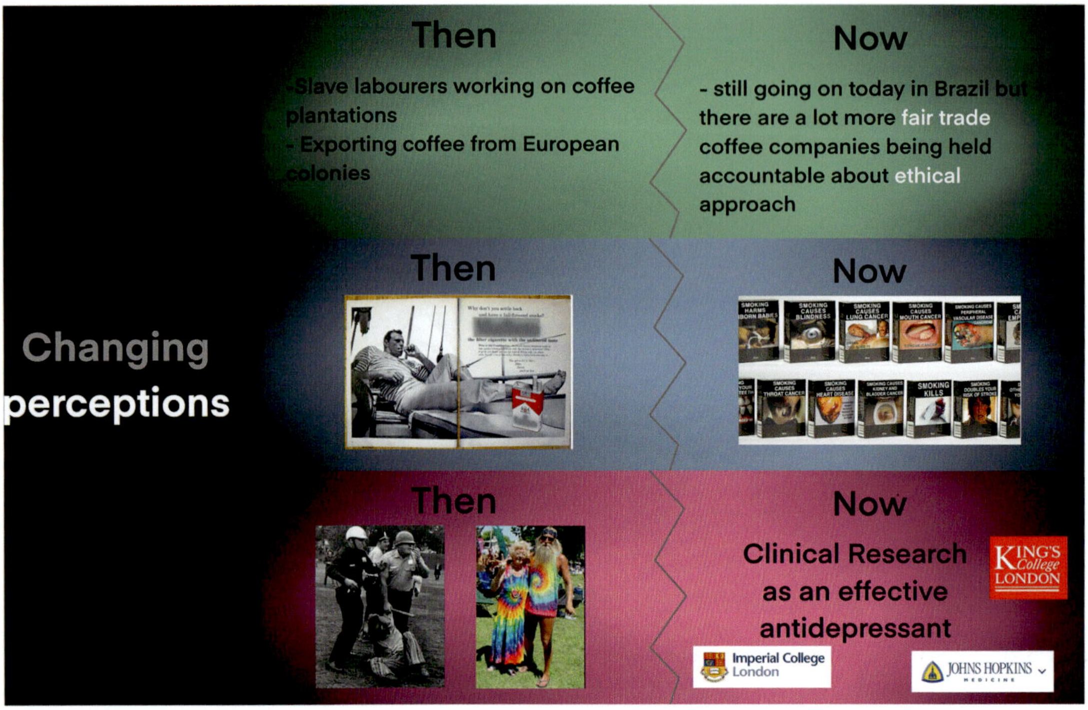

Fig. 6.22 Set of sequences comparing legislation and views on each of the substances in the past and nowadays

Fig. 6.23 Transition screen summarising all the knowledge presented and preparing the viewers for the "big question"

(Fig. 6.22a) was and still is associated with exploitation in how it gets procured. This is improving in recent times through fair trade initiatives. The nicotine in tobacco products (Fig. 6.22b) used to be associated with the consumption of cigarettes as having a high social and satisfactory status (with brands associated with a fun and lavish lifestyle) versus nowadays depicting graphic tissue and organ damage in advertisements and tobacco packaging. The psilocybin in 'magic mushrooms' (Fig. 6.22c) used to be and to some extent still is associated with other class A hallucinogenics. They were illegal and often associated with 'hippie' movements (Smart 1970). Whilst they continue to be outlawed in many places, research into their antidepressant effects has emerged and shows promising medical results (Postránecká et al. 2019; Dudysová et al. 2020). On this note, the story slowly comes to an end.

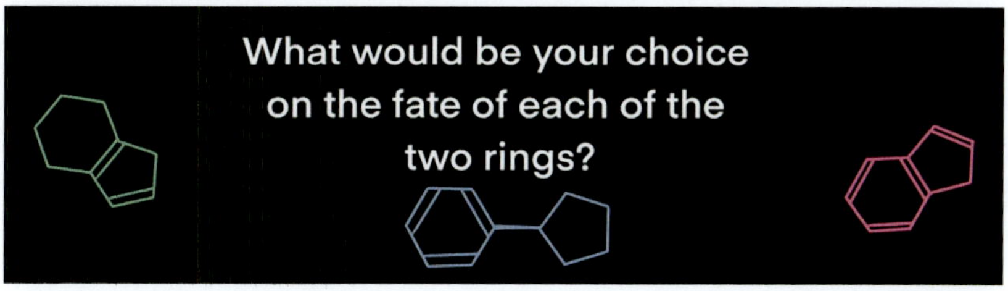

Fig. 6.24 The final screen with the big question challenges the viewer to process the information they were given about the substances and their background and formulate an informed opinion about the future of these substances

The sequence shows a transition screen summarising and preparing for the "big question" to the audience (see Fig. 6.23). The narrator summarises how these very similar-structured submicroscopic entities have a huge impact on society, health, and how we think as humans about what is 'right' and 'wrong'. Is this the correct way of deciding what is 'right' and 'wrong' in relation to these substances and their use and legality? What should change and what should stay the same? As the narrator speaks, the ring structures coloured in their specific colours emerge, the person from the very beginning of the story, collecting plants appears, and a symbol of balance and justice, enters the picture.

This sequence is then followed by the final sequence (*see* Fig. 6.24) with the take-home question: 'What would be your choice on the fate of the two rings?' The message is simple and clearly displayed across the whole screen, accompanied by simple two-ring artistic representations of each of the three substances. This story aims to provide objective information and explanations about these substances by provoking the viewer to think about the biological mechansims, legality, use, and future of the three chemical actors.

6.4 Discussion and Implications

The discussion is framed by first describing the evaluation and feedback generated when colleagues viewed and responded to our completed prototype presentation. We then consider what the prototype contributes to visual storytelling for science communication, and close by offering avenues for future work.

6.4.1 Evaluation and Feedback

The evaluation of this project consisted of qualitative feedback from expert reviewers. In the context of this chapter, an 'expert reviewer' is a person whose professional career and academic research focus on visual communication and/or design. In our case, these backgrounds were associated with interaction and information design, data sciences, information visualisation, and science communication design. Table 6.2 provides a breakdown of the expert reviewer comments. Approval for use of any expert feedback provided during the online conference was given at the beginning of the feedback event. In January 2022, the conceptualised project was presented as a part of the 'Visual storytelling and immersive media day'. This was the culmination of the 6-month-long doctoral Visual Storytelling course hosted by Linköping University, Sweden. The presentation was viewed and evaluated by experts in visual storytelling who organised the programme, keynote speakers with a focus on scientific data presentation and analysis, and other students of various backgrounds taking part in, and working on visualisation projects. Two expert reviewers in the field of effective data visualisation and a fellow student specialising in design for scientific communication provided critique on *Effect of the Rings.*

The perceived overall strength of our prototype was the content, which was commented on by the viewers. The critique that the commentators pointed out included elaborating on the specific ways of interacting with the audience. There was also a suggestion to use a 2D format of animations to make the creation of the

Table 6.2 Evaluation of the visual story prototype design that compares three chemical characters describing the evaluators' expertise background, examples of obtained positive and negative comments, and potential areas for future improvement

	Reviewer Expertise	Examples of feedback comments	+/−	Area of improvement	Potential future improvement
1	Interaction and information design	'Fantastic journey from culture to molecules and back up'.	+	Content	
		'The presented story has a very clear structure'.	+	Structure	
		'How is it seen from the point of view of the audience? Is it the audience interacting with it? Or is there a presenter who makes the choices? How do you see the actual process of the audience appropriating this material?'	−	Interaction with audience	Consider having a speaker, as with some current productions at Visualization Center C.
		'Since you tell three stories in parallel, have you considered how to handle the cognitive load? Will it be too demanding to follow three parallel stories?'	−/+	Cognitive load	The story was purposely communicated in a way to simultaneously show differences and similarities.
		'I'll take that but in the end, only testing would show…'	−	Testing	Conduct an evaluation experiment, described at the end of this chapter.
2	Data analysis and scientific visualiszation	'Inspiring journey'.	+	Content	
		'I loved the moment when you had an animation of 3D receptors'.	+	Content	Perhaps include more 3D animations.
		'When you are telling such a broad story, is it possible to have a project where you use 3D animations and other 2D illustrations which are very intuitive? Long story short, 3D is time-consuming, is it doable to do it in a non-consuming way, do you have a process/ would you consider doing it all in a 2D format?'	−	Time effectiveness, technique	2D animations could be created in order to make the project more time effective, however 3D animations are useful learning tools and their alternation with 2D animations could naturally segment the whole project.
3	Science communication design	'I had to wait until late in the story to understand what it is about. Maybe that's the intention, but for people not familiar with scientific terms they may ask themselves what am I watching and why I am watching?'	−	Content	Present the content in a more understandable manner to make it clear what viewers are about to watch. Can be done with a title with a strong explanatory subtitle. In the case of presenters, the presenter can state that the audience will watch a comparative story about three substances and their effects in order to arrive at their own opinion on the situation.

product simple and time effective. Several qualitative responses are detailed in this section and summarised in Table 6.2.

The first expert reviewer with a focus on interaction and information design commented that the strengths of the prototype were the content, clear structure, and the overall experience of the project described as a fantastic journey. The critique consisted of focusing on the experience of the audience from the presentation type. This question was aimed at how exactly the interactive parts would be controlled: how many audience members would control the interactive components of the prototype, or would the audience communicate with a designated speaker who would control interactivity based on the audience's consensus? What would the measure of consensus of the audience be? One of the possible solutions could be a guided immersive experience, such as the current production used in a full-dome theatre at Visualization Center C in the piece 'Vast 3D: The Cosmic Journey Through Space and Time'. In the guided immersive experience, the audience engages with what is being shown, whilst a designated speaker expert walks the audience through the experience of explaining details through a narrative complementary to the visual stimuli. In a small intimate setting, such as the VR Arena designed for approximately 20 viewers, this experience could also include interaction with the audience where the viewers can tell the designated speaker how to respond to the interactive parts of the story.

Another point that was raised was regarding cognitive load (Plass et al. 2010) and the possibility of information overload, since effectively, three stories are told simultaneously. The tri-parallel model was created on purpose and with care to allow the audience to view the similarities and differences between the three chemical species, how they occur in nature, are used throughout history, are consumed, metabolised, affect the body and mind, and their current perception. The design of this specific model of the presentation was purposefully chosen and developed in a way to prevent the visuals overwhelming the visual sense. This included the use of simple drawings and models that are bounded with narration. The similar and the same occurrences used 'forking' where the display(s) merged into one, the differences were in tri-parallel mode with as many similarities identical on each screen and with the differences made obvious and salient. However, it is possible that some parts of the prototype may be more suitable for this type of presentation than others, for example the written information about the substances. Whilst these sections were reduced to as little writing as possible in this prototype to minimise cognitive load, they could also be converted to a pictorial format. As suggested by the respective expert, our research and implications would have to be to formally evaluated as to whether this novel tri-parallel and forking presentation of information is an effective way of conveying information.

The second expert reviewer described the prototype as 'inspiring', a positive and welcomed response since the *Effect of the Rings* was intended as an educational tool and aimed at young adults. In this sense, this is a confirmation that the prototype could potentially achieve its aim. Another strength to an extent was the use of prototype 3D animation showing receptor binding (Fig. 6.4). Our prototype intended to use 3D models and animations in the final product because 3D models have advantages in showing data that require more cognitive processing, such as data that has to be visualised by the viewer (Solomon 1994). In contrast, creating 3D animations is very time consuming when one considers creating the animation itself, rendering, and post-processing, which is a point that was raised later in the critique. We agree with the expert's suggestion that 2D animations could be used throughout the entire prototype to make the project more time and cost-effective. However, 3D animations have potentially beneficial attributes as already mentioned previously, in addition to being an attractive medium (Mansor et al. 2020) which is a desirable trait in educational initiatives. We also believe that using a combination of 2D and 3D animations could help naturally segment the information into smaller, more digestible perceptual chunks for the audience, which would help reduce cognitive load and increase information retention (Mayer and Moreno 2003). Using the altering types of animations could also point out the more detailed information and direct the viewers' attention to focus more when certain information is presented with more detail in a more attractive way. For example, in Fig. 6.8 the viewer needs to be aware that the chemicals have a similar ring arrangement, yet different structure, without having to know the specific organisation of atoms in each molecule. The short 3D animation of molecules binding in the synaptic cleft (Fig. 6.4) is a process that should be understood in more detail as different substances with different ligand binding affect neurotransmitters differently, which is one of the main differences that lead to different physical and psychological effects. We also do appreciate the need for a more time/cost-effective suggestion and agree that 2D illustrations could be used more extensively.

In the case of the third expert reviewer, the prototype was critiqued as not being clear about what the story presented from the very beginning. It was stressed that this issue was more pronounced for people who do not have a scientific and biochemical background. We agree with this statement and suggest that if in the future the story was to be presented by a designated speaker as mentioned in response to the first expert critique, the speaker could introduce the audience to the topic, the tri-strand presentation, and guide the audience during the overall journey. A more complete production would be clearer and more suitable for people without an in-depth scientific background, given that the target audiences are teenagers.

6.4.2 Contributions of the Design as a Visual Story for Communicating Science

The Effect of the Rings aims to combine art, technology, and science to create an educational platform about real-life issues. It tackles two frontiers, one technological, and one informational. The technological aspect is the use of state-of-the-art technology at Visualization Center C, Norrköping, Sweden, that allows wide-screen viewing and thereby an immersive group comparison experience. This is suitable and meaningful as the target group consists of young adults that could discuss the information presented and form informed opinions even after the storytelling segment has finished. The specific format was also chosen because the wide-screen platform is ideal for comparison presentations. Our comparative video parts use three screens for tri-parallel and forking techniques to present information that overlaps in similarities (forking) but also has significant differences in the same processes (tri-parallel). To our knowledge, this technique has not been yet applied in education.

The second novel concept explored by this prototype is the comparison of the specific chemical actors presented: nicotine, caffeine, and psilocybin, despite being three natural substances extensively used throughout human history. Therefore, it is interesting and important to note their status in legislation based on measurable medical data, rather than isolated cases and stereotypes. This project aims to educate the public about the effects of these substances. The target group was chosen as a specific age because they can make an informed decision about the future of the mentioned substances as they continue to grow older. This refers to their personal ingestion and use of substances and their use in research, attempting to provide an objective overview on the subject rather than persuade viewers to take specific action for or against specific substances.

Our prototype of interactive screening and tri-parallel comparison of three chemical actors is intended to inform younger generations of the use of these substances, so that they can contribute to informed decisions about the use of these substances. Whilst our target group is aged 14–18 years, they are not necessarily expected to make immediate decisions, but to have informed and unbiased access to information about potentially life-altering substances. One of the limitations of our study is that this prototype was not tested on our target group, but only evaluated by experts in visualisation, data analysis, and science communication. A future iteration of this study could also involve a functioning prototype beyond design alone, which could include data collection from actual target audiences.

6.4.3 Conclusions and Future Work

The described design process resulted in the conceptualisation and production of a digital visual story that communicates the science of the biological effects of psilocybin, nicotine, and caffeine using innovative display technology. It links human body processes across multiple levels of scale and interweaves historical, societal, value, and judgement dimensions, which is targeted for public engagement. In a simultaneous comparison of the biochemical effects of the molecular actors, the visual storytelling prototype transcends the submicroscopic and macroscopic making data about these three species more accessible to the public.

Future work could evaluate how the prototype might be validated as an educational or communicative intervention with students and citizens. The feedback from experts during the presentation of this project was meaningful and informed future ideas for improvement. The introduction of the parallel simultaneous tri-display approach shows promise for setting the scene for the viewers from the beginning and allowing for comparative reflection. In conclusion, the presented design has the potential to harness visual storytelling for providing novel ways of communicating and teaching new scientific content interspersed with implications for humans and society.

Acknowledgements We would like to acknowledge and thank the other members of the project group *Robin Skånberg* and *Yuxuan Zhuang* who contributed to the conceptualisation, design, and presentation of the visual design prototype. We thank the teachers and facilitators from the course at Linköping University from which this chapter was inspired, and Norrköpings Visualiseringscenter C for the opportunity to test the designed visual prototype in the VR arena. We also thank the expert evaluators and course peers for their feedback on our visual story design prototype.

References

Akiba H, Wang C, Ma KL (2010) Aniviz: a template-based animation tool for volume visualization. IEEE Comput Graph Appl 30:61–71. https://doi.org/10.1109/MCG.2009.107

Bedrina O (2016) Teaching with animation: from theory to practice–ICT in practice. ICT in Practice. Available at: http://www.ictinpractice.com/teaching-with-animation-from-theory-to-practice/

Berg A et al (2019) Representational challenges in animated chemistry: self-generated animations as a means to encourage students' reflections on sub-micro processes in laboratory exercises. Available at: https://pubs.rsc.org/en/content/articlehtml/2019/rp/c8rp00288f

Bertzen C, Basjuka J (2022) Designing for a multiple screen setup: interactive storytelling and attention guiding for a perceivable and engaging experience of UTM explore. https://liu.diva-portal.org/smash/get/diva2:1667543/FULLTEXT01.pdf

Botsis T, Fairman JE, Moran MB, Anagnostou V (2020) Visual storytelling enhances knowledge dissemination in biomedical science. J Biomed Inform 107:103458. Available at:. https://doi.org/10.1016/j.jbi.2020.103458

Crocq M-A (2007) Historical and cultural aspects of man's relationship with addictive drugs. Dialogues Clin Neurosci 9(4):355–361. https://doi.org/10.31887/DCNS.2007.9.4/macrocq

Cropper A, Luna RE, Mclean EL (2015) Scientific storytelling: from up in the clouds to down to earth a new approach to mentoring. In Proceedings of the IEEE integrated STEM education conference (ISEC), Princeton, NJ, USA, pp. 252–257. https://doi.org/10.1109/ISECon.2015.7119934

Delp C, Jones J (1996) Communicating information to patients: the use of cartoon illustrations to improve comprehension of instructions. Acad Emerg Med 3(3):264–270. https://doi.org/10.1111/j.1553-2712.1996.tb03431.x

Demian P, Fruchter R (2009) Effective visualisation of design versions: visual storytelling for design reuse. Res Eng Des 19:193–204. https://doi.org/10.1007/s00163-008-0051-4

Dudysová D et al (2020) The effects of daytime psilocybin administration on sleep: implications for antidepressant action. Front Pharmacol 11:602590. https://doi.org/10.3389/fphar.2020.602590/full

Erickson T (1995) Notes on design practice: stories and prototypes as catalysts for communication. In Scenario-based design: Envisioning work and technology in system development, pp. 37–58. https://doi.org/10.5555/209227.209231

Gable RS (2006) Acute toxicity of drugs versus regulatory status. In: Fish JM (ed) Drugs and society: U.S. public policy. Rowman & Littlefield, Lanham, Maryland, pp 149–162. Available at: https://psycnet.apa.org/record/2005-14597-007

Hadzigeorgiou Y (2021) Biographical profiling of Nikola tesla for the creation of an engaging story. Educ Sci 12(1):12. https://doi.org/10.3390/educsci12010012

Ma KL, Liao I, Frazier J, Hauser H, Kostis HN (2012) Scientific storytelling using visualization. IEEE Comput Graph Appl 32:12–19. https://doi.org/10.1109/MCG.2012.24

Mansor NR., Zakaria R, Rashid RA, Arifin, RM., Rahim BHA, Zakaria R., Razak A. (2020). A review survey on the use computer animation in education. IOP Conference Series: Materials Science and Engineering, 917 https://doi.org/10.1088/1757-899X/917/1/012021

Mayer RE (ed) (2014) The Cambridge handbook of multimedia learning, 2nd edn. Cambridge University Press. https://doi.org/10.1017/CBO9781139547369

Mayer RE, Moreno R (2003) Nine ways to reduce cognitive load in multimedia learning. Educ Psychol 38(1): 43–52. https://doi.org/10.1207/S15326985EP3801_6

McCurdy M, Connors C, Pyrzak G, Kanefsky B, Vera A (2006) Breaking the fidelity barrier: an examination of our current characterization of prototypes and an example of a mixed-fidelity success. In: Proceedings of the SIGCHI conference on Human Factors in computing systems, pp 1233–1242

Plass JL, Moreno R, Brünken R (eds) (2010) Cognitive load theory. Cambridge University Press. https://doi.org/10.1017/CBO9780511844744

Postránecká Z, Vejmola Č, Tylš F (2019) Psychedelic therapy in The Czech Republic: a theoretical concept or a realistic goal? J Psychedelic Stud 3(1):19–31. https://doi.org/10.1556/2054.2019.003

Pugh S (1981) Concept selection: a method that works. In: Hubka V (ed) Review of design methodology. Proceedings international conference on engineering design. Heurista, Rome. Zürich, pp 497–506

Pugh S (2009) The systems engineering tool box. Available at: http://www.burgehugheswalsh.co.uk/uploaded/documents/Pugh-Matrix-v1.1.pdf

Schönborn KJ, Anderson TR (2010) Bridging the educational research-teaching practice gap: foundations for assessing and developing biochemistry students' visual literacy. Biochem Mol Biol Educ 38(5):347–354. https://doi.org/10.1002/bmb.20436

Smart RG (1970) Some current studies of psychoactive and hallucinogenic drug use. Can J Behav Sci 2(3): 232–245. https://doi.org/10.1037/h0082727

Solomon G (1994). Interaction of media, cognition, and learning. Hillsdale, NJ: Erlbaum. https://doi.org/10.4324/9780203052945

Teravainen-Goff A (2019) Literacy and financial capability: an evidence review. National Literacy Trust Research Report. National Literacy Trust. Available at: https://files.eric.ed.gov/fulltext/ED603666.pdf

Tong C, Roberts R, Borgo R, Walton S, Laramee RS, Wegba K et al (2018) Storytelling and visualization: an extended survey. Information 9(3):65. Available at: https://www.mdpi.com/2078-2489/9/3/65

Weinberg BA, Bealer BK (2001) The world of caffeine: the science and culture of the world's most popular drug. Routledge

Więckiewicz G, Stokłosa I, Piegza M, Gorczyca P, Pudlo R (2021) Lysergic acid diethylamide, psilocybin and dimethyltryptamine in depression treatment: a systematic review. Pharmaceuticals 14(8):793. Available at: https://www.mdpi.com/1424-8247/14/8/793

Williams WR (2019) Attending to the visual aspects of visual storytelling: using art and design concepts to interpret and compose narratives with images. J Vis Lit 38(1–2):66–82. https://doi.org/10.1080/1051144X.2019.1569832?cookieSet=1

Xiao Y (2013) Animation trends in education. Int J Info Educ Technol 3(3):286–189. Available at: https://web.archive.org/web/20170812012116id_/http://ijiet.org/papers/282-JR112.pdf

Ynnerman A, Ljung P, Bock A (2020) Reaching broad audiences from a science center or museum setting. In: Foundations of data visualization. Springer, Cham, pp 341–364. https://doi.org/10.1007/978-3-030-34444-3_19

Diabetes Year One, *Random Fluctuations*, and *Mental Goblins*: Auto-Ethnographic Pathographies and Medical Comics

7

Tony Pickering

Abstract

Three comics; *Diabetes: Year One* (2018), *Random Fluctuations* (2020), and *Mental Goblins* (with Dave Pickering) (2020) were produced as a result of being diagnosed with illness, through engagement with the medium of comics. Using the mechanics and potential of the comics medium within the growing field of graphic medicine, this chapter discusses the ways in which the creation and reception of 'pathographies' can help consider notions of patient agency—the ways in which being a patient is interpolated within individual identity, and cement patient understanding. In addition, I review *Covid Chronicles: A Comics Anthology* edited by Boileau and Johnson (Graphic Mundi, 2021), *The Bad Doctor* by Ian Williams (University Press, 2014), and *Coma* by Slatterty (2021) to explore the ways in which graphic medicine explores the patient–practitioner interface.

Keywords

Pathography · Medical comics · Graphic medicine · Diabetes · Auto-ethnography · Metaphor

7.1 Prelude

The curtains are drawn around you.

The space is marked off—delineated, barriers are put up. Slowly you are sealed off from the world. Outside there are looks, and hurried conversations. Charts are shown, and eyebrows are raised. You have begun the process of identifying as a patient—though of course, you don't know this yet.

The pace of the decision is both fast-moving and glacial, slowly building a dissonance between the world you exist in, and your certainty about what came before. Meaning is communicated with the words from professionals—words that are new to you, yet familiar in their mouths, words which should make sense, but seem to glide over the weight of the moment. Hushed tones are used—to create empathy. Eye contact is made, to check understanding. You nod, a lot, it seems the right thing to do. There are leaflets that are handed out—leaflets are good right? Something to read—to explain everything. Anxious to please the word 'Oh' returns frequently.

Becoming a patient does not happen through explanation, actions, performance, or diagrams; it happens through the swirling mass of all these things—over time.

T. Pickering (✉)
Graphic Illustrator, Greater Manchester, UK

M. Roughley (ed.), *Approaches for Science Illustration and Communication*, Biomedical Visualization 4, https://doi.org/10.1007/978-3-031-41652-1_7

7.2 Introduction

Illness and illustration are synonymous for me, as my entry point into illustration practice was when I was diagnosed with type 1 diabetes. In pursuing ways to illustrate the unknown and unseen, my work moved more and more towards concept art and illustrating ideas. I am an illustrator and comics creator, with a practice situated in Graphic Medicine, cartooning, conceptual art, and mapping spaces through narrative and folklore. Much of this stems from my decision to explore a field that straddles the divide between auto-ethnographic reflection and the understanding of medical and scientific concepts—and find ways to communicate and fuse both impulses.

7.2.1 Why Comics?

Presented with a mass of information that I couldn't grasp easily and emotions that I could not fit into my established paradigm I worked my way to auto-ethnography through journaling, poetry, and sketching. In seeking to weave together multiple media I found that comics worked as the ideal vehicle for me to process and express my experiences of living with type 1 diabetes. This stemmed from the way that comics are '*a form that is not only about but is also the thing itself*' (Sousanis 2015, p. 59); that is, in drawing or creating a comic it is necessary to conceptualise the narrative in ways that are spatial and chronological, a necessity demanded by the comic medium.

Scott McCloud defines comics as '*juxtaposed pictorial and other images in deliberate sequence, intended to convey information and/or to produce an aesthetic response in the viewer*' (McCLoud 1993, p. 9). This definition is useful and identifies the idea of sequence at the heart of comics as a medium. At the heart of Comics is the structure of frames and panels that allows you to improvise an order of sequence on the page—and to delineate time by the arrangement of panels and pages. This includes the potential of fluid pages and one-shots that express the idiosyncrasy of thoughts and emotional flow. In this way sequential patterns become expressive of time. McCloud and Sousanis point to the use of juxtaposition across the page, and the ability to communicate through different channels in a way that is both 'simultaneous and sequential' (Sousanis 2015, p. 62). This capacity to explore an issue both inside and outside time make comics uniquely placed to explore the patient experience—able to consider the out-of-body experience of re-building an identity, but also to re-integrate that identity within a narrative of existence.

McCloud (1993) prioritises the visual nature of comics—noting that words are not necessary to make a comic, and that letters are visual symbols, which, when collected, form words—examples of 'other images' if you will. The arrangement of text follows not only the laws of the language it is written in, but also can be affected by placement, panels, speech/thought balloons, font, colour, size, and so on. Narrative can be written as prose, staccato-like dialogue, or depicted as special effects—where sound is visualised in a dynamic word. The ability to create a visual flow or pattern with the text spoke to the strands of poetry that had become increasingly prominent in my reflections on type 1 diabetes.

The definition also highlights the dual impulses of comics—to 'convey information' and to 'produce an aesthetic response'—that is to say, that the intention is to impact on the reader both in communicating ideas, but in impressing some form of visual and visceral response. In this way, comics can be seen as situated on the fault line between communication and artistic exploration—or in other words as demanding from the comics-creator an approach to imparting information (whether factual or narrative in nature) held within an aesthetic language. In this way comics have developed tropes and shorthand that can be used, or subverted to dissect the subject matter imparted. Thus, drawing on a scale from realism to cartoon can be utilised, or varied, in order to bring the reader closer or further away from the subject or narrator (McCloud 1993, p. 36).

Though not specified as an essential part of the comics medium, it cannot be denied that the use of these 'other images' provides a multi-media strand to the comics medium. That is to say, there is the potential to use narrative, dialogue, thought and images in overlapping, conflicting or harmonious arrangement to isolate a moment of crisis or a series of events. Seeking to interrogate lived experience the grammar, structure and tropes of comics helps to break down and organise simultaneous and conflicting thoughts. Sousanis (2015, pp. 41, 32) argues comics provide the '*tools for navigation*' imagining the geography of this experiential terrain, through what he terms '*Unflattening*', the: '*simultaneous engagement of multiple vantage points... to engender new ways of seeing*'. In this sense, comics provide an excellent vehicle through which to examine experiences of identity in transition, in order to bring the individual elements into focus. My confidence in using comics in this manner was boosted through my discovery and engagement with the field of graphic medicine.

7.2.2 Graphic Medicine

Graphic medicine is a movement, increasingly world-wide that explores the intersection of comics and medicine. Kendra Boileau summarises graphic medicine as follows:

> The term [graphic medicine] was coined in 2007 by UK physician and comics artist Ian Williams...Graphic medicine began as an area of study for scholars, educators, practitioners, and artists who saw in the subversive power of comics the ability to challenge prevailing attitudes towards the disabled, the ill, the dying, and those who care for them. In time, and quite rapidly, graphic medicine grew into a movement and a diverse community that includes not only scholars, educators, and practitioners but also people who create comics, people with illnesses and disabilities, family caregivers, medical students, librarians, and publishers. The movement became as much about the creation and dissemination of comics as about the study of comics... (Boileau and Johnson 2021, p. ix)

The field, thus: explores the work of medical practitioners in communicating medical conditions to patients and the personal experience of providing healthcare; it also covers the practice of creating comics as a way of research and understanding—in some cases, the process is as, or more important than, the result; finally, it includes the sphere of pathography—the experiences of patients as they become patients. It is this last strand where I come in, using comics to merge '*the personal with the pedagogical, the subjective with the objective*' (Czerwiec et al. 2015, pp. 2–3).

Recent strategies in healthcare have identified patient involvement as, *a key determinant in informing commissioning decisions and in shaping healthcare delivery*' (National Institute for Clinical Excellence 2004, p. 12). Traditionally, the patient–practitioner interface has been practitioner orientated—centred on treatment goals or practitioner education, reducing patient experience (and the lived experience of the practitioner) to information. Green and Myers (2010) identify comics as a tool for use at the patient–practitioner interface, noting the dual perspectives of the professional and the patient. The professional perspective, whilst recognising the nuance of comics as a form, emphasises the practical outcomes—the: '*connections between new information and existing knowledge [which increase] recall of health information*'. This approach is illustrated by a practitioner-developed comic *Type 1 Origins* (Kar and Patel 2016), using the analogy of superheroes to discuss the diagnosis of type 1 diabetes with the intent to provide a "'*soft foundation' [for]harder medical literature*" (Revolve Comics 2016). In this sense, the comic is designed as an entry point, but does not seek to provide a wider intersectional and immersive understanding of the experience for the patient (Squire, Czerwiec et al. 2015, p. 48).

Expanding on the field of graphic medicine, Ian Williams (Williams, Czerwiec et al. 2015, p. 132) posits that this wider understanding, depicted in graphic pathographies: '*create [s] valuable new knowledge, which informs the iconography of illness*'; knowledge that ultimately can be dispersed within a patient community that can re-negotiate and challenge our relationships to the patient–practitioner interface and thus impacts on medical practice.

In this sense, the field of graphic medicine can be seen to cover the traditional elements of medical treatment and communication. However, in seeking to understand the individual and societal elements that are bound up with the delivery and effectiveness of graphic medicine by default moves into areas such as public health and patient experience. I would suggest that though the purpose of examples of graphic medicine could be loosely categorised as either practitioner focused or patient focused—and thus informative or immersive, by the necessity of production and reception within the wider context of medical practice, all comics straddle these categories to a greater or lesser extent.

The sequential elements of comics are key to the way in which graphic medicine seeks to explore medicine and the patient-practitioner nexus. There is a concern to arrange time through space, to find the '*new knowledge*'. This ability of comics, as graphic pathography, to access both synchronic and diachronic time (Squire, Czerwiec et al. 2015, p. 46) is a framework that can enable the patient to '*slow-down*' (Myers, Czerwiec et al. 2015, p. 102), and become immersed within their experience. In this sense comics can be used to judge the scope and entry point of/into the patient's narrative; exploring the way in which imagining that narrative, and giving voice to it, can redefine access to healthcare, and establish a position of patient autonomy within their individual treatment.

Although there is no prescribed style for graphic medicine, we should note the teaching of Lynda Barry whose *Making Comics* syllabus influenced many of the group's founders to consider a sense of the aesthetic (Barry 2014, 2019). Barry emphasises that writing is a form of drawing, and that the two have not always been considered different disciplines. Her focus is on drawing for storytelling, drawing that has 'aliveness' (Barry 2019, p. 1)—and she devised comic-making activities that are designed to take away the fear of drawing and find ways to express stories. In this sense graphic medicine looks to the idea of the power and potential of creating comics, and in particular, a guerrilla approach emphasising a style that is accessible to all—a methodology that is concerned with the working through of delivering and receiving treatment. The result is a variety of styles—from naïve through to complex—with the emphasis on finding ways to tell 'your' story—to communicate an individual perspective of wider issues. This is particularly so in the patient, or auto-ethnography strand mentioned above. Whilst texts more focused on information may, on the surface, apply a more static approach, this is to ignore the pathway that may be needed by the patient or recipient to move from knowledge to understanding—that information may best be imparted through a visual language that is able to articulate the personal perspectives that imbibe these narratives, and contextualise information.

This begs the question who is the audience for graphic medicine? And as suggested earlier—this is both specific and varied. Graphic medicine can be used as an educational tool within the medical profession, spreading awareness of new developments or conditions that are unfamiliar. There is the potential that this form of graphic medicine drifts towards infographics in style and submits narrative to information, though it cannot omit the context of its production. Additionally, it can be used as a vehicle for communication of complex (or even simple) medical ideas, and to communicate with those who are unable to access more text-based information systems.

In these cases, the emphasis is on the visual layout—often using the tropes of cartooning and established comic symbols to help communication, an approach that assumes familiarity with the traditions of comics, an assumption that can be problematic as comic traditions across cultures can differ wildly—for example, Japanese Manga is read right to left, not left to right in western traditions. In this sense the creation of a visual language that both engages and communicates is important, and a key part of the creative process.

Lastly graphic medicine works as an object to receive—as a book, comic, or file for patients who have experience with specific illnesses, but also for those around who wish to know and understand more of the experience. In this case, the aim is to connect to and disseminate the experience with those who seek to understand and empathise. These auto-ethnographic comics or Pathographies—Graphic novels/comics concerned with a 'narrative of illness' (Squire, Czerwiec et al. 2015, citing Anne Hunsaker Hawkins, p.45), consider the lived experience of patients, or of people who have experienced the illness of the person involved and are focused at those in a similar medicine situation—who have some experience of the specific illness, or those who want to know about it and those who enjoy the narrative potential of the move from person to patient—who are invested in the fluctuations of identity that the topic considers. In this sense it is harder for these books to 'target' an audience as the skill of the narrative 'hook' is as important as its relevance or use. Auto-ethnography thus has the potential to make the difficult or traumatic relatable—to explore the shades of illness and treatment providing the capacity to infer and imbibe information or an approach through the exploration of another's experience. In this case, the auto-ethnographic approach has the capacity to provoke an empathetic response, which allows information to pass through discovery and reflection, rather than by a didactic thrust.

7.3 A Review of Graphic Medicine

To explore how Graphic Medicine straddles patient and practitioner perspectives, and to consider auto-ethnographic approaches, I would like to look at a few examples from *Covid Chronicles: A Comics Anthology* edited by Boileau and Johnson, *The Bad Doctor* by Ian Williams—including a close reading to explore the detailed workings of comics as a medium and *Coma* by Zara Slattery.

7.3.1 *Covid Chronicles: A Comics Anthology* Edited by Boileau and Johnson (2021)

Covid Chronicles: A Comics Anthology was collected from comics creators during the pandemic and covers many different personal perspectives. As Boileau covers in her introduction:

> The comics here run the gamut in perspective and style. Some are true, deeply personal stories; others are invented ones, either based on real events or inspired by a vivid imagination. They are documentary, memoiristic, meditative lyrical, fantastic and speculative, offering a view onto the countless ways the Covid-19 pandemic has changed lives...In short, these comics reveal the pure fear, anxiety, and grief so many of us are experiencing these days – feelings that will no doubt be with us for years to come (Boileau and Johnson 2021, p. ix).

In this opening, Boileau points to the ways in which comics creators respond to a subject—with varying forms, and in doing so raises the question does a pathography have to be a journal or documentary? Comics provide many visual pathways to communicate—and the decision on how to portray an avatar of the authors can range from deciding how to draw yourself, to creating another persona to voice your thoughts and feelings. Any approach requires you to think yourself from the outside, and puts your own personal thoughts, feelings, and actions under a microscope—which can be both revealing and challenging, and part of an examination of illness and treatment. In this sense, though pathography may appear as a journal—I would suggest that this is only one way to create a narrative of illness. As Hillary Chute argues, comics make '*readers aware of limits, and also possibilities for expression in which disaster, or trauma, breaks the boundaries of communication, finding shape in a hybrid medium*' (Chute 2017, p. 34). These 'possibilities for expression' are the reason for the potential that comics have in exploring medicine in new ways.

Joe Decie's *It is a Laugh a Minute Round Here* (Boileau and Johnson 2021, p. 74) presents as a comic strip—using a four-panel situational joke to explore the gap between action and meaning, moving from a cough to the explanation of the sort of cough it is—inserting two extra panels that drag out time, accompanied by the diminishing sound of 'tap' as realisation dawns about what a cough could mean. The change of perspective draws focus from the other person in the room to the 'cough-er' (presumably Joe), who realises the implication of the pregnant pause and explains to offer a reassurance—sort of. Decie's drawing style makes these static scenes energetic, creating a paradox whereby the framing of the panel clearly separates each from each, and the gutter space builds up the slowing of time; yet the drawing creates the impression that the images in the panels are moving vigorously whilst not moving. The images suggest that a lot is happening in the still moments—beyond the only real action—that of typing on the keyboard. This is an example of the ways comics communicate the multiple understandings of a simple act. Within these panels we see an innocuous cough reveal the medical and societal implications of the pandemic; we move away from day-job priorities to the wider social context, and the slow realisation that the future may not be the same as yesterday.

In panel three the sound is reduced to a solitary 'tap' as we focus on Joe's face which is in thought—attempting to self-diagnose his symptom and categorise himself as dangerous or safe—a decision that is rife with problems either way. The strip's pay-off, revealing that it was 'just a normal phlegmy cough' and drawing the response 'oh good', plays with the idea of a last-panel punchline. Here instead of a gag, the pay-off is laced with awkwardness and bathos—a veneer of reassurance that the two people crave and hope for but cannot really access as they both know they cannot be truly sure. Yet they also fear to check—aware of what that could mean. In this way, Decie's comic strip serves to present a key duality of the lockdown—the desire to do the right thing for public health, and the hope that you do not have to.

Pandemic Precarities (Boileau and Johnson 2021, p. 211) by Kay Sohini takes the form of a memoir—straddling Sohini's experience in New York with memories of her grandmother from Calcutta who dies during the pandemic. The comic begins in a documentary style—visually moving the reader from an academic conference in Boston back to New York through emptying streets until we are inserted into Sohini's apartment. This travel sequence is depicted using a god's eye view perspective, creating a sense of the reader as the observer—viewing the first signs of lockdown on the busy streets of New York. This perspective also serves to form a distance, a sense of the barrier with society—working as a metonymy for social distancing and the numbing to the world around that was created by lockdown.

Sohini leaves the constraints of the panel to stride across her two cities—New York and Calcutta, a separation that feels so much more with her inability to attend her grandmother's funeral. Here, thoughts are depicted through panels of pure text and reflect that: 'so much of the grieving process involves exchanging stories with your loved ones' (Boileau and Johnson 2021, p. 212) and segues into an anecdote about her grandmother giving Sohini's favourite food to travel with back to the USA—refusing to accept that it will not be permitted to travel into the country with her.

Pandemic Precarities develops from this initial memoir to recollections of isolation alone with the presence of her memories, moving to polemic in the consideration of the ways in which the pandemic has been handled in both India and the USA—considering issues such as discrimination through inequality, religion, race, and issues such as mask wearing and pandemic denial and pay for the most vulnerable in society. Visually the images move from situational to metaphor, as we see the foetal Sohini underlaid beneath text that recounts her experience alone in her apartment dwelling on her inability to mourn her grandmother fully. As we move into polemic recreations of news photos are used to ground the argument—and footnotes are used to cite sources and indicate veracity—both floating in different coloured text boxes and inserted in contrasting colour at the bottom of panels. Finally, the comic finishes with Sohini against the background of the world, contrasting the COVID statics pointing to the position of the USA and India as of the time of writing—both in the top three worst-infected nations. This direct address to the reader serves as a challenge, a move from personal experience to wider societal conclusions.

The Dance of Death (Boileau and Johnson 2021, p. 173) by Peter Dunlap-Shohl presents a pictorial sequence of allegories updated from the Middle Ages to the COVID present. Dunlap-Shohl prefaces the sequence by noting *The Dance of Death* was a 'popular meme... where Death is depicted as the great equalizer' (Boileau and Johnson 2021, p. 173). The 12 images take contemporary moments—moving between the every day and nationally important and insert death into the image in the form of a skull—either in the background or as an active participant. Death appears in a coffee shop, at the grocery, walking a dog in the street, at airport security, in the beach, in an evangelist church, beside Donald Trump at a press conference, at an anti-lockdown protest, in a hospital, at a wedding and in prison. These juxtapositions speak to the ability of COVID to spread easily anywhere, and to the randomness of infection, and death. Dunlap-Shohl engages with the range of society, and how the normal so quickly becomes the site of the obscene, how death becomes a risk to factor in—a simple part of life.

7.3.2 The Bad Doctor by Ian Williams (2014)

The Bad Doctor explores the General Practice of Iwan James and his own experience of OCD. Though not explicitly the avatar of the author, Iwan as the protagonist serves as the narrative voice throughout, and it is through his own anxiety and sense of doubt that we are shown the day-to-day experience of his work in general practice. In this way, the graphic novel serves to chart patient and practitioner from a psychological perspective as opposed to a didactic or informative approach.

The narrative of *The Bad Doctor* juxtaposes the presence of Iwan's OCD with his role as a medical authority within the GP surgery. His OCD manifests through feelings of guilt and responsibility, which then take the form of anxiety about the mistakes he will make, or the harm he has caused, or will cause, others. Visually his OCD is shown through insert panels that suggest his own suicidal thoughts—specifically him putting a gun in his mouth and pulling the trigger (p. 22). There is the constant use of circular motifs to emphasise the patterns of behaviour which Iwan is drawn to, and how small things are able to trigger larger issues (p.121)—moving from a need to say goodnight to his stuffed animals and toys in order, to distress over the effect of failing to do this task properly—effects that in *The Bad Doctor* are seen to be satanic in nature (influenced by the cultural memory of his upbringing). These circular motifs can be seen in the use of cycling as a constant theme—how Iwan processes his anxieties, and in reference to the benefits of exercise in treating mental health issues. Visually we see the presence of circles throughout—appearing in gears, wheels, and cogs (p. 96) during some of the chapter break images, and in the organisation of panels across pages as the influence of the satanic plays on Iwan's mind—taking the form of sigils or symbols associated with satanic worship (p.167).

In working through Iwan's feelings about his OCD, and his worry that it might again be overwhelming him, we see him dealing with a patient who exhibits OCD tendencies. This 'mirroring' device allows him to externalise his own feelings—although it is made clear that Iwan feels comfortable treating the issues of others, but does not feel that he can take his own advice. In his encounter with this patient, Iwan blurs the patient–practitioner line, by revealing his own experience with OCD, something the patient finds helpful, as he begins to see that he is not alone. In doing so the graphic novel asks the question about how the experience of being a patient can inform the process of treatment, or indeed help to question the paradigm of what we expect from our GPs.

The way in which the professional and personal elements are woven together and the impact of features of the comics medium can be illuminated through a close reading of the opening chapter. This blurring of the professional and personal is seen immediately from the beginning of the book. The opening image depicts an almost pastoral setting (p.7). The GP surgery is a modern building framed by a tree and with grass and a shrubbery bordering the driveway out front towards the reader. There is the figure of a woman running across the grass towards the building—at first glance you might take her as a child in play, but on closer inspection she is carrying a shopping bag and breathing hard. Further up the picture, there is a bird in silhouette—resonant of Edgar Allan Poe (though NOT a Raven).

The woman, it turns out, is running to report that someone has collapsed at a nearby bus stop. But first we are introduced to Iwan's office (p. 8)—with a staccato panel to set the scene of his desk, before slowly breaking down the action of picking up the phone to call the next patient—another panel indicates the distance between the Iwan's room and the waiting room by spreading the dialogue between his room and completing the dialogue from the speaker in the waiting room. The space created by the gutters (the spaces between the panels) gives weight to each panel so that the reader feels these are moments of time that have been stolen within the rush of the day. This is supported by the voice-over narrative of Iwan in the first person that articulates how he has been working on automatic—having omitted to go to the toilet when he had the chance.

The urgency of the request for help is juxtaposed with the slow deliberate journey of Mrs. Davies, a patient waiting, from the waiting room to 'Room 4'. Her odyssey on pages 8–9 is told through panels that utilise perspective to emphasise the slowness and difficulty of this journey. The first panel is shown from a floor angle, so we see the pressure involved in standing. The arch of her back as she moves, inching forward suggests both her pain and vulnerability. On page 9, the perspective moves to overhead—as if caught by CCTV, with the gutter used to show how long it takes to round the corner. On arrival, her face is clouded and gaunt indicating her pain or distress. Having made it this far Mrs. Davies' consultation is interrupted by news of the person who collapsed at the bus stop—a surprise as we have seen the receptionist depart at a quicker speed than Mrs. Davies can muster, but this is soon clarified as the receptionist explains she has already asked Dr. Smith—who claimed he was too 'BUSY' (the capitals suggesting that this response is typical) (p. 10).

At once, in the opening pages, we have the urgency and mundanity of General Practice. We have the way in which time moves at different speeds, and we can see the ways in which treatments overlap in ways that prioritise some at the expense of others; we as the reader are still unaware of Mrs. Davies problem as her dialogue is depicted through sequences of vertical lines—suggesting a rhythm or pattern that is familiar, but not specific. Additionally, we can see that personal relationships in the practice are strained and subject to a hierarchy through Dr. Smith's delegating the emergency to Iwan. This opening serves to show medicine within a personal context, and asks the reader to consider medicine within a holistic framework.

Exploring the way the opening transitions into the emergency on the bus allows us to consider the way in which panels work to enhance meaning within comics, and point to the power of graphic medicine in fleshing out pathography. Pages 8 and 9 use a six and seven panel structure to pace out the action—using the extra panel to slow the speed of the events so that the emergency storyline can be contrasted with the movement of Mrs. Davies. When Iwan agrees to attend the panel sequence is reduced to speed up events, but also to allow for a panel in wide-shot that allows us to see Iwan and Sister Wendy in action rushing to the patient with emergency equipment. The page ends with a touch of bathos as we see James—now still, asking 'where's the patient?' only to receive the reply 'Got on the bus'. This moves the reader to anti-climax—but there are suggestions that we are premature in our assumption, as behind Iwan are splashes on the pavement. Turning the page—an act that works as a reveal for the reader, we are presented with the splashes in close-up—they are now more obviously blood, and we can see the frequency with which they are falling. The reveal of the page turn allows the reader to feel the fluctuation between the adrenalin of the call out—the seeming relief that the patient is okay, and then the realisation that they are still hurt, and walking around untreated. In this way we are beginning to get a sense that medicine does not work on one level, that the professionals involved are subject to the whims and caprice of moods and the thousands of thoughts that we are subject to as individuals in the network of the moment.

On page 12, we follow the blood trail, and once again find the panels adapting to the narrative direction—with borderless panels of blood splashes interrupting the tracing of the patient onto the bus, passed the annoyed bus driver, to where the patient is sitting with an obvious head wound. This passage is dialogue free—allowing the reader to follow Iwan's point of view in assessing the scene—taking in the patient, but also the reactions of the bus passengers—this treatment will be a public show. When the conversation starts on page 13 Iwan's work in diagnosis is interspersed with the reactions of two teenagers who are videoing the events on their phones—they will later upload the video to social media. The panel structure varies through the encounter—again showing the staccato rhythm of patient interaction: we move through the attempts to create a rapport; the details of medical practice—putting on gloves; probing questions to find out what has happened; the confusion and resistance of the patient; the public gaze—building towards another reveal on the page turn as the patient becomes aggressive in a potentially horrific moment depicted in close-up with blood dripping down the patient's face and smeared on a raised hand. Iwan's attempts to calm the situation and persuade the patient to go back to the surgery are again interrupted by a violent outburst.

Iwan's attempts at treatment are interrupted by Sister Wendy informing him that the patient will not cooperate, by the driver who is visibly annoyed and demands that the patient 'Mad Mog' is bandaged as he cannot keep 'bleeding' over his bus. Leaving Wendy to 'patch' up Mad Mog Iwan returns to the surgery. In the meantime, another patient faints—presumably because of the blood, much to Wendy's consternation and annoyance. Here panels are used to quicken the action—formed into strips to speed the collapse in the manner of animation cells (the more cells the smoother the movement).

Iwan's return to the surgery takes us through a glimpse of his professional relations—the suggestion that he has feelings for Lois Pritchard, a fellow GP, and his antipathy for Robert Smith—the practice senior partner, who it appears to be concerned with funding, and is oblivious towards Iwan's annoyance that he was left to deal with the incident. The sequence utilises the semiotics of icons and objects: a heart appears on Iwan's sleave as he greets Lois—she in return tucks her hair behind her ear—her gesture and body language suggest there is a connection between the two; Robert is reading a newspaper with the headline 'More cuts to GPs income'; posters around the surgery proclaim the dangers of the cold virus and show the Bristol Stool Chart—props that build the mis-en-scene of the surgery. The texture of this setting accentuates the personal relations of the practice, enabling the reader to see the personality of the doctors outside the medical experience they provide to others, and so enable us to see Iwan as a patient.

The chapter finishes with Iwan finally managing to go to the toilet, before returning to his office. This action is again slowed by dividing it into panels, forcing the reader to acknowledge this is a moment of calm within the chaos of the previous events. Back in the office, Mrs. Davies has seemingly left, and Iwan audibly expresses his relief and checks on the next patient—only for her to be admiring photos on the wall behind the door. This pay-off to the chapter again uses the reveal of the page turn to travel between Mrs. Davies' dialogue, so that we 'hear' her before turning the page to reveal a one-panel page that conveys the humour of Iwan's awkwardness and embarrassment by the completion of the dialogue, but also through Iwan's stressed pose, turning abruptly from the computer—and movement and emotion lines, suggesting the speed of his movement, and the alarm at being caught in an unprofessional moment. In this sequence, we see Iwan as the reluctant hero, as socially awkward, as empathetic, as callous, and as the comic protagonist of a series of unfortunate events.

The Bad Doctor uses a professional context—the work of a GP to turn the tables and place the focus on the doctor as patient. As the book develops it becomes clear that Iwan's career as a GP can be seen as an extension of his experience living with OCD. Through flashback sequences we can see the origins of his pathology stems from a belief that he is responsible... for everything. His OCD behaviours are rooted in trying to prevent bad things happening. These flashbacks are portrayed in softer focus, with a greater range of tone. The style is both romantic and other from the main narrative visual style which is more linear with flat tones. It should be noted that the style is connected—so that we see more texture added to a recognisable present in the flashback sections—so that objects and places seem to have more intensity. A stylistic choice that echoes the idea that Iwan's OCD is not a constant—but something that can be worked through and treated. Throughout the flashbacks, we see the motifs of heavy metal and satanic symbols developing—growing from a biblical concern with sex and sin. This is at least in part the impetus behind the title—where 'The Bad Doctor' is a moral categorisation, not a professional one. A labelling that is bound up with notions of personal responsibility, and a hyper-awareness of how our actions can impact those around us—notions that are amplified to pathology and codified through spiral thinking into behavioural paralysis.

7.3.3 Coma by Zara Slattery (Slatterty 2021)

Coma by Zara Slattery explores the author's experience of falling into a coma as a result of Necrotising Fasciitis—a deadly bacterial infection, which ultimately leads to a leg amputation. The graphic novel takes the form of alternating chapters which take the viewpoint of Slattery in her coma, and the others which use the diaries kept by her husband and the ward nurses to explore the experience of those around her, and how they deal with the impact of her coma. Slattery varies the two strands visually, with the coma sequences in vivid colour, and full of a sense of movement that is not grounded in conventional physics—enabling her to move in all directions for dramatic and emotional impact. The coma sections have panels without borders—suggesting that they are uncontained, giving the impression that the panels spread out in space—suggesting a roaming psychological landscape filled with fantasy and horror tropes with swirling skies, ominous corridors, plunging seascapes, and surreal creatures. The images are depicted using a fluid brush or pen line and filled with tonal washes that are both emotive and atmospheric.

In contrast, the diary sequences are drawn using blue pencil with tonal elements added through shading. The panels are clearly delineated to reflect the effort of structuring these days of balancing waiting and carrying on. There is an emphasis on detail and texture in these sections—an almost overwhelming sense of the presence of ordinary things—food, birthday parties, traveling to and from the hospital—carrying on, feeling helpless. Whilst the diary deals with the weight of the every day, Slattery engages in a surreal quest through her prison—she questions herself, feeling guilt at abandoning her family, blaming herself for what she is experiencing she falls through realms of her own purgatory and faces a Bacon-esque creature—a representation of the bacteria that chases her through an increasingly uncertain landscape. The use of two narrative voices enables Slattery to contrast the immediacy and dream-like feel of the coma, with the perspectives of those around her, and the focus on the practical elements like the implications of the amputation and the emotional moves from hope for a simple operation to the acceptance of the size of what is now in store for the family. This structure allows the reader to appreciate the events as they happened, but also as experienced.

Coma like *The Bad Doctor* and *Covid Chronicles* seeks to find ways to express the uniqueness of illness within the necessity of existence. In all texts, visual languages are created that point to the ways that both are present at the same time. Pathography finds ways to be both 'fantastic' and 'stoic'—in the sense that it provides ways for illness to enter the imagination and for it to be reconciled (if uneasily) to the world in the illness exists—a world that is not always accommodating of its need. In these texts, we also see the psychology—the stresses and strategies that patients find to aid this interpellation of illness into concepts of the individual—or the 'Self'.

7.4 Case Studies—My Comics: *Diabetes: Year One* (2018), *Random Fluctuations* (2020), and *Mental Goblins* (with Dave Pickering) (2020)

7.4.1 Diabetes Year One (2018)

The impetus for my illustration and subsequent graphic medicine practice was my own diagnosis of type 1 diabetes late in life—I was 41 years. Though subsequent research (talking to other people with type 1 diabetes) has revealed that this is not unusual (indeed people have been diagnosed much later than me), nonetheless the popular perception is still that type 1 diabetes is 'juvenile diabetes' and thus is predominately identified in childhood and teenage.

My diagnosis was contemporaneous with the beginning of my Master's degree in Fine Art: Illustration so logically my research in the two arenas dovetailed. At once I was exploring the medical science of type 1 diabetes, experiencing the impact of changes in lifestyle and the implications to daily behaviours, and interrogating my artistic practice—specifically what was unique about what I could offer—and how design influenced my illustration choices. I built towards comics through the process of journaling and sketching around this research. This auto-ethnographical impulse was developed through refining sketches and thoughts into poetry and working up illustrations. My illustrative work in particular explored ideas of frustration—the inability to express the unseen, and my reaction against the NHS visualisation in pamphlets and posters—which I felt created a distancing relationship with patients who needed to not just be informed about, but to understand their own experiences. I was interested in automatic drawing, ways in which I could generate expressions of ideas that were in process—scribble patterns created through writing, immediate responses to music or text that sprawled outward—indicating marks to be isolated or developed. These ideas were pursued through working in traditional media—specifically coloured pencils, leading to experiments with colour and usefully with the ideas of erasing space—an excavation of the image, which lead to an approach that unearthed the psychology of the experiences. Colour led to a move away from solid black outlines, replacing them with colours to create a tonal harmony across the page—using the weight of mark making for contrast where necessary.

Having discovered graphic medicine and coming to the comics medium, my process for *Diabetes: Year One* involved a simultaneous approach to text and image. On an open spread of my a3 sketchbook, I would arrange snippets of text and associated or derivative images. In this sense, the interaction between text and image was designed—initially with the idea that one should not duplicate the other. Though as time went on the capacity to extract more from the text through images became apparent. Text in this sense took on a visual impact that was originally hand lettered—ultimately becoming a handwritten font that I created to consider the intimacy and specificity of the narrative voice.

Diabetes: Year One began with an exploration of how to depict the unseen. The comic is, to some extent, an experimentation of how to solve this problem as articulated by Susan Sontag: '*Illness is not a metaphor, and... the most truthful way of regarding illness – and the healthiest way of being ill – is one most purified of, most resistant to, metaphorical thinking'* (Sontag 1991, p. 3). However, *one cannot think without metaphor* (Sontag 1991, p. 91).

In an unused page. I began with animals as my metaphors—but felt that this was a crude attempt to find qualities that mirrored the broader experiences of type 1 diabetes. Some animals do make it into the comic—specifically a parrot and bees—metaphors chosen as much for their aural qualities as their visual specificity. The humming of the bees suggests the nagging presence of type 1 diabetes throughout everyday tasks especially—a quality akin to the squawking repetition of a parrot's obsession with a learnt phrase.

An octopus is used as a visual aid—from an association that I made when I experienced my first ECG during my diagnosis—the physical presence of the creature adds psychological weight to the sense of the electric sensors as tentacles sprawl across my body (*see* Fig. 7.1).

As the comic progressed my attempts to visualise type 1 diabetes became more focused on the equipment that accompanied the experience—the shapes and forms of vials, testing strips, monitors, injection pens, needles, and the general detritus of medical waste that begins to surround you. This developed into the creation of landscape settings in which I placed the avatar of myself—a wonderland or dystopian realm in which I inhabited. In this sense *Diabetes: Year One* moves between the internal experience of type 1 diabetes—exploring symptoms such as changes in sight, and hypoglycaemia from a first-person point of view, to an external perspective, which though still in first person, attempts to allow me to observe myself, to put myself in the position of observer or watcher, and to unpick externally the impulses I experienced. This attempt to use comics as a medium for untangling thinking is also seen in the way that my avatar moves between 'icon' and 'real'—changing the way in which the reader interacts with it—at times seeing themselves in the 'everyman' icon, and then choosing to see the 'other' real depiction.

This switching also provides the capacity to explore the science of type 1 diabetes—allowing the comic to present symptoms, the science of (what we know about) type 1 diabetes—and to consider how this can impact lifestyle—or be managed. I engage with science, and the history of diabetes, as part of an ongoing motif in the comic—the idea of movement: whether that be walking as thinking or processing; or travel across distance—climbing, moving house, going on holiday, or simply the idea of the title for the final section 'the journey'. This sense of progression is important, as the comic is presented as in an episodic structure—it has a sequence of moments rather than a narrative, so the comic does not so much have an end but a coda. Though the comic takes time to reflect on the psychological impact of becoming and being a person living with type 1 diabetes, there is also a sense of progression—of achieving, with one eye on embedding being a patient into my sense of self.

This moment is reflected in the progression to a more fluid and experimental use of panels away from the more structured approach (influenced by Dave Gibbons in *Watchmen* 1986) that the comic begins with. We see the use of more panels to slow time, and wider panels for a more immersive experience (both used to show varying aspects of hypoglycaemia), smaller panels to convey the psychological pressure of everyday situations in an exploration of diabetes rage (caused when blood/sugar levels spike or drop quickly), but also floating panels to create a more upbeat feel and sense of freedom in the travels around Italy.

7.4.2 Random Fluctuations (2020)

Random Fluctuations is a spin-off from *Diabetes: Year One*—inspired by Chris Ware's *Building Stories* (Ware 2012) where the comic is presented in a box and the story is combined through the readers' arrangement of the comic which is delivered in various formats—board, broadsheet newspaper, booklet etc. With that in mind I considered a similar approach for *Diabetes: Year One*—but was unable to complete the concept due to time constraints. One residual idea from this concept I had that I did want to see through was the notion of a comic that came in the form of the diabetes diary I was given when diagnosed. In this diary, I had to record the amount of carbohydrates I consumed at each meal, along with the dose of insulin I took (this would vary depending on the ratio for the time of day). The result of this was pages, and pages of numbers—*Random Fluctuations* would be a visual interpretation of these numbers.

Random Fluctuations was produced using digital media, and is more design focused in the sense that each page is a panel in itself, and each panel is designed in response to a line of a short poem that condenses the experience of living with type 1 diabetes. The format of the booklet—longer and narrower than panels I normally use, forces the composition into contortions: squeezing in clocks, forcing type around objects (*see*

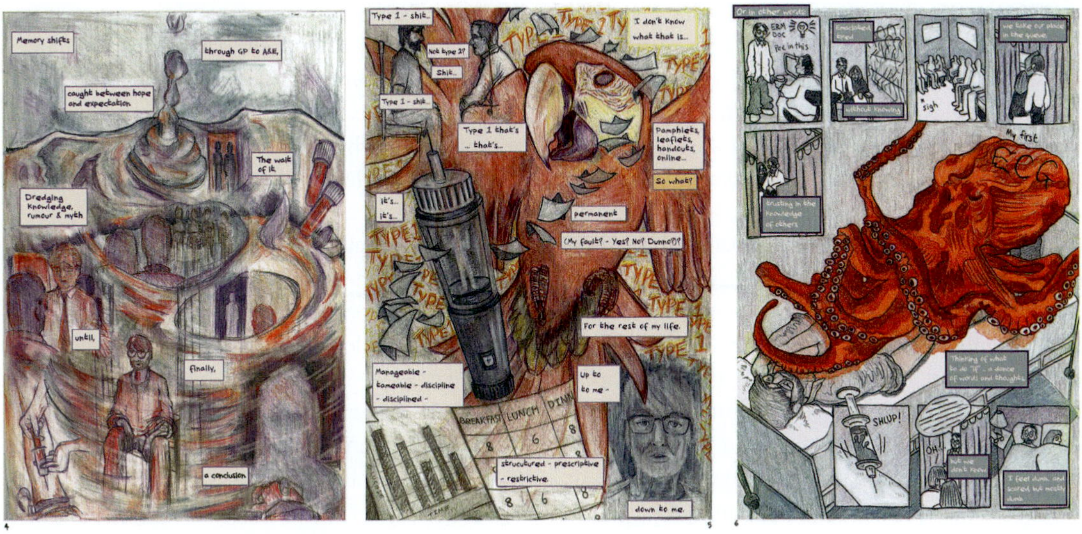

Fig. 7.1 Pages 4,5 and 6 from the comic *Diabetes: Year One* by Tony Pickering

Fig. 7.2), cascading images down the page, extending the flow of chemistry experiments down graph axis, and surreally extending limbs and breaking bodies around corners.

Words in *Random Fluctuations* vary in type and font—all hand realised. How the words are written becomes a key part of the way in which images are formed. From blocks to handwriting, to mechanical, to classic, the text embeds itself as part of the image—so that the poem is fleshed out and its links to type 1 diabetes exemplified. This is especially noticeable in the way that the text is given thrust—pages where the layout provides the drive to the next page, and the theme of progression moves the reader from page to page.

The integration of text as image, and the movement of the comic build with the use of layering in the image—backgrounds of faded information that lie beneath the main image (the greyscale lying behind a black and white foreground), so that we feel the idea of type 1 diabetes as a calculation that is always in the background—a constant presence that needs to be managed for the every day to be accomplished.

Fig. 7.2 Pages 4, 12, and 19 from the comic *Random Fluctuations* by Tony Pickering

Various motifs repeat in the comic: images of precariousness—hanging over cliffs, skulls, the wobble of a hypo, the charge of the cavalry (as insulin—which if not taken carefully can lead to hypoglycaemia), and the shaky grasp of reality in a diabetic rage; geometric shapes—a desire to find shapes that compliment or harmonise, and the need to force situations to fit these shapes; and linked to the geometry the motifs of maths and science—time, grids, equations, and experiments. These all speak to notions of patient experience and specifically to my type 1 diabetes. Finding a way to live between coma 1 and coma 2—fitting how I want to live and how I need to live (and recognising the privilege I have to do that—for example access to insulin), and my need to understand aspects of science and maths that previously I had no engagement with—and the importance of being able to do so.

My avatar in *Random Fluctuations* is consciously cartoony—learning from my experience in *Diabetes: Year One* I wanted my narrative voice to be accessible as an 'everyone' too. It allows for elements of Harold Lloyd, for the image of myself as an ineffective boy scout, the surreal stretching of my body and for myself as an embodiment of rage.

Though *Random Fluctuations* can be said to engage with many of the themes in *Diabetes: Year One*—as indeed it was originally conceived to do so, it is in finding a new—or alternative, visual language that compliments *Diabetes: Year One* that I find rewarding. Especially the way in which cartooning and typography combine to find a route through the structured stress of the imposed format.

7.4.3 Mental Goblins (2020)

Mental Goblins is a departure in my graphic medicine work. The script was written by my brother Dave Pickering and explores his experiences with anxiety through the trope of personal goblins—which in the comic surround and impact on his life. In contrast to my personal work *Mental Goblins* uses an extended metaphor throughout—the goblins that stand in for anxiety. We see the goblins as selfish, manipulative, destructive, and ridiculous. They have the capacity to appear from everywhere—and choose mundane and unexpected places such as cereal packets from which to surprise Dave. They have a capacity to blend into the background but are still present even where they cannot be seen. Their faces mock, as they force Dave to question what he thinks he knows, and to worry about what may or may not be. It is this constant pressure that slowly grinds down Dave's sense of who he is within society—at one point he can be seen hiding beneath the table, unable to deal with the constant bombardment of goblins (*see* Fig. 7.3).

Now working digitally, I consciously interpreted the artwork in a handdrawn style—emphasising pencil strokes and the element of erasing that I had employed in *Diabetes: Year One*. Drawing for another person allowed me to develop a more consistent avatar for Dave and to explore the character design of the goblins—balancing mischief, malevolence, and the slapstick required by the script. They have elements of the 'Blue Meanies' from the film *Yellow Submarine* and the gremlins from *Gremlins* and appear on almost every page—inserting themselves into everyday life—growing in annoyance and numbers until the tension breaks into conflict. The colour palette of *Mental Goblins* is more restrained than *Diabetes: Year One*—allowing for the colour of Dave's avatar and the goblins to stand out—showing the way goblins insert themselves into events and how this makes him feel displaced in everyday life.

The script is written in first person discussing Dave's experience with goblins and informing or warning the reader about how unpredictable goblins can be—providing a strategy for understanding and coping with goblins. Usually, this text is contained in a text box at the top of the page—allowing the rest of the page to move between floating panels over a large image that goes from wide angle to close-up on detail, more structured pages that emphasise the passage of time, disembodied panels that allow for more conceptual transitions—such as the evolution of the goblin on page 14, and dynamic panels that vary in position and shape to accentuate action—

Fig. 7.3 Pages from the comic *Mental Goblins* by Dave and Tony Pickering

such as page 15. A key difference in this project was the way in which text and image related—whilst some pages use the text as part of the images—as voices for the goblins, in most cases, the visuals expand on the story rather than standing alone, and in one case an elaborate image is used as background for a passage of extended text—with the image of goblins squashed together placed, faded down, behind the text as an ecosystem for the words which expand on the nature of goblins and the guidelines that can help (p. 10).

In the role of an artist, I was able to look at the narrative both as someone invested but also distanced—to bring out elements of humour, as well as the stress, tension, and frustration that forms much of the experience of anxiety. It was my responsibility to find uniqueness in the artwork, which as I have discussed previously forms the nexus of what we think of as graphic medicine pathography. So, although Dave's avatar is cartooned, it is still identifiable as him, and his clothing/costume can be seen to express his identity as he grows in confidence—such as when he begins to gain the upper hand against the goblins, and he is wearing a dress.

Mental Goblins provided a much more linear structure than *Diabetes: Year One* or *Random Fluctuations*—something that is seen in the fluidity of structures discussed above—and in this way suited a homogenisation of style across the comic. An awareness of the rhythm and pattern of the narrative allowed for my concern being to 'stage' the text—to find ways to express the energy and importance of each stage. The movement of the images works in contrast to the direct address of the text—fleshing out a metaphor or extending a simple word like 'fight' (p. 15) to a whole page. In this way, my role was a hybrid between theatre director and designer—interpreting the stress of the text through lettering and typography and creating the mis-en-scene and action of the comic.

7.5 Discussion

Since the establishment of my practice in graphic medicine I have focused on pathography—and especially on finding ways to convey individual experience. The more I have been able to apply aspects of comic theory, the more I have come to appreciate the ways the comic medium can work and bend to new situations. My personal works have been as much about comic thinking as comics practice. In *Diabetes: Year One* and *Random Fluctuations* the form has worked for me to interrogate my experience, to understand feelings and experiences I've had as elements of becoming and being a patient, of accepting these qualities into my own sense of self.

In this sense, the act of comic making has been both an active understanding, and an attempt to articulate my experience—to make it accessible and interesting to others—looking for ways to help, but also ways to engage people outside the type 1 diabetes experience. Type 1 diabetes is a strange illness—it requires the patient to self-manage—to take an understanding of type 1 diabetes and apply regular diagnosis and treatment protocols to the maintenance of your own life. In this way, type 1 diabetes is a useful condition to explore the ways comics straddle the patient-practitioner interface. Comics can be utilised as both process and product—appealing to different learning styles, providing the stimulus for conversations and questions that may not come up in the hustle and bustle of everyday medical practice. The slowing of time that comics provide can be a useful way for a patient to think themselves into their illness—negotiating their relationship with their sense of self, and the design of text can provoke a patient to grasp details they never thought they could. For practitioners, comics can provide a site for patients to engage with their situation, and for them to engage with the individual before them.

Working with the experience of someone else's pathography draws upon elements of my experience through the conduit of empathy—so that design choices are negotiated, and emotions and specifics are discussed and realised. Working with another's stimulus asks me to apply my experiences to the context of another illness. In this case, the mental health issues of *Mental Goblins* do cross over with my approach to type 1 diabetes, where much of my work has been concerned with the psychological impact of living with an unseen and constant illness—and so aspects of the feelings of stress, isolation, anger, and frustration were areas I could speak to and use these experiences in the design—especially in considering aspects such as gesture and movement—allowing body language to speak where text may have other things to say.

7.6 Conclusion

In exploring graphic medicine as a field and creating my own pathographies I think it is clear to see ways in which this type of graphic medicine plays a role in the treatment of patients, but more in the way in which the patient comes to understand how they wish to be defined as a patient. Comics provide changes of perspective and reconsider time so that the patient is able to move from knowledge of what they have to an understanding of how they want it to impact their lives. Seeing comics both as a process and product adds another layer to this pathway—that of active understanding, a practice in which the patient can apply knowledge in an accessible field, and in doing so reclaim elements of autonomy and agency that the moment of diagnosis can destroy. Pathography, as an auto-ethnographic process can help us reconsider the role of the patient in treatment, and bring back a sense of control so that the patient can see themselves as human again and start to think of their situation in the context of friends, family and community. With agency shored up, responsibility for, and engagement in treatment can be re-energised, and defined on the terms of the patient themselves.

The model of pathography—the importance of engaging with narratives of illness, narratives that can be considered spatially as well as chronologically, can be extended as a paradigm for considering the nexus of the patient–practitioner gateway. Finding ways to guide patients into a personal understanding of the medical science they are often overwhelmed by. Looking to support healthcare professionals in recognising patient priorities are not always medical, and that through a combination of media—such as the comics medium, it is possible to work towards a sense of agency that can be disconnected at diagnosis.

References

Barry L (2014) Syllabus: notes from an accidental professor. Drawn & Quarterly, Montreal, Quebec

Barry L (2019) Making comics. Drawn & Quarterly, Montreal, Quebec
Boileau K, Johnson R (2021) Covid chronicles: a Comics anthology. Graphic Mundi, Pennsylvania
Czerwiec MK et al (2015) Graphic medicine manifesto. The Pennsylvania State University Press, Pennsylvania
Chute C (2017) Why Comics? From underground to everywhere. Harper, New York
Green MJ, Myers KR (2010) Graphic medicine: use of comics in medical education and patient care. [Online] https://doi.org/10.1136/bmj.c863. Accessed 16 January 2018
Kar DP, Patel DM (2016) Type 1: origins. revolvecomics.com: https://revolvecomics.files.wordpress.com/2016/09/diabetes-t1-comic-low-res1.pdf. Accessed 18 January 2018
Moore A, Gibbons D (1986) Watchmen. DC Comics, New York
McCLoud S (1993) Understanding Comics: the invisable art. Kitchen Sink Press, Northampton, MA
Myers KR (2015) Graphic pathography in the classroom and the clinic. In: Czerwiec MK et al (eds) Graphic medicine manifesto. The Pennsylvania State University, Pennsylvania, pp 87–114
National Institute for Clinical Excellence (2004) Improving supportive and palliative care for adults with cancer: the manual. National Institute for Clinical Excellence, London
Pickering T (2018) Diabetes: Year One. Self, Manchester
Pickering T (2020) Random Fluctuations. Self, Manchester
Pickering D, Pickering T (2020) Mental Goblins. Manchester: Self
Revolve Comics (2016) Revolve comics. revolvecomics.com: https://revolvecomics.com/creative-services-2/. Accessed 18 January 2018
Sontag S (1991) Illness as metaphor & aids and its metaphor. Penguin Books, London
Slatterty Z (2021) Coma. Myriad Editions, Oxford
Sousanis N (2015) Unflattening. Harvard University Press, London
Squier SM (2015) The uses of graphic medicine for engaged scholarship. In: Czerwiec MK et al (eds) Graphic medicine manifesto. The Pennsylvania State University Press, University Park, Pennsylvania, pp 41–66
Ware C (2012) Building stories. Jonathan Cape, London
Williams I (2015) Comics and the iconography of illness. In: Czerwiec MK et al (eds) Graphic medicine manifesto. The Pennsylvania State University Press, Pennsylvania
Williams I (2014) The bad doctor. Myriad Editions, Oxford

Playing with Science: Games as a Medium of Science Communication

8

Pawel Stachyra and Mark Roughley

Abstract

By metrics, such as average play hours and money spent (as measured per capita or market gross volume), gaming is at the forefront of cultural content consumption; being the most popular form of entertainment in the developed world. While the most common association with gaming would be related to digital gaming, traditional games, such as tabletop roleplaying games and boardgames, are undergoing their own renaissance. All the while, classic traditional games like chess or classic deck card games remain popular. For science communicators, it is important to acknowledge games as effective learning and communication tools.

To date, there is no broad-perspective evaluation of the gaming medium published for science communicators. There exist singular studies dealing mostly with digital games, but their particularities have not been explored comparatively. This chapter defines a game; outlines a brief history of games, the parameters, and genres of gaming, and discusses a selection of games that act as science learning and communication tools. Furthermore, our own experience designing '*Belly Rumble: a microbiota board game*' will contextualise the use of tabletop board games as a biomedical visualisation learning tool.

Keywords

Science communication · Gaming · Ludopedagogy · Game design · Serious gaming · Transdisciplinarity

8.1 Introduction

To have any discussion on the potential applications of gaming for science communication, it is important to establish what a game is. There exist many published works analysing the phenomenon of gaming through different methodological lenses; however, they often bemoan the lack of commonly accepted and/or clearly delineated taxonomy of games (Ruggill and McAllister 2011; Shubik 1983).

8.1.1 Definition of a Game

Huizinga (1938) outlines many qualities of play rather than gaming that include engrossment with the activity (both that of the spectators and the participants), disinvolvement from the realities of daily life, participants experiencing 'fun', and repetitiveness; play being able to be completed and then performed again in its designed form

P. Stachyra (✉)
Włodawa, Poland

M. Roughley
Liverpool John Moores University, Liverpool, UK
e-mail: M.A.Roughley@ljmu.ac.uk

M. Roughley (ed.), *Approaches for Science Illustration and Communication*, Biomedical Visualization 4,
https://doi.org/10.1007/978-3-031-41652-1_8

somewhere and sometime else. Other traits include time constraints and specific space in which the play takes part, ritualised rules, layout of the game play and lastly the community-building effect of play activity. Huizinga (1938) also extends the principles of play to include diverse phenomena such as performance art, adjudication, warfare and other areas of human experience (Tanghe 2016). This is indicative of how the experience of play is integral to the human condition and, by extension, so is that of gaming.

Huizinga's approach is philosophical and reflective, in contrast to more modern, academic models. For example, in 2010, Arnetta established a set of directives to recognise a 'game'. There are commonalities between the two approaches, and these are listed in Table 8.1.

Furthermore, there is a great deal of nuance when considering what passes for a game. One could define a game as any activity that involves one or more players and has defined rules of victory that can be resolved by the end of said activity's sequence. Such a definition would stretch to board games, most videogames, card games, and sports, but would exclude roleplaying games, free-form play, and some videogames that might include cinematic choice-based stories (Arrambide Figueora 2022). Using the term 'rules of play' as opposed to 'rules of victory' helps accommodate said exceptions, but further opens the definition to more diverse aspects of daily life.

Traditionally, sports including football and the Olympic disciplines all meet the aforementioned criteria, but there exists a pervasive cultural perception of sports as a separate phenomenon; it seems to be implicitly understood that playing football is participating in a sport, however, playing chess is considered gaming. There are arguments to be made on both sides and Kobiela (2018) argues for the inclusion of games in sports under the term 'mental sports'. The opposite argument could be made by naming football an 'athletic game'. The boundaries become even more unclear when one considers the phenomenon of E-sports. In this case, competitive video gaming is considered by some a veritable sporting activity (Jenny et al. 2017). It is a controversial space and public discourse presents arguments for both the inclusion and exclusion of e-sports from the purview of sportsmanship proper. Given, that E-sports garner remarkable spectatorship and funding, perhaps the cultural perceptual difference between the 'games' and 'sports' may be predicated upon profitability and public engagement (Jenny et al. 2017). Does a game become instead an 'E-sport' after generating a large and profitable enough competitive scene?

The common quality of activities commonly considered 'sports' is physical exertion. To focus the scope of this chapter, the integral element of physical exertion has been elected as an exclusion criterion, an approach shared by others, including Parry (2019). None of the games considered within this chapter necessarily involve physical exertion. Even though it can be present as an element of Live Action Role Playing (LARP) games or megagames, it is rarely a requirement for playing them and does not significantly affect their resolution. However, some people choose to include simulated physical activity in their games, for example, playing virtual sports (e.g. FIFA and Madden) while remaining sedentary (Jenny et al. 2017).

Considering the criteria mentioned above and informed by Arnetta's (2010) model, for the purpose of this chapter a 'game' will be defined as having these characteristics:

1. Feeling the activity can be successfully completed.
2. The player can concentrate fully on the activity.
3. The activity has clear goals.
4. The activity provides fast feedback.
5. The player is deeply involved in the activity.
6. A sense of control over the actions is necessary to perform the activity.
7. Self-awareness disappears during flow.
8. There is an altered sense of time.

8.2 Brief History of Gaming

Playing games is arguably an activity integral to the human experience. In archaeological sites

Table 8.1 Analogous concepts between Huiznga's and Arnetta's Frameworks

Analogous concepts between two theories of gaming	
Huizinga (1938)	Annetta (2010)
Engrossment	Immersion
Time, space, and set-up specificity	Identity
Disinvolvement with reality	Immersion
Repetitiveness and replicability	Interactivity
Community building	Identity

spread across the continents and cultures, one often discovers dice, gameboards and other gaming paraphernalia (see Figs. 8.1 and 8.2). One of the oldest games with a traceable history is the Ancient Indian game of Chaturanga—a known direct descendant of chess, played possibly as far back as 3000 BC and no later than sixth century AD. This game has been argued to be an early example of a boardgame, a wargame and a roleplaying game (Ewalt 2013).

In his landmark publication, *Homo Ludens* (1938), Johan Huizinga argues that play participation is as intrinsic to the human experience as reasoning, speech, or toolmaking. He also draws a compelling case for the universality of play, citing dogs engaging in play fights with the appreciation of boundaries, conventions, and rules (e.g. pretending to attack the throat but not biting hard). In fact, many species of mammals and birds engage in forms of play, analogous to that observed in humans (Tanghe 2016; Pellis and Pellis 2017; Gray 2019). An idea presents itself, that perhaps play behaviour is an evolutionary adaptation that lets the animal boundary test and simulate important, yet dangerous activity in a safe, inconsequential environment (Gray 2019).

Arguably, games that take place in theatre of the mind such as roleplaying games, may be the oldest form of organised human gaming. At their very basic core experience, they require an ability to tell a story; an arbiter (being a subjective person) or impersonal game mechanics or a mix of both, and to resolve any conflicts within the story (Ewalt 2013). Analysis of ancient Greek accounts of gaming by writers such as Herodotus or Plato (Austin 1940), not only describes many types of board games recognisable to the modern audience including racing dice games similar to chutes and ladders, and war games, but also mentions the prevalence of such games amongst geographically distant cultures such as the Celts.

Spanish conquistadors found the Mesoamerican native peoples playing a boardgame of 'Patolli'—a game of strategy and chance played over material stakes (Erasmus 1950). The game's description and layout are functionally near identical to a known Indian game of 'Pachisi' (Fig. 8.3) dating back to circa 1100–800 BC (Finkel 2004). Given the historical migrations of populations, the game has either predated the prehistoric land migration to North America or was developed independently in a near-identical form. A similar game is marketed contemporarily in Western markets as 'Ludo' and was introduced by British sailors travelling between Britain and India (Bell, Richard 1979). This point serves to illustrate the cultural pervasiveness and longevity of gaming.

An important historical employment of 'serious gaming', which is gaming for an explicit purpose beyond enjoyment (Giasolli et al. 2006), is war games. Going back to the seventeenth century, there are confirmed historical instances of military forces using game pieces and board to simulate combat encounters in order to inform the future strategy (Glick and Charters 1983). Given, that both the evidence of warfare and tabletop gaming predate that, and the relatively low technological requirement for producing such simulations, such games could have existed for far longer. Moreover, if one considers any sort of war training scenarios involving real-world drills and simulated combat that are structurally reminiscent of LARP or megagames, they can be observed as games (Eng 2020).

Fig. 8.1 Ancient Egyptian icosahedral dice. The purpose of it is unclear and is theorised to be either used for divination practices or gaming. Nate the structural similarity to a later, modern example (Open Access)

A major development in modern gaming is the introduction of digital games, with the text based 'The Oregon Trail' being recognised as the first commercially available digital game in 1971, although the game 'Tennis for Two', also known as 'Pong', was available to play on the oscilloscope in the year 1950 (Smith 2019). A device that a digital game operates on serves as

Fig. 8.2 Modern retail available icosahedral (d20) dice included for comparison. Commonly used in Roleplaying games like Dungeons and Dragons. (Creative Commons)

Fig. 8.3 An example board and pieces of the game of Pachisi. Note the structural and gameplay similarities between this game and both the Pre-Contact Mesoamerican Patolli and the modern marketed game of Ludo (Creative Commons)

a rule interpreter but is also an output for a more detailed graphics and a more complex gameplay experience (Wankel and Blessinger 2012). The introduction of the digital component to gaming has served to lower the entry-level barriers for the player as the ruleset and the set-up are both managed by the software. The scope of the games available also increased, with the development of new technologies, especially 3D graphics, video games have achieved levels of immersion previously impossible to their analogue predecessors. Simulation programmes have been not only employed as tools of pure pastime enjoyment but also as learning tools for the military (Yao and Huang 2020), medical staff (Beal et al. 2017) and machine operators (Allerton 2010).

With the Internet being a ubiquitous phenomenon, it opens virtual games to the mass market. In the year 2016, the revenue of console and the mobile gaming industries combined exceeded the combined revenue of movie sales across all platforms. In developed countries, nearly every household possesses at least one device capable of running videogames (Sweeney 2021; Witkowski 2021). With greater access and popularity of digital online gaming, the focus of gaming scholarship has shifted drastically from other forms towards online video games (Annetta 2010).

While understandable, the distinction is needless, as video games are essentially a further development within the space of gaming at large and need not be considered separate. Given that gaming follows a similar set of rules and principles regardless of platform, this chapter will remain platform-agnostic and the findings that pertain to one type of game are considered applicable to the others unless stated otherwise. However, in the final analysis the nature of different types of gaming for effective science communication and knowledge transfer will also be taken into consideration.

8.2.1 A Note on the Cross-Culturality of Gaming

Gaming is not a phenomenon unique to any particular culture, which makes them useful vehicle for science communication. Games with contemporarily-familiar structures can be found through distant ancient cultures (Erasmus 1950; Finkel 2004). Many of the theoretical frameworks discussed in this chapter point to gaming behaviour as a function of human cognition and decision-making, which can also be observed outside of gaming proper (Huizinga 1938; Kuhn 2021; Pellis and Pellis 2017). Traditional games are found in virtually every known human civilisation, and digital games are ubiquitous in developed societies, with massive global revenue and a lively international e-sports scene (Jenny et al. 2017). Different gaming cultures are present within different world locales (Cartlidge 2022; Kikkawa 2022) but overall, the media of gaming remain near-universal and this inquiry found no contriving levels of locale differences. As such, the findings of this chapter should be considered applicable regardless of the geo-cultural context.

8.3 Types of Games

There are multiple ways to categorise games (Ruggill and McAllister 2011). The exact method and paradigm differ depending on the goals of one's inquiry. Games can be perceived through the broad logical/mathematical lens of 'game theory', where most human choice-based phenomena are objectified as games (Shubik 1983). It is an approach reminiscent of the broad scope of *Homo Ludens*. An example of a game that is understood by such a framework is the 'Prisoner's Dilemma'. It is a hypothetical scenario, in which two subjects are acting without the knowledge of one another's choices trying to minimise harm, while maximising their own possible gain (Kuhn 2021). There is the experiential framework of how the player engages with a game, and what type of experience it best emulates to them. This approach is more open-ended and prone to subjective classifications. For example, the game 'Darkest Dungeon' later discussed is commonly referred to as a strategy game, a roleplaying game and a 'rouge-lite' based purely on the reviewer's experience of the game and similar games. 'Operational gaming', which Shubik (1983) further discusses, has more to do with the operations of economic systems rather than games as understood in this chapter.

As this inquiry considers games' suitability for science communication purposes, it will consider game design as an artform, with distinctions based on its medium. This is the most salient and practical approach as it is based on an empirical signifier (the game medium) and avoids the subjective experiential classifications, while also steering away from the detached, mathematical view of 'game theory'. It is an approach designed to help a maker make informed choices for their practice.

8.3.1 Traditional Games

The first (chronologically speaking) branch of gaming is 'traditional gaming'. Traditional games are most broadly games that do not involve an electronic device in their play. This is a broad classification but helpful no less for an aspiring game designer. Traditional games have a lower technological entry barrier for both the player and the creator. While requiring more investment of time and attention to learn the rules and keep track of the physical/mental play pieces, they also near universally encourage human interaction (Jore 2011).

Traditional games can be mediated through a variety of commonly used items. Cards, dice, boards with miniatures, and rulebooks are most common, as demonstrated in Fig. 8.4. The same pieces can be used to play different types of games, for example, a French-suited deck of playing cards can be used, among others, to play 'Blackjack', 'Bridge' and 'Solitaire', which can be classified as (in order) a game of chance, a cooperative tactics game and solitary logic puzzle game. The rules of the play are most often contained within the rulebooks, although may be contained within the pieces, or be passed down through oral tradition. The space of play may be a table in case of a card game, a board in case of a boardgame, the theatre of the mind for a roleplaying game or an open outside environment for a live-action roleplaying game.

Traditional games may be games of luck, skill, strategy or even physical reflex. Adversarial, competitive games are often referred to as war games. A quintessential example of such a game is chess (von Hilgers and Benjamin 2012). A competitive element is present within most games, even in cooperative games that do not include open-player rivalry, players compete

Fig. 8.4 Full 4-player set-up of the game 'Belly Rumble' created by Pawel Stachyra. It serves to illustrate typical elements of traditional games: cards, dice, board and pieces. Photograph by Pawel Stachya

against the structure of the game itself (Zimmerman and Fortugno 2016).

8.3.2 Digital Games

Conversely, the term 'digital games' refer to any game played through a medium of an electronic device. Videogames, compared to traditional games, have a higher technological and financial entry-level requirement for their creators. On the other hand, the entry level is lower for the players. The computerised engine of the game handles most of the set-up and ruleset of the game, and there is no intrinsic need to require multiple players to be in the same location (Rot et al. 2022). Single-player digital experiences are more common with the digital gaming medium and the multiplayer game is most often facilitated through an Internet connection.

There is a wide range of scope between individual digital games available. Large game development studios produce 'triple A games' that are large-scope projects with budgets and credits often exceeding these of major Hollywood films (Iskarius 2020). Examples could include the 'Witcher' (Culture.pl 2018) or Senua's 'Sacrifice' (NinjaTheory. 2017). Conversely, smaller studios tend to create smaller, more focused games like the 'Darkest Dungeon' (Red Hook Publishing 2016; Cartlidge 2022). That being said, party games and local multiplayer games are also present and enjoy some popularity. An example of a digital party game is '*Kahoot!';* a fast-paced custom quizzing platform, which has been adapted for educational purposes on multiple occasions (Castro et al. 2019; Neureiter et al. 2020). Digital party games are an interesting example of continuity between different gaming media. While they are mediated by digital platforms, they are often played in a group of people within the same physical space same way traditional board games and card games are played. Often a digital game emulates the experience of a traditional game (e.g. Chess simulators or card games played through an electronic device).

While many games played on digital devices can also be played traditionally, the digital medium enables the use of real-time gaming. Digital games often test the player's coordination and reaction time as well as any more mental attributes. Digital games can achieve a faster pace to that of the traditional games. It is possibly partly responsible for the emergence of the e-sports phenomenon, where the audiences can spectate the game in real time, as large team's efforts are mediated, streamed and captured by the software (Jenny et al. 2017). Another, distinctive functionality of digital games is record keeping—a game can automatically record and store data of any player's action—as such digital games somewhat lend themselves to the objective observation of human behaviour.

8.3.3 Serious Gaming

There exists a pervasive cultural perception of gaming being antithetic to productivity or even morality itself (Binde 2007; Elequin 2020; Laycock 2015). Coming back to Huizinga's and others' theories of mastery of play, it is a harmful assumption. High-stakes tasks are commonly simulated through games, as a means of training. Wargames were used to visualise and simulate combat since at least the Napoleonic era (Glick and Charters 1983; von Hilgers and Benjamin 2012). Simulators are used in training of medical professionals and machine operators (Allerton 2010; Beal et al. 2017; Yao and Huang 2020). The concept of games produced for purpose other than enjoyment, or using pre-existing games for such causes is referred to as 'serious gaming' (Giasolli et al. 2006).

Less often, the process of teaching through the use of gaming is referred to as 'ludopedagogy', although it seems to be a recent term and mostly concerning formal education (Wareham Morris and McGowan 2018). In this chapter, 'serious gaming' is defined as gaming intentionally for purposes beyond enjoyment. The opposite of a 'serious game' becomes a 'pure game', which is one played mainly for enjoyment. 'Serious games', however, are not as clearly delineated from the 'pure games' as 'traditional games' are from 'digital games', as games created for serious purposes may be found enjoyable by a wider audience, which may be unintended, or intentional on behalf of game's maker. Marketability and attractive gameplay are factors driving interest in game's serious subject matter (Brody 2019). Table 8.2 is a table summary classification of types of games based on the medium.

8.3.4 Summary Table

8.4 Gaming as a Medium of Science Communication

Science communication is an important skill for future scientists in disseminating their research as well as informing both public opinion and policy (Donev 2013). For the purpose of this discussion, 'science communication' is considered to be predicated on two values: 'fidelity' and 'coverage'. 'Coverage' can be defined as a combined amount and demographic of people engaged with the scientific material. 'Fidelity' refers to quality of the communication; it must be clear, approachable for the target audience, and not contrived. Approachability is predicated on both the delivery of the material and voluntary engagement with it on the part of the audience.

Science communication can be described broadly as exposing the target audience to scientific concepts and information (Donev 2013; Illingworth and Allen 2016). The audience can be reached through magazines, art exhibitions, content creation and broadcasting/podcasting, and public engagement programmes. With the advent of common internet access, science communication is also increasingly open to private individuals, rather than educational organisations, both to produce and consume (Fox et al. 2021). The aforementioned vectors of public engagement are used to gain coverage.

While scientific knowledge is a field of open enquiry and debate, scientific facts may be defined with a high degree of objectivity, especially within the STEM fields (Kanyapasit and Srisawasdi 2014; Tanghe 2016). As such, a vector of science communication is required to be precise and accurate in its ability to portray the concepts and at the very least not to contradict the science with its messaging (Taddicken and Reif 2020). An interesting dynamic emerges, where science communicators must balance fidelity and coverage. Questions arise: Can one make a scientific concept approachable? Does the popular appeal compromise the scientific value?

Scientific material can be provided through detailed verbatim exposition, but it can also be hinted at. The target audience must be specified and understood (Illingworth and Allen 2016), and the target of the scientific communication outreach may instead be signposted towards the primary sources for their own further reading. The values of *pathos* (emotional appeal), *logos* (logical argumentation) and *ethos* (appeal to one's credentials) need to be measured and balanced for an effective approach (Braet 1992). This can

Table 8.2 Example classification of games based on the medium

Virtual reality game	Augmented reality game	Video game	Megagame	Live action roleplaying game	Miniature game	Card game	Boardgames	Roleplaying game	Game type
Digital	Digital syncretic	Digital	Traditional syncretic	Traditional	Traditional	Traditional	Traditional	Traditional	Technology
Electronic device with a virtual reality set	Mobile device or an AR headset. Physical location	PC, Console or Mobile device with required Input devices (pads, keyboards, mouse, touchscreen)	Real-world locations, game boards and maps, tokens and counters, attributes of a LARP. a wargame and a boardgame.	Real-world locations, dress-up, accessories, often faux weapons	Maquette board, miniatures, dice, measures	Cards, counters, tokens	Board, board pieces, tokens, dice	Theatre of the mind, character sheets, rulebooks, dice	Common setup
Disconnect from the surroundings, long history of experimentation, often used for educational purposes	Overlaying the digital imaging over a real-world location, often GPS tracking, often used for educational purposes	Great variety and a large modern market, very dependent on the current state of technology, often emulating traditional games	Syncretic elements of LARP and boardgame. Often played over long time within large or outdoor venues, may include digital l elements	Cooperative or competitive roleplaying action, may or may not involve a form of codified combat and environmental putties, no clear victory condition	High material/ time investment, open modular play within the rule set, set on a 3D board; collectible/ hobby aspect	Chance or skill focus, high variety, often collectible, portable	Great variety of structure, approach, and timing, chance and skill elements	Collaborative storytelling, creative licence, no defined victory terms, open structure	Main Qualities
Flight simulators; VR chat;	Pokemon GO, Kahoot! (arguably), Zombies, Run	World of Warcraft, Fallout 1, Candy Crush Saga, FIFA series, Niche, Darkest Dungeon	Extemporaneous	Typically based on modified Roleplaying game franchises	Warhammer 40 kg; Bolt Action, Carnevale, Rumbleslam	Poker, Solitaire, Magic the Gathering, Rummy, Coup, Yu'Gi'OH!	Chess, Mah-jong, Entaglion, Monopoly, Pandemic	Dungeons And Dragons, Shadowrun, World of Darkness	Examples

(continued)

Table 8.2 (continued)

Virtual reality game	Augmented reality game	Video game	Megagame	Live action roleplaying game	Miniature game	Card game	Boardgames	Roleplaying game	Game type
Single player to massively multiplayer	Most often massively multiplayer	Single player to concurrent millions of players	Multiple						
Competing teams, numerous support staff and game- runners	Multiple with multiple support staff and game-runners	Two to multiple	One to multiple	One to multiple	At least two	Number of players			

be somewhat simplified as: *pathos* drives popular interest; *ethos* uses authoritative sources to legitimise the message, and *logos* can be used to sway more expert and inquisitive audiences (Higgins and Walker 2012). Overall, good science communication needs to include a balance of approaches predicated upon the knowledge of the audience (Illingworth and Allen 2016).

Additionally, a distinction can be made between 'inward-facing' and 'outward-facing' science communication (Illingworth and Allen 2020). The latter is congruent with the definition thus far and describes specialists communicating expertise to the often less knowledgeable broader public. 'Inward facing' science communication occurs between qualified scientists, through dialogue, conference and publications. A peer-reviewed journal article can be seen as an example of 'inward-facing' science communication.

In relation to the 'coverage' of science games it would be wise to consider: how many people can a game reach? What demographic will the game be more attractive towards? Will people actively reach out and seek to play the game? (Illingworth and Allen 2016). A game is a participatory art form, often demanding a high time and emotional investment form its audience. Through a mixture of game design, attractive artistic presentation, gripping narrative or competitive thrill, games can captivate and often retain players. Often, a 'flow-state' may be observed in a gamer, where the person involved in the activity loses the sense of self-awareness and time instead of becoming fully engrossed in the activity at hand (Nah et al. 2014).

A science communication game has to not only appeal to a large audience, but also has to involve each player directly and personally to a point of them making a time investment assuming the gamer identity (Annetta 2010). The gamer identifies and is identified as a part of the community of fellow gamers. They also invest their time and effort to increase their mastery of the game. As such, a game must have a toolset to appeal to a diverse enough selection of potential players each directly, accounting for different tastes and preferences. Some may be drawn by the aesthetics of the game, some by the narrative, and some by the competitive experience. Likely every single part of the game will never be fully appreciated by a single individual but is necessarily included to attract the multitude.

8.5 Games as Educational Tools

One of the major loci, where science communication occurs, is within educational institutions. While the exact approaches will vary by location, culture as well as school type and the jurisprudence of the teacher, overall communication of scientific knowledge, impartment of skills and cognitive development can be described as consistent. Cross-culturally, a student is challenged to acquire skills and information to better function in society as an adult.

The process of learning occurs within or without the institution, however, within the formal setting examination and accreditation are introduced to track student's progress. The accreditation and examination results are imparted upon a student not unlike an initiation ritual (Illich 1971). By progressively challenging the learner of any age, facilitating their cognitive development, and equipping them with skills, the learner can be progressively exposed to more demanding tasks and material, as described in the scaffolding theory. Here, the learner learns things in their proximal zone of development and expands their knowledge outwards towards erstwhile distant subjects (Vand Der Stuyf 2002). Within such framework, a structure can be outlined where a learner progresses from cognitive development and through skill acquisition into being able to critically engage with scientific subject matter presented to them (Walker 2003). Eventually, a learner being knowledgeable and competent enough can seek out and interrogate the information they require (Illich 1971).

There is a constant theme of dissatisfaction with the modern educational system, with it being reported to be antithetical to creativity and individuality of the student (Illich 1971; Collins 2004; Day-Black et al. 2015). In a traditional classroom, students are obliged to learn at a pace that might compromise the needs of some (Wong and Csikszentmihalyi 1991). Such compromises tend to leave either party with their needs unmet (Illich 1971), conducting and directing their own research based on priorly obtained knowledge (andragogy) (Lage, Platt and Treglia 2000; Betihavas et al. 2016) and squandering the potential for transformative teaching approaches. This might unfairly disadvantage those outside of a privileged socioeconomic position (Darbyshire 1993; Sandlin 2005).

The search for new experimental learning techniques implies the active role of the teacher (pedagogue) (Darbyshire 1993), with in-demand University degree courses calling for new innovative methods of learning (Day-Black et al. 2015). Moreover, approaches transcending the pedagogy-andragogy dichotomy between leading the learner and leaving them to lead themselves are required.

Multiple benefits of a gamified approach to learning exist. An example would be problem-solving focus, where instead of memorising the material, students actively engage in learning as the educational subject matter becomes crucial to the competitive success of the player. Additionally, as games are a structured experience, they offer better conceptual convergence. In their conference proceedings, Kanyapasit and Siraswasdi (2014) report an educational issue, where students would all claim to acknowledge and understand the exact sequence of the prophase mitosis within the cell cycle, yet a significant portion of them would give divergent answers incongruent with the instructional material. The paper later goes on to propose a digital game learning framework to help ameliorate the issues. An organised, sequential gameplay is proposed as a framework to which to relate the exact steps of the cell cycle sequence. Games, often relying on a particular turn order and a particular sequence of moves required to win, can lend themselves well to intimately simulate complex processes in a way that makes the player familiarise themselves with the process in question.

8.5.1 Ludopedagogy

Combining the derived definitions of gaming, science communication and learning, the concept of 'ludoeducation' (Giasolli et al. 2006) can be introduced. Arnetta (2010) proposed a framework for designing video game and virtual reality experiences (Fig. 8.5). This model concerns multiple facets of gaming experience nestled within a hierarchy from the most fundamental being *identity*, to the *instructional* being the broadest. The 'external' facets are predicated on the success of the 'internal' facets and only with the successful implementation of the more intrinsic elements of gaming, the educational purpose may be met successfully.

'Identity' is a crucial factor, that may not seem obvious. The player becomes a subject within the game, either within oneself or within a group. As such the experience becomes personal to the student and promotes higher engagement. At the same time, every player having played the game gains a shared experience that has been resolved within the same rules and parameters. Players with otherwise divergent processes and backgrounds can all relate to this standardised experience. There emerges a development of a learning culture—the players become a community with its internal loyalty and accountability. In Strickland and Kaylor's (2016) experimental study, this theoretical framework seems to be validated by nursing students involved in a megagame reminiscent of a scavenger hunt. The students not only were motivated to perform and learn a variety of practical and theoretical tasks but were further encouraged to aim higher to 'win' the game as members of competing teams.

Another facet of note that runs through all gaming experiences is competitiveness. 'Increasing Complexity' is an expression of challenging gameplay. The player is encouraged (again, through their identity as a student or a group member) to tackle progressively more challenging tasks or opponents (Castro et al. 2019; Vorderer et al. 2003). It is not unlike the very nature of learning, although the motivation does

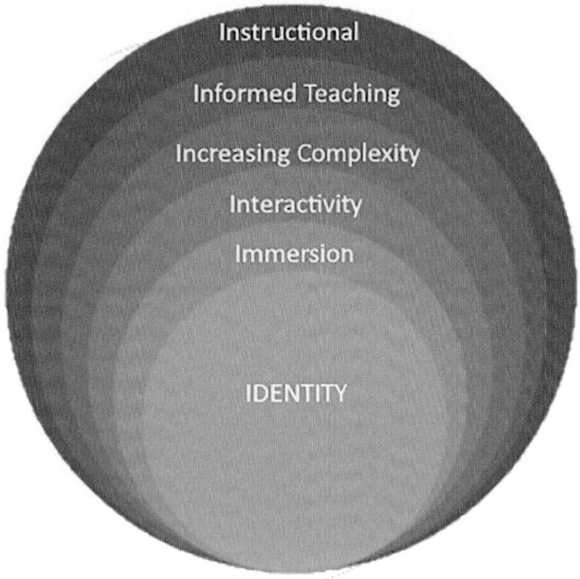

Fig. 8.5 Diagram repurposed from Anneta (2010) by Pawel Stachyra. The model argues building upon the core gaming elements of (Identity to Increasing Complexity) to support the subsequent implementation of educational content

seem to occur more intrinsically rather than through the external means used in classic education (grades, deadlines, standardised testing). Moreover, the increasing challenge may be beneficial for gearing the challenge towards students' individual levels. It is the student themselves, who seeks out harder challenges and opponents, once mastered their current level. This seems to be congruent with the observations of Kanyapasit and Srisawadi (2014), where an emphasis on critical problem solving is brought up as a benefit. When engrossed within a game, the player is encouraged to involve their higher critical learning faculties beyond memorisation and recall (Davidson and Candy 2016).

In previous research projects, first author, Pawel Stachyra has repurposed the Anneta model (Stachyra 2021) and added some nuance to allow for non-digital games, as well as delineating the pure game from a serious game within the framework's boundaries. This is presented in Fig. 8.6. With terminology introduced and definitions established, case studies of gaming for science communication can be discussed and analysed.

8.6 Science Games Case Studies

When considering science communication through game, the most obvious games to discuss are those made explicitly for such purpose. Games are often used as an attractive object to drive attention for both students and the public alike. Additionally, the structure of a game is usually fixed, thus lending itself to explain scientific procedures clearly. In relation to games as science communication tools, the educational outcomes of the subsequent case studies will be judged on their achievement of skill impartment, cognitive increase, and communication of science.

8.6.1 'Niche': A High Fidelity Science with Enjoyable Gameplay

An example of a game with high-scientific fidelity that achieved full implementation of 'pure' gaming design values is the video game 'Niche' (Stray Fawn Studio 2017). The game was originally developed as an honours degree project but

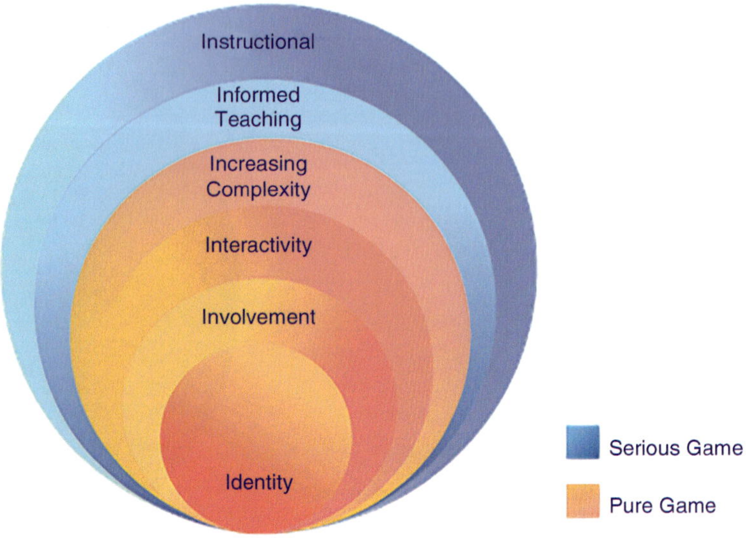

Fig. 8.6 Annetta's model repurposed and updated by Pawel Stachyra. Distinction is made between 'pure gaming' principles and the 'serious gaming' principles. Serious gaming effort needs to be based on a well-implemented pure gaming core

moved on to become a popular standalone title as the author employed a team of developers. The core principles of ecology, including Darwinian selection, genetic drift, and adaptations, were implemented with a high degree of fidelity into the game experience (Stray Fawn Studio 2017). The player assumes control over a species of animal, called '*nicheling'* trying to adapt and survive within a changing environment. Despite being a marketed game with a popular appeal, the game remains accurate to the core tenets of its scientific subject matter. The main aim, however, is providing an enjoyable experience. As such, 'Niche' may be considered a 'serious game', depending on the intention of using it. The game holds favourable reviews with both pedagogues (Bristol 2021) and gamers (Sinclair 2017; Valve Software 2021).

8.6.2 'Entaglion': A Traditional Game about Quantum Computing

Another example of a well-implemented serious game is a quantum computing boardgame titled 'Entaglion'. It is a simple traditional boardgame dealing with an otherwise high-scientific concept that helps to explain the process of quantum computing, met with generally favourable reviews (Brody 2019). 'Entaglion' can be classified as a 'serious game' as it was created with a goal of explaining a scientific concept, but it can also be seen as a 'pure game' in the manner it is being marketed to the public.

Interestingly, 'Entaglion' reverses the commonly used approach of using narrative rather than play elements to communicate a scientific process. The narrative setting of the boardgame is a rather generic science fiction universe, with the players tackling the issues of space navigation. The scientific material is instead contained within the gameplay system, and the success of a player is predicated upon mastering the choice sequences present on the game board. Said sequences are representative of the *Boolean* logic gates used within the field of quantum computing. The scientific accuracy was rated positively in Ashoori's (2018) study, which saw stringent playtesting and iterative design rounds of the game, under the supervision of experts within the field of quantum computing, and by experts who field tested the game (Brody 2019). Additionally, the game has been met with enthusiastic reviews and some commercial success (Brody 2019).

Game design can lend itself well to communicating high-fidelity, high-specificity hard science. The sequential, structured nature of play lends itself well to making the player intimately follow along a complex process (Kanyapasit and Sirsawasdi 2014). The enjoyment factor of the game keeps the player motivated to do so. Some caution must be taken by science enthusiasts making games so as to not underdeliver the gaming aspect of them, least the game becomes more of a study help than an actual ludic experience.

8.7 Gaming within a Science Education Setting

Ludopedagogy within nursing education is common with research studies dating back to 1960s. Nursing education relies on both teaching practical skills as well as theoretical knowledge. As such different aspects of ludo pedagogical approach can be tested. Day-Black et al. (2015) cited problems within the nursing classroom including learning materials being considered unappealing and difficult to relate to, with a desire for alternative, interactive approaches being expressed. Given that nursing is an in-demand occupation, efforts have been made globally to facilitate and optimise the development of future workforce within the sector using games.

8.7.1 Hijacking 'Kahoot!' for Education Purposes

The first example comes from Spain, in the University of Valladolid (Castro et al. 2019). The faculty took it upon themselves to improve both the engagement and the educational attainment of the students. Concerns were expressed regarding increased intake of students and decreased funding. As such an educational tool of high efficiency was required and the game, '*Kahoot!*', was repurposed for the task. '*Kahoot*!' can be described as a customisable mobile phone online quizzing platform, where players track their progress by 'gaining experience' for correct answers. As such, without being purpose-built for nursing education, it can be described as a pure game 'hijacked' for the purposes of learning.

The study found a slight improvement in test results but more interestingly, the games' structure let the researchers easily measure response times for each question and compare it against the accuracy, providing additional data points for analysis. The questions that took less time to answer were near-universally found more often answered correctly. The authors conclude that gaming can indeed prove a useful tool for education as it increases engagement and interaction, as well as makes for more effective testing (Castro et al. 2019).

The findings of the study were later expanded upon and confirmed in another study (Nerueiter et al. 2020), which not only implemented the very same game in an educational setting but also managed to produce a robust experimental study of the game's efficacy. The reported time for answering decreased by roughly 50% and the median attainment of correct exam answers increased from 45% in the control group to 76%. It is a stark, marked improvement and the gamified approach seems to not only better command student attention but improve teaching outcomes overall.

In case of the Nerueiter's 2020 study, it was used on a medical student cohort and suggests '*Kahoot!*' and similar games can be successfully repurposed for multiple divergent fields of study. The instant feedback is provided not only to the students but also to the faculty and can be used to inform both the staff and the students themself on what questions seem to be a struggle. It may point to a phrasing error, or a particularly complex, perplexing part of the material (Neureiter et al. 2020). Overall, quizzing platforms seem to provide near-ready solutions to lots of educational problems. The parameter of a game as an activity providing instant feedback (as in Arnetta's 'Interactivity' value) is exemplified here. Additionally, as Castro et al. (2019) report, the increased engagement between students, discussing answers and consulting one another. This community-building quality, in turn seems to reflect Arnetta's 'Identity' value, also remarked by Huizinga (1938), and Kanyapasit and Srisawadi (2014).

The described uses of '*Kahoot!*' for science education represent a use of a pure game for educational purposes. Interestingly, a similar study with similar results was performed using the open rules of the quizzing card game '*Jeopardy*' (Boctor 2013). As such, by making the quizzing environment more interactive and providing the player identity to the learners, a group of learners may be better engaged through the use of traditional gaming mediums. The only loss, in that case, would be the loss of direct digital feedback to the educator. A similar approach was implemented with a game of '*Deal or no Deal*' for the mathematics students, for the results of better engagement and retention (lChow, Woodford and Maes 2011). Overall, there seems to be a strong toolset of pure gaming titles, that may lend to an educational instruction.

8.7.2 'Transcription' Boardgame: A Game as a Biomedical Visualisation Learning Tool

'Transcription' is a board game focusing on the subject matter of RNA transcription, and was a cooperative experience with no competitive element between the players (University of Dundee 2012). The game was tested over a few years and was fabricated in multiple copies for internal use in tutorials. First author, Pawel Stachyra, being a student at the University at the time, had an opportunity to play this game. In teams, players took turns to 'move' the cell cycle forward. There was no single or team victor by design. On the other hand, the progression of the game sequence was closely analogous to the sequence of the RNA transcription. Different subcellular elements were represented by a variety of custom tokens, counters, and board pieces. The game is resolved by skill once the entire sequence has been completed as represented by a desired end-configuration of board pieces. The boardgame thus became more of a simulation than an actual competitive experience and was not further casually used by students.

To our knowledge, there was no use of the game outside of the tutorials, and the game was not released to the public or mentioned in any scholarly publication other than the university website. It is an interesting example, as on one hand the structure of a traditional game lent itself well to portraying a scientific process with high granularity. In Anneta's model, the game provides 'Informed Teaching' and 'Instruction'. The board with multiple representative pieces also provided 'Interactivity' in a general sense, but the parameters, the moveset of the game did not allow for experimentation with the gameplay. Here, 'Increasing Complexity' was not achieved in the game design. The lack of a competitive element has removed any stakes for a player to assume the gamer identity and the same starting layout always led to the same ultimate end-state, meaning that the game did not allow for increasing mastery within the game.

Overall, the game did not meet most of the core (pure) gaming parameters, while achieving the distal, serious parameters. It is an example of a game with high-accuracy scientific material but missing some of the basic game design considerations—a common error present in games made by scientists and not game designers (Zimmerman and Fortugno 2016). From personal experience playing the game, it could be improved by adding an explicitly competitive element (either vs other players or a timer), as well as adding a degree of randomness to the set-up to add a dynamic challenge.

8.8 Indirect Science Communication: Designing a Killer Plague to Save Lives

Sometimes, games not necessarily intended for a serious purpose, can also serve as science communication devices. One such game is 'Plague Inc.' (Ndemic Creations 2021). It is an example of a very popular pure game available on mobile devices and PCs. The player takes control of a deadly pathogen and undertakes a somewhat accurate epidemiological simulation to wipe out humanity. It is remarked that the transmission vectors and the evolutionary adaptations of the virus are very reminiscent of those in the real world and the creator was even invited to speak before the American healthcare authorities (Microbiology Society 2016). The game became very popular early in the SARS-CoV-2 pandemic and arguably helped propagate correct attitudes towards disease spread and control. A very interesting observation is to be made in case of those three games.

Through the 'Identity' principle, the gamer is put in unusual, unrealistic positions (controlling a disease or a species) but because of the implied unreality and detachment of the experience, the suspension of disbelief is present and helps mitigate the problem (Seraphine 2018). Despite this, 'Plague Inc.' is credited with a positive impact on social attitudes regarding epidemiology globally, both during the Ebola crisis from 2013 to 2016 and in the early stages of SARS-CoV-2 pandemic (Jiang et al. 2021; Mitchell and Hamilton 2017; Peng and Bai 2021).

Through gameplay, the player would observe the humans modeling scientifically sound attitudes. The user would creatively engage with the subject matter. Both qualities of the gameplay experience are cited as educational benefits of using the game (Peng and Bai 2021). 'Plague Inc.' represents how a game may use its popularity as a vector of correct scientific terminology and attitude, as long as it references sound scientific material. It is perhaps an underused avenue of science communication, where informed efforts can launch off the momentum of a popular marketed title. The public is primed with the correct terminology and introduced to key concepts that may be better suited to engage with scientific outreach efforts in the future.

8.9 Driving Interest in Science Ethics through Gaming

Narrative is one of the most powerful tools available to a science communicator. The introduction of in-narrative profound ethical choices in games is not unlike the established educational practice of discussing case studies to promote critical thinking (Popil 2011). Games indeed seem to display continuity with works of literature, particularly science fiction, that take their long-form nature to slowly and thoroughly explore difficult scientific subjects within a casual, entertaining setting (Barnett et al. 2006).

The video game 'Alpha Centauri' (Firaxis Games 1999), through its minimalistic narrative design, points the player towards ethical questions that include the value of individual vs collective, the ethics of independent AIs, biomodification and cloning of humans. The short, often ominous cutscenes do not provide answers, but rather phrase and rephrase the question of profound ethical implications. Given, that games often include complex, well-referenced narratives, it is possible that video game carry the function previously covered by science fiction novels that popularised the subjects of science. In this aspect, a science communicator, artist and a game designer may not only reach but also over time create an audience ready to engage with the hard science topics (Illingworth and Allen 2016). A study of prescriptive narratives in science communications seems to further build upon the point of the narrative's potential to build and model behaviour, on top of promoting dialogue (Downs 2014). The most promising aspect of the employment of such narratives is that they can remain independent of the game mechanics, thus offering more of a freedom of expression for the author (Seraphine 2018).

8.10 Upstream Interactions of Games and Science: Inward-Facing Science Communication

Lastly, games can also be used in inward-facing science communication. Having previously discussed accidental and/or indirect science communication, this section will conclude with some impactful examples of games educating the educators. As previously mentioned, games and especially the online games have an innate ease and reliability of recording player's behaviour. Every single decision made by each single player is traced, noted and stored, often in massively multiplayer games with high millions of concurrent players active at one time. As such, digital worlds provide one thing scientists take an interest in—large datasets. Gaming behaviour of a player is not as divorced from their real behaviour as one could suspect. The player identity is strong as per Annetta's or Huizinga's approaches. The player assumes the role of a subject in game and makes large time, attention and financial investment into their play behaviour. Harm to their online avatar is a thing actively avoided and thus players act in the virtual world with a self-interest not unlike that displayed by them in real life (Toh 2021).

Games can be used to derive highly data-rich models of human behaviour, and the most famous case is likely that of the 'Corrupted Blood' incident in World of Warcraft (Blizzard 2004), one of the most popular digital massively multiplayer online (MMO) games with around 1.5 million concurrent active players. The 'Corrupted Blood' incident occurred to the game developer's surprise and was not an intended part of the game experience. Players realised that a particular 'Non-Player Character' (NPC) raid boss afflicted their characters with a simulated disease, which damaged their character's health and subsequently spread to nearby other characters. The damage was significant for a highly developed character but immediately deadly for any lower-level character. Some malicious actors began to preserve the 'disease' and carry it into large gatherings of players; auction houses, guild halls, town squares, to spread it. What followed was a turmoil within the virtual world, with major population centres abandoned and weaker players fleeing to more remote areas in the game. Eventually, as the virtual pandemic raged on, entire servers became depopulated and hostile to digital life (Smith 2012). Many different behaviour patterns have been observed. Some players started displaying altruistic behaviours and helping the weaker players survive. Others would intentionally keep infecting one another to preserve the plague and release it again upon other players. Self-organised quarantines were imposed, and certain vistas were blockaded. The behaviours were used to model future pandemic and became a primer for the epidemiologists to begin tackling the covid crisis later (Kriz 2020). The significance of this event, is that an unprecedented of unintentional participants have provided data on their movement and reactions within an environment, where their action could be easily catalogued and preserved for the future scientific research. Future observation of online communities became recommended as the effect (Girish 2019; Smith 2012).

Another example of using gaming to inform the scientists is that of the 'Phlyo Project'—a citizen-science platform presented as a puzzle game for voluntary participants (Kawrykow et al. 2012). While the game itself was divorced from the scientific subject matter, it presented the player with pieces of a DNA sequence in the form of a logical puzzle. Players would look for patterns matching the coding regions of the DNA codes and be scored well on the speed and efficiency of finding said patterns. Scientific data produced was considered of better quality than the mechanistically obtained data and was supplied readily at high efficiency (Kawrykow et al. 2012). Given, that players often willingly give their time and focus to games, the player base remains somewhat of an untapped resource that the gaming medium offers. It is a promising avenue to explore, as long as the game design can be used in such a way that the game output becomes sound scientific data, while providing entertainment to the player.

Examples discussed in this section, provide some more in-depth focus on how a game can indirectly promote certain attitudes and drive informed conversations. The stand-out example is that of the 'Plague Inc.', where a game accidentally provided the general public with a primer on epidemiology education, especially communicating the potential downfall of poorly implemented policies and safety measures. The impact of the game on the public response to the SARS-CoV-2 pandemic may never be truly explored, as such studies would face a strong survivor bias. A game narrative is free to be shaped, however, the author sees fit and carries a potential of subtle, non-confrontational presenting of ideas to the public, although a manipulative, almost-propagandistic approach should be used responsibly, with caution and humility (Kikkawa 2022).

8.11 Developing a Science Board Game: 'Belly Rumble: a microbiota tactics'

Having suffered screen fatigue and low social interaction after 2 years of SARS-CoV-2 pandemic lockdowns, first author, Pawel Stachyra opted to create a traditional game to help others learn about the gut microbiota. He was initially inspired by an existing boardgame, 'Quantum' (Zimmerman 2013). In particular, he enjoyed how the game re-imagined the use of dice as counter and board pieces as well as random number generators, and envisioned a game dealing with the subjects' gut microbiology, where dice value spread over the game board would signify living organism density area. Given the possible adversarial interactions between players, he settled for a strategy boardgame using cards and dice as props.

Initially oblivious to much of the terminology within the field, Stachyra approached the subject as a gamer himself, focusing more on game design and the experiential terms of play. Reading Zimmerman and Fortungo's (2016) developers' reflections, the idea of making the game a competitive experience was reinforced, while driving interest, pointing towards sources and not contradicting the science (Brown 2012; Taddicken and Reif 2020). The gameplay needed to be 'fun' and engaging; therefore a simple game play was initially devised. It comprised an interplay of unit count on the dice (representing bacteria density) with different evolutionary pressures, adaptive strategies and random events operated by cards. The game's artwork was highly stylised with anthropomorphised bacteria (Fig. 8.7) and stylised, colour-coded modular board pieces (Fig. 8.8) to represent the transverse sections of the small intestine. The game cards produced for the Belly Rumble also included QR codes to relevant scientific and gameplay information to encourage fact checking (Fig. 8.9).

In the development process, an AGILE approach was used, meaning a minimal functional product was created, tested with participants and then improved upon based on feedback in a cyclical manner (Dingsøyr et al. 2012). There were at least two major iterations of the game made after issues were identified through initial playtests with students from the 2020–21 cohort studying for an MA in Art in Science at Liverpool John Moores University. For example, the dice count was uncapped initially but later capped because it needlessly slowed down the gameplay, required unfeasible amounts of dice for a physical copy and contradicted the limiting nature of ecological niches (Qin et al. 2010). Game was designed to accommodate between two and four players, however, it seemed to have the best balance/dynamics in the two-player variant.

The game play used a turn-based gameplay, where players roll a number of dice based on the gut area their 'bacteria' control to use it as biomass. Evolutionary adaptations and strategies are picked by players using cards with both their in-game effects, comic illustration and often QR codes linking to relevant research. The main goal was to communicate science through narrative and structuring the gameplay so as to not give perverse incentives or contradict the subject matter, following the recommendations of Brown (2012). Three aims were developed that sought to:

1. Provide enjoyable gameplay experience that players would seek out in their free time.
2. Drive interest in the subject and link to relevant scientific material for self-study.
3. Encourage and drive conversations about the subject between players.

In the development of the game board itself, there were considerations of board pieces (pre-existing die) fitting, as well as the board size for portability, therefore a modular design and a compact size were opted for. The complete boxed set of the game with the instructions, cards, board and dice was used in five in-person playtests. The final version of the game is available to download, including its ruleset and

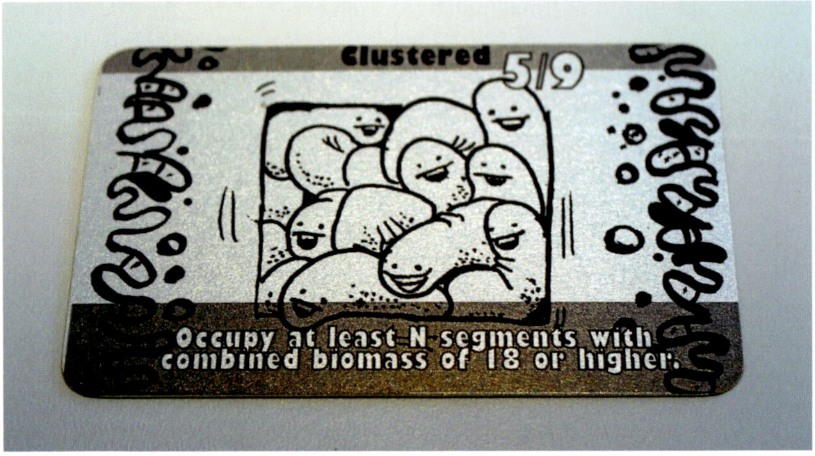

Fig. 8.7 Game card produced for 'Belly Rumble' featuring anthropomorphised bacteria. Photograph by Pawel Stachyra

templates to fabricate the board pieces and cards in the spirit of open-source sharing on the project's webpage (https://pstachyra6.wixsite.com/website/work).

The dice were ordered from online resellers and the smaller 12 mm size was used for the sake of portability as opposed to more common 16 mm. Remaining pieces were fabricated using a Fab Lab—a local maker space. The board pieces were cut from MDF using a laser cutter, after being printed/painted on using a silk printing technique. The card designs were initially hand painted and then digitalised and added text was added in Adobe Illustrator. The card material was 180 gsm card stock and large gantry printers, and laser cutters were used to produce cards off the designs. There were multiple learning curve moments involved. Stachyra had to learn the practicalities of silk printing, operating a laser cutter with the associated software and then problem solve the printing orientation and spacing of cards. What is heartening, was that discounting not insignificant supervision and advice of the maker space and print room staff, the game was successfully fabricated within 2–3 weeks from the first physical prototype (Fig. 8.10). Moreover,

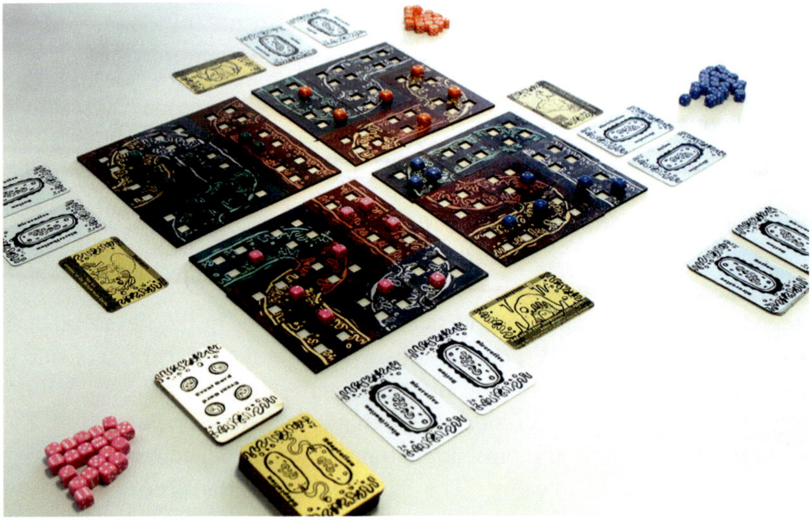

Fig. 8.8 'Belly Rumble' game set-up (third prototype). Dice represent different bacterial strains (players), and coloured modular board represents the intestinal sections. Each player can gain up to two gold 'adaptation' cards and has two secret victory conditions (silver). Photograph by Pawel Stachyra

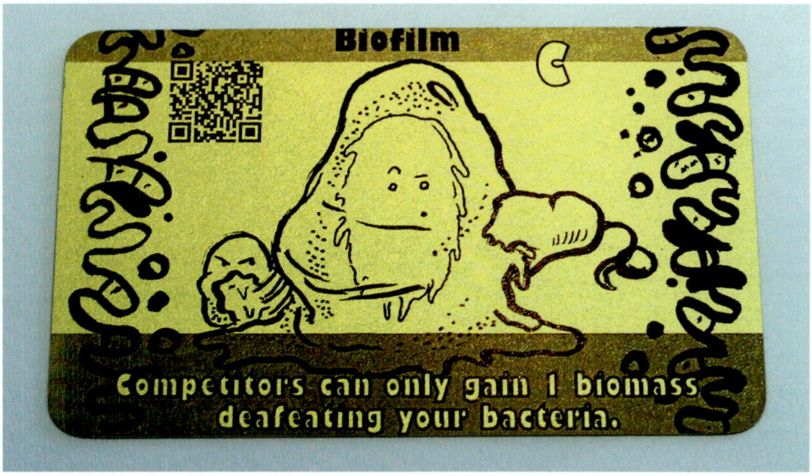

Fig. 8.9 Game card produced for 'Belly Rumble' displays an interplay of the narrative and ludic elements of a game piece. After including all the game-relevant information any type of story can be conveyed through artwork, wording and in this case, QR codes. Photograph by Pawel Stachyra

these designs were easily recorded and saved for further reproduction and iteration. It should be promising for any single practitioner or a team thereof that considers developing a traditional game—it is achievable, relatively simple and cost-effective in any locale with a functioning maker space.

8.11.1 'Belly Rumble': Playtests and Reflections

During the game's development phase, closed playtests were performed online with peers during the 2020–21 UK lockdowns who gave informal verbal feedback. Between April and August 2021 playtests in an experimental setting with participants face-to-face in social public venues such as coffee shops took place (Fig. 8.11). Ethical approval was obtained from Liverpool John Moores University's Research Ethics Committee (21/LSA/013) and qualitative feedback was gathered. Participants comprised different demographics and experience levels. Five playtests ran with 10 different participants and combinations thereof. The shortest complete playthrough took 24:37 min to resolve (a two-player variant of the game), while the longest (four-player variant) took 1 h and 47 min. In every game played, it successfully

Fig. 8.10 All the board pieces, dice and cards assembled. The entirety of the game fits in an a5 footprint box. For reference each MDF board square has a side of 5.5 mm. Photograph by Pawel Stachyra

Fig. 8.11 A two-player playtest of 'Belly Rumble' at Melodic Distraction, Liverpool UK, during the summer of 2021. Photograph by Pawel Stachyra and included with the consent of Dr. Lee Haines (pictured)

concluded. The players left each playthrough commenting on the gameplay and discussing their strategies for the next playthrough. This hints at the successful implementation of the 'Increasing Complexity' principle discussed by Arnetta (2010), as the players were able to grasp the game rules and assume their gaming identity with their goals and possible moves.

An anonymised online questionnaire was distributed to gather feedback. While the results were overwhelmingly positive, the sample size is small with limited diversity. For example, only three out of ten participants identified as women. It is uncertain, how large the effect of the sampling bias was, but overall, some observations can be made based on the feedback and the author's own experience running the game tables.

The game did successfully conclude in each testing run, for the second and third prototype versions. The parameters and rules were simple enough for the players to grasp and build upon. The initial apprehension about engaging with the rules, would eventually dissipate as the players tested the boundaries of the gameplay. It was observed that novel, unpredictable strategies devised by the players and the discussions would ensue after the game regarding the course of action in possible future playthroughs. The competitive element was well received and seemed to drive the enjoyment of the game. The player's gender did not seem to affect the experience, regardless of the preconceived notions of competitiveness between said genders. The game would become challenging and engrossing around the 2-3rd round where the conflicts would begin occurring and players would attempt to close their victory conditions.

The audience that was more knowledgeable about the subject matter of gut microbiota, would remark that the gameplay is reminiscent of something they would imagine the gut ecology to unfold as and would not only recognise scientific references but also propose some of their own into the future versions of the game. The subjects would universally report increased awareness and knowledge of the subject, although a less subjective measure should be included in future investigations to avoid biases. Further testing in a more neutral environment with a wider demographic range of participants could help improve the data reliability. While an online survey is appropriate for the purposes of the study, it sacrifices the free-form feedback previously gained in the closed development tests. For further testing, the two methods should be cycled and supplemented. Lastly, the game had some balancing issues for the four-player variant, where the turns would take long for the players to conclude and some victory conditions were much simpler to achieve than others. Such issues could be only addressed with continuing playtesting and iterations, and the authors are confident they can be addressed without compromising the core gameplay.

Overall, the game has managed to fulfil or show a promise of fulfilling the stated objectives.

The robust theoretical framework of Anneta's model is credited, but also free-form reflective journals of Zimmerman, with guiding the authors through the process. The fabrication itself was also relatively simple with the help of the maker space staff and modern consumer technologies and hence was more of an afterthought to the actual game experience and messaging.

An interesting sidenote was the mystique associated with me making a game. Play testers would usually display some awe at the process of my designing, producing and delivering a game. For the authors, this process was deemed to be relatively easy, given their prior knowledge of the theory and practice within the field. It should be a promising prospect for the future ludopedagogists—as long as the core gaming values are delivered, the scientific expertise can be layered freely and creatively on top of them and the enjoyment factor refined and improved enough to make an attractive, engaging communication tool as will be discussed below.

8.12 Discussion

From analysis of the case studies in the sections prior, this section will discuss five promising qualities and three pitfalls of games for science communication to advise those considering games as their mode of engagement.

8.13 Promising Qualities of Games for Science Communication

8.13.1 Dynamic Complexity

A game meets the learner at their level and provides a dynamic scaffolding, where the player raises up to new challenges posed by the game, and in turn seeks further challenges while their mastery of the subject increases (Annetta 2010; Vand Der Stuyf 2002). The progression pace becomes personal and perhaps more reflective of the player's skill level.

8.13.2 Feedback Stream

Performance within a game can be easily tracked both by the player and the facilitator (Castro et al. 2019; Nerueiter et al. 2020). The player is constantly made aware of their progress and performance throughout the game (Kiili 2005) so they either are reassured in their performance or notified to immediately seek assistance. More interestingly, feedback in a digital gaming setting can also be collected and tracked by a facilitator or educator, to help further improve the learning experience and knowledge transfer and identify the areas in need of improvement. As such, some digital games can create complex databases of traceable information as was exemplified in previous examples of MMO games (Belaza et al. 2020; Girish 2019). A feedback stream can be used by scientists themselves to derive insights, and the player can be nudged and guided towards correct attitudes through a game system.

8.13.3 Identity Commonality

A game builds a community and commonality between the players, with them participating in a self-same structured experience within the same parameters (Huizinga 1938; Annetta 2010; Kanyapasit and Srisawasdi 2014; Tanghe 2016). It also provides a neutral exploratory space perceived as detached from real-world consequences. As such it promotes more bold, risk-taking and cooperative approaches together with well-natured competitiveness (Zimmerman and Fortugno 2016). This helps demystify and reduce stress in learning a new subject, which becomes enclosed within the parameters of the game (Annetta 2010). The 'commonality' value becomes important when dealing with complex scientific processes as the gameplay can be made analogous to the process itself—thus giving the player a practical example of a highly complex process as it was achieved with both the 'Entaglion' and 'Niche' games.

8.13.4 Incidental Exposure

Games with rich narratives expose a player engrossed in the pure game experience to whatever subject the narrative may include (Davidson and Candy 2016; Liz 2003) (Liz 2003; Davidson and Candy 2016). As such, the players may become exposed to complex scientific ideas in a non-confrontational manner over the span of hours. This promotes self-directed inquiry and helps shape the future of discourse by equipping the player with the vocabulary, motivation, and knowledge on the subjects (Illingworth and Allen 2016). As such, designing scientifically informed game narratives may come to the forefront and build a future audience rather than identify an existing one, as with the 'Plague Inc.' game that informed about large player base about the concepts of epidemiology (Peng and Bai 2021).

8.13.5 Cognitive Involvement

While games are considered a pleasant experience, and through their identity they are somewhat easier approachable than real-world tasks they still pose a challenge (Huizinga 1938; Kanyapasit and Srisawasdi 2014; Zimmerman and Fortugno 2016). Such challenges promote the player to cogitate at their peak capacity, including problem solving, critical thinking and states of flow. More complex tasks can be broken down and challenged within a game. Through direct feedback players build confidence in their independent thinking and problem solving. It also provides a scaffolding structure, where the players can practice their critical faculties before tackling practical tasks or exams (Vand Der Stuyf 2002). A detailed example of a proposed use of games and puzzles to prepare for introduction of more complex subjects, such as cryptography and quantum computing, is discussed in a guideline paper by Moursund (2006).

8.13.6 Equity

Having a shared experience, which tailors itself towards the player's needs and means, promotes and enables a more inclusive learning experience, especially for players with learning disabilities or divergent learning styles. Gamified interventions have proven successful in assisting disabled students (Sholz et al. 2008) to perform at higher levels than they would otherwise. Additionally, traditional games can be developed and delivered in settings that may not afford fast broadband internet due to their low technological requirement. This potentially serves to improve education in less developed countries.

8.14 Shortcomings of Games as a Science Medium

8.14.1 Overreliance on Gaming

A ludopedagogical approach is tempting to facilitate learning of complex science information, however, it may not be the most appropriate vehicle for knowledge transfer. Sometimes the game overshadows the learning (Christians 2018; Wan and Chiou 2006; Mauri et al. 2011), and successful implementation of a gamified approach should be seen as an additional tool and not a replacement for other modes of learning (Sholz et al. 2008; Chow, Woodford and Maes 2011; Boctor 2013; Kanyapasit and Srisawasdi 2014; Castro et al. 2019; Nerueiter et al. 2020).

Any form of science communication should lead the audience towards further reading (Illingworth and Allen 2020). It is simply unfeasible to expect the audience to derive a fully nuanced understanding of the subject from just one instance or medium of communication. Games can be used to prime and energise the audiences but there needs to be further reading and clarification presented. A well-intended game may become a pastime and a distraction to knowledge transfer if not supplemented with instruction, information and further progression.

8.14.2 Poor Game Design

Another pitfall is poor gaming design. Through too narrow focus on the learning material or foundational mistakes in the designer approach

(Zimmerman and Fortugno 2016; Gallegos et al. 2017) a game may fail to engage players. As observed with unstructured math learning, different students may develop wholly different processes, which prove valid and functional learning (Carraher, Carraher and Schliemann 1985; Saxe 1988), unless regarding a process-oriented subject (Kanyapasit and Srisawasdi 2014). Otherwise, the engrossment of a player within a competent gaming experience may be enough to expose them to the principles of the science being communicated and build interest in the subject. Some games are not deemed as successful as they do not offer a 'fun' experience and might meet harsh criticism for not delivering on its gaming promise (Gallegos et al. 2017).

8.14.3 Sensationalising Science

Lastly, is the potential to communicate distorted, sensationalised science. Brown (2012) argues the real risk of sensationalised science is not the framing itself, but the lack of critical faculties of the public to engage and evaluate the content (Brown 2012). The extreme expressions of a public unequipped to critically engage with science are behaviours of Scientism—uncritical semireligious perceptions of science, risking stifling the debate and promoting irrationality (Pigliucci 2018). As such, the principles of efficient gaming education in the form of involvement of higher thinking faculties can be employed to provide a scaffolding structure for the members of the public to practice and develop their critical thinking (Walker 2003; Kanyapasit and Srisawasdi 2014). A brief note must be made regarding the proliferation and spread of fake or distorted science information. Formal logic structures can possibly be taught through gameplay mechanisms and case studies explored in some game narratives can help the public build resilience to unfounded information (Pennycook and Rand 2021).

8.15 Reflection on the Arnetta's Models

In keeping with the structuralist (Robey 1973) approach of this work, it is important to build up one's understanding and practice upon relevant theory and heuristics. Overall, the 6 'I' model by Arnetta has proven promising for delivering said heuristics for any would-be game designer. Apart from the nestled hierarchy discussed previously, the paper also presents a useful list of signifiers that can be used to decide whether a game fulfil its promise as a game. These parameters are listed below and use the example of chess to show how they might apply.

Anneta's (2010) signifiers of a game with an example of chess:

1. Feeling the activity can be successfully completed—Chess is a balanced two-player game with clearly readable outcomes.
2. The player can concentrate fully on the activity—Chess is a self-contained experience that can be resolved within its own parameters.
3. The activity has clear goals—There are codified, standardised rules for chess.
4. The activity provides fast feedback—The success of a player can be tracked by analysing the board layout.
5. The player is deeply involved in the activity—Players are focused on the game itself and their interactions with and through it. Intense analytical and logical faculties are employed by the players.
6. A sense of control over the actions is necessary to perform the activity. Game is balanced, and the conclusion is predicated upon players' actions within it.
7. There is an altered sense of time. Arguable, as that would be players' internal experience. Moreover, competitive chess often involves turn counters.

Moreover, chess as a game meets the first 4 "I's" of Arnetta's model. As a game designed purely for the game design's sake, chess does not meet the two outward "I's" of Informed Teaching as well as being Instructional. It does attest to specificity of the Arnetta's framework, however, considering the reported learning benefits of chess play (Sholz et al. 2008), a further blind spot of the model is exposed—it does not possess the non-direct and unintentional learning. It is proposed, as such with pure gaming, its previously hinted potential for providing educational benefits should be outlined and given recognition (Bart 2014; Gallegos et al. 2017; Gobet et al. 2004; Nerueiter et al. 2020).

As it can be observed, the framework can retrospectively recognise a game and break down the elements required to fully realise a gaming project. Moreover, having said parameters pointed out helps identify areas where the game needs to improve or pinpoint where it went wrong. For example, the Dundee University Transcription Boardgame failed to provide point 6—the game was set up to conclude eventually at its one correct arrangement eventually. Including a competitive (e.g. racing) element could help mitigate that and provide better Interactivity and Identity value.

For *Belly Rumble*, the framework still applies:

1. Feeling the activity can be successfully completed—The victory conditions were achievable and players were aware of the mechanisms of achieving and declaring victory.
2. The player can concentrate fully on the activity—All resources and knowledge required to engage with the game were contained within the box set. The instructions, the board and the pieces.
3. The activity has clear goals—As in point one, the first player to achieve both their randomly drawn objectives twice is declared the winner and the game stops.
4. The activity provides fast feedback—The success of a player can be tracked by analysing the board layout.
5. The player is deeply involved in the activity—After learning the rules the players were happy to engage with the game and to strategise/negotiate in a way that demonstrates understanding.
6. A sense of control over the actions is necessary to perform the activity—As above.
7. There is an altered sense of time—Playthroughs lasted between 26 min and over 90 min based on the number of players but the flow of time was not apparent.

The game also is designed to provide informed teaching through ludo narrative and instruction linking to relevant literature within the subject of gut Microbiota.

There are things the 7-point model would not be able to track: the game balance needed fine-tuning and it still does at the time of writing this work. Playing with 4 players would also make the experience longer and less dynamic. Thus, the model can detect whether an activity is or is not a game but cannot point to qualities of it. Other fields in which this framework falls short, is the lack of inclusion for non-digital gaming. There is no explicit argument within the text against them, but it is an unrealised potential not to recognise that the principles of gaming seem to apply regardless of the delivery medium.

Point 7 relies on the attainment of a flow-state as described in cognitive psychology (Csikszentmihalyi 1990). Such states of perception are characterised by partial dissolution of self-awareness, feeling of competence/mastery, improved performance as well as partial obliviousness to time and the immediate environment outside of the activity. Interestingly, it also has been observed that the moment the self-awareness (identity) is restored, the flow-state becomes disrupted. Arguably, the flow-state exists in the state of default network (Mason et al. 2007; Buckner, Andrews-Hanna and Schacter 2008), where the mind idly wonders as a subject 'goes through the motions' of daily activity.

Flow-states are rare outside of high-performing activity of a skilled subject. It is often observed that video games produce an altered sense of time, and are highly engrossing, thus producing a flow-state when the players' identity becomes temporarily obliterated outside of the game (Nah et al. 2014; Murch et al. 2020) (which is also one of proposed models for developing a video game addiction). There are, however, also alternative findings with long-time video game addicts displaying a state more akin to the default network when gaming (Wan and Chiou 2006). As argued by other papers (Kanyapasit and Srisawasdi 2014), there are other important aspects of learners' cognition that become engaged by gaming: problem solving and critical thinking. Moreover, while a considerable amount of scholarship explores the potential of flow-state in education (Nah et al. 2014), it is not apparent if a cognitive phenomenon so widely observed is of clear benefit to learning. Flow-states are, for example heavily implicated in addiction to things like gaming (Wan and Chiou 2006), gambling (Murch et al. 2020) or social media (Fox et al. 2021; Mauri et al. 2011; Nah et al. 2014). A reader, whoever experienced spending hours fully engrossed in scrolling their social media feed, may relate to the fact. Flow-states appear to be powerful cognitive phenomena, that can be harnessed for good or ill. On the other hand, it is widely argued that critical thinking and problem solving are crucial not only to learning but also to performance in the professional world at large (Annetta 2010; Kanyapasit and Srisawasdi 2014). As such, we can postulate modifying the Immersion quality to be considered within the parameters of Involvement [of higher thinking faculties. Overall, with this slight modification, Annetta's model is probably useful to a ludopedagogist in conceptualising what a game is for the sake of educational use, prioritising design elements and discussing the results.

8.16 Importance of Narrative Crafting

When considering the potential of communicating complex science ethic issues through gaming two games have been mentioned. Interestingly, both Alpha Centauri and S.O.M.A are video games and can be considered pure games. The issue of consciousness is in no way intrinsic neither to a gameplay of a 4X game (Alpha Centauri) or a survival horror (S.O.M.A). Instead, the core experience remains that of a pure game. Keeping the audience captive in a state of flow; focusing their attention on the successful resolution of the game; the game narrative exposes them to ongoing debates within the science philosophy, often quoting terminology and primary sources, so one can follow up and engage a debate within the community of players (Identity/Community-building principle). Exercises in cooperative storytelling like LARPs and storytelling games (e.g. DnD) can be used to incorporate mature philosophical themes. These in turn can be explored within the simulated/imagined reality without the adversarial character of an open academic debate.

Interestingly, in adult population, psychological indicators between different gaming mediums, indicated that content is less appealing than the interaction itself—to summarise the experience of the game seems to matter more in the test subjects (Marston 2013). Perversely, it should be considered an opportunity, as it indicates the pure gaming experience may be able to keep the player entertained for long enough to explore some more challenging science material. Narratives present themselves as a very powerful tool in the science communicator's toolkit. Games like Senua's Sacrifice, the Darkest Dungeon, Alpha Centauri or S.O.M.A, stripped of their narrative content, become common representatives of their respective genres. Again, the gaming element carries the narrative content. The game can be used to expose the player to the material they did not explicitly come to seek out, which may be outside of their comfort zone or intellectual awareness at the time (Gowler and Iacovides 2019). It is explicitly a manipulation technique, one can use to any end—be it positive or nefarious (Kikkawa 2022).

8.17 Conclusion

Despite modern preconceptions, gaming is not a new phenomenon. The understanding of what makes play and how to partake in it seems to be hardwired into not only humans but even other higher animals. The digital medium has merely served to elevate the complexity and marketability of gaming. As such, even as science communication practitioners we are intrinsically equipped with a working understanding of how to make a game no less than the target audience is primed with motivation and understanding on how to play it.

In the spirit of transdisciplinary cooperation (Nicolescu 2014), the medium of game development may become a 'third space' in which scientists, artists and game designers may meet to bridge divides within the body of epistemological knowledge as well as push further our cultural understanding and participation within arts and sciences. The syncretic nature of game design opens itself up to large collaborative projects, where game designers, artists and science communicators contribute uniquely and valuably through their available tools—both ludic and non-ludic. While the authors believe that the narrative exposure may be the single most powerful tool in a ludopedagogical approach, it cannot fully manifest without engaging gameplay design and becomes purposeless or harmful and bereft of sound scientific content.

This chapter serves three main purposes: it is a primer on the use of games in scientific communication aimed at those considering such an approach in their practice; it facilitates through practical distinction between pure and serious gaming as well as the digital and the traditional media, and it assesses and balances their potential and difficulties for the reader to better inform their approach. Many projects exist in isolation partly because of lack of dialogue between the different specialists. As such, dialogue must be built and maintained to facilitate change. Lastly, this chapter is a call to action—ludic approaches can be used to effectively promote science, improve the quality of learning, and aid in scientific information transfer, however, the potential is still mostly unrealised and that is good news for prospective serious game designers.

References

Red Hook Publishing (2016) Darkest dungeon. https://www.darkestdungeon.com/darkest-dungeon/

Allerton D (2010) The impact of flight simulation in aerospace. Aeronaut J 114:747–756. https://doi.org/10.1017/S0001924000004231

Annetta LA (2010) The "I's" have it: a framework for serious educational game design. Rev Gen Psychol 14(2):105–113. https://doi.org/10.1037/a0018985

Arrambide Figueora MK (2022) Player agency, decision-making, and morality in cinemaitc choice-based adventure games. [University of Waterloo]. http://hdl.handle.net/10012/18204

Austin RG (1940) Greek board games. Anquity 14:257

Barnett M, Wagner H, Gatling A, Anderson J, Houle M, Kafka A (2006) The impact of Science fiction film on student understanding of Science. J Sci Educ Technol 15(2):179–191. https://doi.org/10.1007/s10956-006-9001-y

Bart WM (2014) On the effect of chess training on scholastic achievement. Front Psychol 5:762. https://doi.org/10.3389/fpsyg.2014.00762

Beal MD, Kinnear J, Anderson CR, Martin TD, Wamboldt R, Hooper L (2017) The effectiveness of medical simulation in teaching medical students critical care medicine: a systematic review and meta-analysis. Simul Healthc 12(2) https://journals.lww.com/simulationinhealthcare/Fulltext/2017/04000/The_Effectiveness_of_Medical_Simulation_in.7.aspx:104

Betihavas V, Bridgman H, Kornhaber R, Cross M (2016) The evidence for 'flipping out': a systematic review of the flipped classroom in nursing education. Nurse Educ Today 38:15–21. https://doi.org/10.1016/j.nedt.2015.12.010

Binde P (2007) Gambling and religion: histories of concord and conflict. J Gambling Issues 20:145–165. https://doi.org/10.4309/jgi.2007.20.4

Blizzard (2004) World of Warcraft. https://worldofwarcraft.com/en-us/

Braet AC (1992) Ethos, pathos and logos in aristotle's rhetoric: a re-examination. Argumentation 6(3):307–320. https://doi.org/10.1007/BF00154696

Bristol J (2021) Niche a genetics survival game. https://www.commonsense.org/education/game/niche-a-genetics-survival-game

Brody J (2019) The hidden structure of the Entaglion game board. Medium https://medium.com/qiskit/the-hidden-structure-of-the-entanglion-game-board-3d09bd6ebbd

Brown P (2012) Nothing but the truth: are media as bad at communicating science as scientists fear? Sci Soc 13(11):964–967

Buckner RL, Andrews-Hanna JR, Schacter DL (2008) The Brain's default network. Ann N Y Acad Sci 1124(1):1–38. https://doi.org/10.1196/annals.1440.011

Carraher TN, Carraher DW, Schliemann AD (1985) Mathematics in the streets and in schools. Br J Dev Psychol 3(1):21–29. https://doi.org/10.1111/j.2044-835X.1985.tb00951.x

Cartlidge J (2022) Becoming afflicted, becoming virtuous: darkest dungeon and the human response to stress. Games and Culture 15554120221084450:170. https://doi.org/10.1177/15554120221084450

Castro M-J, López M, Cao M-J, Fernández-Castro M, García S, Frutos M, Jiménez J-M (2019) Impact of educational games on academic outcomes of students in the degree in nursing. PLoS One 14(7):e0220388. https://doi.org/10.1371/journal.pone.0220388

Collins J (2004) Education techniques for lifelong learning. RadioGraphics 24(5). https://doi.org/10.1148/rg.245045020

Culture.pl. (2018) CD Projekt red reveals how they made the Witcher 3. Culture.Pl/En. https://culture.pl/en/article/witcher-3-how-game-made-CD-projekt-red

Davidson SJ, Candy L (2016) Teaching EBP using game-based learning: improving the student experience. Worldviews Evid-Based Nurs 13(4):285–293. https://doi.org/10.1111/wvn.12152

Darbyshire P (1993) In defence of pedagogy: a critique of the notion of andragogy. Nurse Educ Today 13(5): 328–335. https://doi.org/10.1016/0260-6917(93)90072-A

Day-Black C, Merrill EB, Konzelman L, Williams TT, Hart N (2015) Gamification: an innovative teaching-learning strategy for the digital nursing students in a community health nursing course. ABNF J 26(4): 90–94

Dingsøyr T, Nerur S, Balijepally V, Moe NB (2012) A decade of agile methodologies: towards explaining agile software development. J Syst Softw 85(6): 1213–1221. https://doi.org/10.1016/j.jss.2012.02.033

Donev D (2013) Principles and ethics in scientific communication in biomedicine. Acta Informatica Medica: AIM 21(4):228–233. https://doi.org/10.5455/aim.2013.21.228-233

Downs JS (2014) Prescriptive scientific narratives for communicating usable science. Proc Natl Acad Sci USA 111(Suppl 4):13627–13633. https://doi.org/10.1073/pnas.1317502111

Elequin M (2020) A dungeon Master's guide to the satanic panic: the history, sociology, and rhetoric of conservative Christianity and dungeons and dragons [Baylor University]. https://baylor-ir.tdl.org/handle/2104/10909

Eng D (2020) What are Megagames? https://www.universityxp.com/blog/2020/10/22/what-are-megagames

Erasmus CJ (1950) Patolli, Pachisi, and the limitation of possibilities. Southwest J Anthropol 6(4):369–387. http://www.jstor.org/stable/3628563

Ewalt DM (2013) Of Dice and men: the story of dungeons and dragons and the people who play it. Scribner. https://books.google.co.uk/books?id=AJJqBAAAQBAJ&pg=PA36&redir_esc=y&hl=en#v=onepage&q&f=false

Finkel I (2004) Round and round the houses: the game of Pachisi. In: Mackenzie I, Colin F (eds) Asian games: the art of contest. Asia Society, pp 46–57

Fox MP, Carr K, D'Agostino McGowan L, Murray EJ, Hidalgo B, Banack HR (2021) Will Podcasting and Social Media Replace Journals and Traditional Science Communication? No, but. . . . Am J Epidemiol 190(8): 1625–1631. https://doi.org/10.1093/aje/kwab172

Gallegos C, Tesar AJ, Connor K, Martz K (2017) The use of a game-based learning platform to engage nursing students: a descriptive, qualitative study. Nurse Educ Pract 27:101–106. https://doi.org/10.1016/j.nepr.2017.08.019

Giasolli V, Giasolli R, Giasolli M, Girasoli A (2006) Serious gaming–teaching science using games. Microsc Microanal 12(SUPPL. 2):1698–1699. https://doi.org/10.1017/S1431927606061149

Girish N (2019) "Corrupted blood" and public health. Berkley Scientific Journal 24(1). https://doi.org/10.5070/BS3241046896. Retrieved from https://escholarship.org/uc/item/4qj303gp

Glick SP, Charters LI (1983) War, games, and military history. J Contemp Hist 18(4):567–582. http://www.jstor.org/stable/260304

Gray P (2019) Evolutionary functions of play. In: Smith K, Roopnarine PJ (eds) The Cambridge handbook of play: developmental and evolutionary perspectives. Cambridge University Press, pp 84–102. https://cdn2.psychologytoday.com/assets/evol.functs.play_published.pdf Acessed at 16.06.2021

Gobet F, de Voogt A, Retschitzki J (2004) Moves in mind: the psychology of board games. In Moves in mind: the Psychology of Board Games. https://doi.org/10.4324/9780203503638

Gowler CPR, Iacovides I (2019) "Horror, Guilt and Shame"–uncomfortable experiences in digital games. Proceedings of the Annual Symposium on Computer-Human Interaction in Play 325–337. https://doi.org/10.1145/3311350.3347179

Higgins C, Walker R (2012) Ethos, logos, pathos: strategies of persuasion in social/environmental reports. Account Forum 36(3):194–208. https://doi.org/10.1016/j.accfor.2012.02.003

Huizinga J (1938) Homo ludens: a study of the play-element in culture. 1–220. https://doi.org/10.4324/9781315824161

Illich I (1971) Deschooling society. Harper Colophon

Illingworth S, Allen G (2016) Outreach and public engagement. In: Effective science communication. IOP Publishing, pp 5–23. https://doi.org/10.1088/978-0-7503-1170-0ch5

Illingworth S, Allen G (2020) Introduction. In: Effective science communication, 2nd edn. IOP Publishing, pp 1–5. https://doi.org/10.1088/978-0-7503-2520-2ch1

Iskarius J (2020) The role of investors in video game development [@article{iskarius2020role, title={THE ROLE OF INVESTORS IN VIDEO GAME

DEVELOPMENT}, author={Iskarius, Joseph}, year={2020} }]. https://www.theseus.fi/bitstream/handle/10024/354807/Iskarius_Joseph.pdf?sequence=3

Jenny SE, Manning RD, Keiper MC, Olrich TW (2017) Virtual(ly) athletes: where esports fit within the definition of "sport". Quest 69(1):1–18. https://doi.org/10.1080/00336297.2016.1144517

Jiang R, Shao B, Si S, Sato R, Tsuneo J (2021) Health communication in games at the early stage of COVID-19 epidemic: A grounded theory study based on plague. Games Health J 10(6):408–419. https://doi.org/10.1089/g4h.2020.0135

Jore H (2011) Not just fun and games: the status-quo of commercial games in teaching. DiGRA Conference

Kanyapasit P, Srisawasdi N (2014) Development of digital game-based biology learning experience on cell cycle through DSLM instructional approach. Workshop proceedings of the 22nd international conference on computers in education, ICCE 2014, pp 857–866

Kawrykow A, Roumanis G, Kam A, Kwak D, Leung C, Wu C, Zarour E, Players P, Sarmenta L, Blanchette M (2012) Phylo: a citizen science approach for improving multiple sequence alignment. PLoS One 7(3):e31362

Kikkawa T (2022) In: Kikkawa T, Kriz WC, Sugiura J (eds) Subtle manipulation in games BT–gaming as a cultural commons: risks, challenges, and opportunities. Springer Nature Singapore, pp 171–183. https://doi.org/10.1007/978-981-19-0348-9_10

Kobiela F (2018) Should chess and other mind sports be regarded as sports? J Philosophy Sport 45(3):279–295. https://doi.org/10.1080/00948705.2018.1520125

Kriz WC (2020) Gaming in the time of COVID-19. Simul Gaming 51(4):403–410. https://doi.org/10.1177/1046878120931602

Kuhn S (2021) Prisoner's dilemma. In: Zalta EN (ed) The {Stanford} Encyclopedia of Philosophy ({F}all 202). Metaphysics Research Lab. Stanford University

Lage MJ, Platt GJ, Treglia M (2000) Inverting the classroom: a gateway to creating an inclusive learning environment. J Econ Educ 31(1):30–43. https://doi.org/10.1080/00220480009596759

Laycock J (2015) Dangerous games: what the moral panic over role-playing games says about play, religion, and imagined worlds. University of California Press

Liz H (2003) Group narration: power, information, and play in role playing games. https://www.darkshire.net/jhkim/rpg/theory/liz-paper-2003/

Marston HR (2013) Digital gaming perspectives of older adults: content vs. interaction. Educ Gerontol 39(3):194–208. https://doi.org/10.1080/03601277.2012.700817

Mason MF, Norton MI, Van Horn JD, Wegner DM, Grafton ST, Macrae CN (2007) Wandering minds: the default network and stimulus-independent thought. Science 315(5810):393–395. https://doi.org/10.1126/science.1131295

Mauri M, Cipresso P, Balgera A, Villamira M, Riva G (2011) Why is Facebook so successful? Psychophysiological measures describe a Core flow state while using Facebook. Cyberpsychol Behav Soc Netw 14(12):723. https://doi.org/10.1089/cyber.2010.0377

Mitchell S, Hamilton SN (2017) Playing at apocalypse: Reading plague Inc. in pandemic culture. Convergence 24(6):587–606. https://doi.org/10.1177/1354856516687235

Moursund D (2006) Introduction to Using Games in Education: A Guide for Teachers and Parents (Vol. 06)

Murch WS, Ferrari MA, McDonald BM, Clark L (2020) Investigating flow state and cardiac pre-ejection period during electronic gaming machine use. Front Psychol 11:300. https://doi.org/10.3389/fpsyg.2020.00300

Nah FF-H, Eschenbrenner B, Zeng Q, Telaprolu VR, Sepehr S (2014) Flow in gaming: literature synthesis and framework development. Int J Inf Manag Sci 1(1–2):83–124. https://doi.org/10.1504/IJISAM.2014.062288

Nerueiter D, Klieser E, Neumayer B (2020) Feasibility of kahoot! as real-time assesment tool in (Histo-) pathology calssroom teaching. Adv Medical Education Practice 11:695–705. https://doi.org/10.2147/AMEP.S264821

Neureiter D, Klieser E, Neumayer B, Winkelmann P, Urbas R, Kiesslich T (2020) Feasibility of kahoot! As a real-time assessment tool in (Histo-)pathology classroom teaching. Adv Med Educ Pract 11:695–705. https://doi.org/10.2147/AMEP.S264821

Nicolescu B (2014) Methodology of Transdisciplinarity. World Futures 70(3–4):186–199. https://doi.org/10.1080/02604027.2014.934631

NinjaTheory. (2017) Helllblade Senua's Sacrifice. Ninja Theory

Parry J (2019) E-sports are not sports. Sport Ethics Philos 13(1):3–18

Pellis SM, Pellis VC (2017) What is play fighting and what is it good for? Learn Behav 45(4):355–366. https://doi.org/10.3758/s13420-017-0264-3

Peng L-H, Bai M-H (2021) How Gameful experience affects public knowledge, attitudes, and practices regarding COVID-19 among the Taiwanese public: cross-sectional study. JMIR Serious Games 9(2): e26216. https://doi.org/10.2196/26216

Pennycook G, Rand DG (2021) The psychology of fake news. Trends Cogn Sci 25(5):388–402. https://doi.org/10.1016/j.tics.2021.02.007

Pigliucci M (2018) The problem with scientism. Blog of the APA. https://blog.apaonline.org/2018/01/25/theproblem-with-scientism/

Popil I (2011) Promotion of critical thinking by using case studies as teaching method. Nurse Educ Today 31(2): 204–207. https://doi.org/10.1016/j.nedt.2010.06.002

Qin J, Li R, Raes J, Arumugam M, Burgdorf KS, Manichanh C, Nielsen T, Pons N, Levenez F, Yamada T, Mende DR, Li J, Xu J, Li S, Li D, Cao J, Wang B, Liang H, Zheng H, Consortium M et al (2010) A human gut microbial gene catalogue established by metagenomic sequencing. Nature 464(7285):59–65. https://doi.org/10.1038/nature08821

aan het Rot H, Kortmann R, Verbraeck A, de Vries G, Kooreman M, de Bruin B (2022) In: Dhar U, Dubey J, Dumblekar V, Meijer S, Lukosch H (eds) Facilitated tabletop games in a mediated environment BT–gaming, simulation and innovations: challenges and opportunities. Springer International Publishing, pp 28–41

Robey D (1973) Structuralism: an introduction. Clarendon Press, Oxford

Ruggill JE, McAllister KS (2011) Gaming matters: art, science, magic, and the computer game medium, pp 1–155

Sandlin J (2005) Andragogy and its discontents: an analysis of andragogy from three critical perspectives. PAACE J Lifelong Learn 14

Saxe GB (1988) The Mathematics of child street vendors. Child Dev 59(5):1415–1425. https://doi.org/10.2307/1130503

Seraphine F (2018) Ludonarrative dissonance: is storytelling about reaching harmony. The University of Tokyo

Sholz M, Niesh H, Sfeffen O, Ernst B, Loeffler M, Witruk E et al (2008) Impact of chess training on mathematics performance and concentration ability of children with learning disabilities. Int J Spec Educ 23:138–156

Shubik M (1983) Chapter II–gaming: a state-of-the-art survey. In: Ståhl I (ed) Operational gaming an international approach. Pergamon, pp 15–22. https://doi.org/10.1016/B978-0-08-030870-8.50008-4

Sinclair B (2017) Biology game niche thrives by going against the flow. Gameindustry.Biz. https://www.gamesindustry.biz/articles/2017-10-12-biology-game-niche-thrives-by-going-against-the-flow

Smith J (2012) Guide to the corrupted blood plague documentation collection. Stanford University Library

Stachyra P (2021) Your knowledge has increased!–gaming in Science communication and higher education. Liverpool John Moore

Stray Fawn Studio (2017). Niche: a genetic Surivial game. Stray Fawn Studio. www.niche-game.com

Strickland H, Kaylor S (2016) Bringing your a-game: educational gaming for student success. Nurse Educ Today 40. https://doi.org/10.1016/j.nedt.2016.02.014

Sweeney M (2021, March 19). Lockdown boredom drives UK video games market to £7bn record high. The guardian. https://www.theguardian.com/games/2021/mar/19/lockdown-boredom-drives-uk-video-games-market-to-7bn-record-high

Taddicken M, Reif A (2020) Between evidence and emotions: emotional appeals in Science communication. Media Commun 8(1):101–106. https://doi.org/10.17645/mac.v8i1.2934

Tanghe KB (2016) Homo Ludens (1938) and the crisis in the humanities. Cogent Arts Humanities 3(1):1245087. https://doi.org/10.1080/23311983.2016.1245087

Toh W (2021) The economics of decision-making in video games. Game Stud 21(3)

University of Dundee (2012) Molecular biology boardgame. https://www.lifesci.dundee.ac.uk/studying/media/videos/molecular-biology-boardgame

Valve Software (2021) Steam online store

Vand Der Stuyf R (2002) Scaffolding as a teaching strategy. Adolescent Learning and Development

von Hilgers P, Benjamin R (2012) War games. The MIT Press. https://doi.org/10.2307/j.ctt5vjpbj

Vorderer P, Hartmann T, Klimmt C (2003) Explaining the enjoyment of playing video games: the role of competition. Proceedings of the Second International Conference on Entertainment Computing, 1–9

Walker SE (2003) Active learning strategies to promote critical thinking. J Athl Train 38(3):263–267. https://pubmed.ncbi.nlm.nih.gov/16558680

Wan C-S, Chiou W-B (2006) Wan, Chin-sheng. Cyberpsychol Behav Soc Netw 9(3):317. https://doi.org/10.1089/cpb.2006.9.317

Wankel C, Blessinger P (2012) Increasing student engagement and retention using immersive interfaces: virtual worlds, gaming, and simulation

Wareham Morris K, McGowan J (2018) Gamified learning: a rationale for ludo-pedagogy and gamified classrooms. Worcester J Learn Teach 12:23–31

Witkowski W (2021) Videogames are a bigger industry than movies and North American sports combined, thanks to the pandemic. Marketwatch.Com. https://www.marketwatch.com/story/videogames-are-a-bigger-industry-thansports-and-movies-combined-thanks-to-the-pandemic-11608654990

Wong MM, Csikszentmihalyi M (1991) Motivation and academic achievement: the effects of personality traits and the duality of experience. J Pers 59(3):539–574. https://doi.org/10.1111/j.1467-6494.1991.tb00259.x

Yao K, Huang S (2020) Simulation technology and analysis of military simulation training. J Phys Conf Ser 1746:012020. https://doi.org/10.1088/1742-6596/1746/1/012020

Zimmerman E (2013) Quantum. Funforge. https://boardgamegeek.com/boardgame/143519/quantum/credits

Zimmerman E, Fortugno N (2016) Soapbox: learning to play to learn–lessons in educational game design. Gamasutra. https://www.gamasutra.com/view/feature/130686/soapbox_learning_to_play_to_learn_.php%0A; http://www.gamasutra.com/view/feature/2273/soapbox_learning_to_play_to_learn_.php

Increasing Microbiology Literacy about the Public Health Threat of Antimicrobial Resistance Through Art-Science Interactions

9

Natasha Niethamer, Mark Roughley, and Raechelle D'Sa

Abstract

In response to the recognised urgent global public health threat of antimicrobial resistance (AMR), the cause of drug-resistant infections, art-science activities are increasingly utilised as public engagement (PE) and public outreach (PO) tools to increase microbiology literacy. Though AMR art-science activities aim to increase public awareness, research evaluating the platform as one conducive to the aims of AMR engagement, including long-term behavioural change, is lacking. The aim of this research study is to explore interactive art-science activities aiming to promote microbiology literacy in AMR. Existing activities lack critical evaluation, and development details are not shared with wider audiences. The SARS-CoV-2 pandemic introduced a global shift to online platforms, driving the necessity to reevaluate current PE/PO methods and integrate interactive tools. Two AMR SciArt activities exploring gamification and interactive fiction are described and analysed as case studies. Both were informed by the existing research and evidence-based theories in this chapter, and operated in pre- and post-SARS-CoV-2 pandemic limitations. These activities' methods show promise, however, the most effective strategy in AMR PE is yet to be defined. By establishing a standardised framework and sharing evaluated activities based on evidence-based learning research, more work can be done to inform and strengthen future activities.

Keywords

Antimicrobial resistance · Microbiology literacy · Public engagement · Science communication · Comics, Health literacy

9.1 Introduction

In 2019, Timmis et al. emphasised the urgent need to improve microbiology literacy in society, stating that 'unlike other subjects having a pervasive impact upon humankind, such as financial affairs, health, and transportation, of which there is a widespread understanding, knowledge of relevant microbial activities, how they impact our lives, and how they may be harnessed for the benefit of humankind– microbiology literacy—is lacking in the general population, and in the subsets thereof that constitute the decision makers'. Antimicrobial Resistance (AMR) awareness is a critical aspect of microbiology literacy. AMR results in drug-resistant infections, and is

N. Niethamer (✉)
Liverpool School of Tropical Medicine, Liverpool, UK

M. Roughley
Liverpool John Moores University, Liverpool, UK

R. D'Sa
University of Liverpool, Liverpool, UK

M. Roughley (ed.), *Approaches for Science Illustration and Communication*, Biomedical Visualization 4,
https://doi.org/10.1007/978-3-031-41652-1_9

recognised as an urgent global health threat currently causing approximately 5000 annual deaths in England and over 700,000 annual deaths globally. This number is steadily increasing (Interagency Coordination Group on Antimicrobial Resistance 2019). The most recent systematic analysis of 2019's global burden data suggests that in 2019, there were 4.95 million deaths associated with bacterial AMR, with 1.27 million of these deaths directly caused by bacterial AMR (Murray et al. 2022). Despite the significant global burden of AMR, there remain misunderstandings in the general population of how an individual can take action to help slow and prevent AMR (World Health Organization 2015).

Public engagement (PE) and public outreach (PO) are considered promising methods to slow AMR rates by increasing public awareness of the topic (National Co-ordinating Centre for Public Engagement (NCCPE) n.d.-a; Illingworth and Allen 2016). PE requires two-way interaction between the public(s) and practitioner while PO involves one-way communication of research and higher education-based topics (Illingworth and Allen 2016). 'Art activism' is a method involving PO and PE showing promise in communicating public health-related topics with the aim of long-term behavioural changes such as AMR. The scientific consensus is that both learning and long-term memory are facilitated by emotional and therefore meaningful connection (Davies 2019). For these reasons, it is important to continue creating AMR engagement projects, strengthening, and pushing the boundaries of current practice.

There are many mechanisms that can influence public knowledge and attitudes, with a wealth of research in which the arts have been used (referenced in Curtis et al. 2012), however, the most effective strategy in AMR PE is yet to be defined. For this reason, there is a call for the exploration of innovative AMR PE approaches, as long as they are published and evaluated to fast track development of a strong framework (Redfern et al. 2020).

To evaluate the effectiveness of interactive art-science activities as a tool for promoting AMR awareness and microbiology literacy in the public realm there must exist a standard evaluation and key aim to compare, contrast, and therefore strengthen activity methods (Redfern et al. 2020). Due to the lack of standardised design framework and evaluation in this field, this chapter explores existing literature that may inform the development of an effective AMR PE and outreach activity design framework. Although research literature on AMR PE and outreach activities are relatively new and not yet well developed, research undertaken in other fields may offer considerable insight into factors contributing to effective AMR awareness activities.

It is important to consider the influence of the SARS-CoV-2 pandemic on public perception of microbiology when considering future AMR outreach approaches. This pandemic unprecedentedly forced the world into collective immersion of microbiology and health education through pandemic social messaging, resulting in drastic, widespread infection prevention behavioural changes, such as mask wearing, hand washing, sanitising, and social distancing. Due to this immersion, the public baseline understanding of SARS-CoV-2 and microorganisms' critical role in public health has increased (Read et al. 2022). However, fear surrounding SARS-CoV-2 infection may drive misconceptions about microorganisms, specifically the belief that all microorganisms are pathogens causing disease that need to be destroyed; less than 1% of microorganisms are pathogens with the rest harmless, helpful, or critical to human and environmental health and need to be understood to inform decisions at a personal and global policy level (Carvalho and Lima 2022; Robinson et al. 2021; Timmis et al. 2019).

Negative perceptions driven by misunderstandings about microorganisms coupled with an increased baseline in the publics' microbiology literacy level, specifically a greater understanding of hand washing's role in infection prevention, may require re-evaluation of key education aims of future microbiology art-science activities. This may allow greater complexity in PE/PO art-science AMR awareness activities, but it also may require practitioners to focus on unravelling additional misunderstandings associated with microorganisms while emphasising positive roles.

Furthermore, the SARS-CoV-2 pandemic transformed communication methods utilised in public education. The lockdown restrictions forced educators to shift delivery from traditional, direct communication to one heavily utilising technology and online platforms. Educators were challenged to create engaging online content that facilitated learning despite the lack of face-to-face interaction. This has driven innovative approaches in online education. A technique utilised to promote greater student engagement and reinforce knowledge includes integrating more interaction through gamification, incorporating game motivation techniques and mechanics into non-game subjects (Joshi 2021; Lee and Hammer 2011). Comics were also increasingly utilised, viewed as a natural medium for public health education during the pandemic, 'as a vehicle to present science in graphic narratives, harnessing the power of visuals, text, and storytelling in an engaging format' and as a medium established as having great accessibility to low literacy levels in both children and adults (Kearns and Kearns 2020).

Despite limitations to online learning, it is a major source of health literacy for the global public and considered a valuable tool to connect with a wider audience and disperse public health information (Jia et al. 2021). With the pandemic driving innovative, interactive ways to better engage online, it is important to learn from these ideas and build upon identified strengths. With face-to-face interaction no longer restricted, continuing to utilise supplementary online platforms may provide a valuable resource to widen PE/PO activities' impact and reach from an isolated event to a globally accessible online resource. This is especially critical in regards to the public health threat of AMR, which, to prevent unneeded deaths and economic burden, urgently requires global, collective efforts and widespread policy changes (Tarín-Pelló et al. 2022).

Two case studies utilising art-science practice to increase AMR awareness will be compared and contrasted to provide insight into the benefits and limitations of shifting methods utilised in art-science activities in response to the SARS-CoV-2 pandemic. One activity was developed and enacted pre-pandemic while the other activity was developed in response to pandemic restrictions.

9.2 Literature Review

This literature review aims to understand why art-science activities are considered impactful, and determine their potential as an effective PE/PO tool that should continue being utilised by antimicrobial resistance (AMR) activists. First, the demand for AMR public art-science activities will be described. Then, art-science activities will be defined to understand how and why this platform developed into an activism tool. Following this, literature relating to art-science activities promoting increased microbiology literacy in AMR will be analysed for their strengths and weaknesses. Finally, two case studies will be evaluated to compare and contrast interaction methods used pre-COVID-19 pandemic versus restricted in response to the pandemic.

Due to the lack of published evaluations relating to AMR PE activity design and impact, this chapter will refer to related literature aimed to increase awareness of similar, activism-based topics requiring global policy changes and collective action. By studying activities based on similar topics, this chapter aims to provide insight into the strengths and weaknesses of the PE platform based on more established research. The literature review will describe research undertaken in other fields including social sciences and public health to gain insight into the factors contributing to effective AMR PE activities. Common themes in successful art-science activism activities will be identified and will contribute to the proposition of a framework yet to exist for those wanting to develop their own AMR PE/PO activities. Finally, the existing art-science activities that ran pre-pandemic and in response to pandemic restrictions, both promoting microbiology literacy in AMR awareness, will be detailed and analysed as case studies.

9.2.1 What Is Antimicrobial Resistance?

Antimicrobial Resistance (AMR) results in drug-resistant infections and is recognised as an urgent global health threat, causing approximately 5000 annual deaths in England and over 700,000 annual deaths globally. This number is steadily increasing (Interagency Coordination Group on Antimicrobial Resistance 2019). AMR occurs when microscopic organisms that can cause infections such as bacteria, viruses, parasites and fungi develop weapon-like abilities allowing them to fight and defeat antimicrobial drugs like antibiotics, normally used to fight infections. These mutated, drug-resistant microorganisms are called 'superbugs'. When they are not killed, they continue to grow and spread until the person they have infected is either treated by a different method, has overcome the infection, or dies (World Health Organization 2018). AMR occurs due to misuse of antimicrobials where antibiotic treatment is stopped early when antimicrobials are overused in humans, animals, and plants, or due to poor hygiene (Interagency Coordination Group on Antimicrobial Resistance (IACG) 2019, p. 3).

It is predicted that without intervention, in 30 years, drug-resistant infections, no longer treatable by antibiotics, may cost 100 trillion US dollars and take the lives of up to ten million people each year (Review on Antimicrobial Resistance 2016, p. 15). This would overturn modern medicine making simple medical procedures like surgery, caesarean sections, organ transplants, and cancer therapy life-threatening with significant economic burden (World Health Organization 2018). A systematic analysis of global burden data suggests that in 2019, there were 4.95 million deaths associated with bacterial AMR, with 1.27 million of these deaths directly caused by bacterial AMR (Murray et al. 2022).

9.2.1.1 Why Is Antimicrobial Resistance Public Engagement and Outreach Important?

In 2016, The United Nations Global Assembly politically declared AMR as an established priority for intervention, securing it within the global political agenda (United Nations General Assembly 2016). In this declaration, 'Member States requested the Secretary-General, in consultation with the Food and Agriculture Organization of the United Nations (FAO), the World Organisation for Animal Health (OIE) and the World Health Organization (WHO) to convene an ad hoc interagency coordination group (IACG)'; the Interagency Coordination Group (IACG) on AMR was appointed to generate common direction among global sectors, industry, and nations (IACG 2019). IACG released an updated global initiative in 2019 urging stronger regulations to be put in place, because AMR rates can only be reduced by collective, worldwide action.

One key strategic objective out of the five proposed by the WHOs global action plan is to reduce AMR rates by raising the general public's awareness and understanding of this issue (WHO 2015 p. 8). This awareness is regarded as critical in driving long-term behavioural changes regarding misuse and overuse of antimicrobials (IACG 2019; WHO 2015 p. 8). Since this plan was proposed, many institutions and organisations have designed and delivered AMR-based activities termed 'public engagement' or 'public outreach' activities (Redfern et al. 2020). Additionally, The Society for Applied Microbiology (SfaM), the oldest microbiology society in the UK whose goal is to use microbiology to address challenges significant to humankind, calls for the 'urgent need for microbiology literacy in society' as it is currently lacking. They argue that 'microbiology literacy in society is indispensable for informed personal decisions, as well as for policy development in government and business, and for knowledgeable input of societal stakeholders in such policymaking', specifically emphasising the importance of microbiology literacy regarding AMR (SfaM 2019; Timmis et al. 2019). This lack of microbiology literacy in society further drives the necessity to analyse and develop strong AMR public engagement activities.

Public engagement (PE) and public outreach (PO) are considered promising methods to slow AMR rates by increasing public awareness of the topic. PE is defined as the 'myriad of ways in which the activity and benefits of higher education and research can be shared with the public' in a way that allows two-way interaction, conversation, and mutual benefit (National Co-ordinating Centre for Public Engagement (NCCPE) n.d.-a). PO is defined as, 'a one-way discourse, in which scientists communicate their research to the general public, with particular focus on schoolchildren and young people' (Illingworth and Allen 2016). Though we refer to the general public as a singular term, it should be noted that the public is commonly referred to as 'publics' due to there existing many different target groups within the 'general public' that may require different considerations when choosing a PE method.

Showing widespread public interest in an issue may adjust long-term behaviour to one of appropriate antimicrobial use, generate more time/funding to support further research in alternative treatments to drug-resistant infections and antibiotic discovery, target important stakeholders such as those in the health profession to adjust prescription behaviour, and push political leaders to set in motion public policies needed to make a significant, lasting change (Anderson et al. 2020 p. 47; Hoffman and Outterson 2015 p. 364; SfaM 2019; Timmis et al. 2019). For these reasons, it is important to continue creating AMR engagement projects, strengthening, and pushing the boundaries of current practice.

9.2.1.2 Art-Science Public Engagement Activities and Art Activism

'Art activism' is a PE/PO method showing promise in communicating public health-related topics with the aim of long-term behavioural changes such as AMR. The scientific consensus is that both learning and long-term memory are facilitated by emotional and therefore meaningful connection (Davies 2019). With this basis in mind, 'artistic activism' developed as a promising PE method in which the aim 'is not to force compliance, which art can never do, but to persuade by creating moving experiences that prompt people to question the world as it is, imagine a world as it could be, and join together to make that new world real' (Guerilla Science 2018). With the mounting evidence supporting science communication as 'emotion work' in which wonder and curiosity is harnessed to facilitate effective science communication, 'art activism' is a term that informs this movement (Davies 2019). Thus, science and art practitioners have collaborated in creating art-science public engagement activities in hopes to adjust the targeted publics' understanding of behaviour in some way.

Art in combination with the sciences has been shown to be effective in communicating complex concepts of science to a broader audience by promoting curiosity in a subject and resulting in viewing a concept in a different way (Grimberg et al. 2019; Jensen and Buckley 2014). Art-science or 'SciArt' activities have also been found a potential catalyst for scientific advocacy and community action as supported in research by Curtis et al. (2012) who described the role of art in engaging the public in communicating 'unpalatable' urgent messages about climate change with the aim of inspiring collective public understanding, behaviour change, and action. Curtis et al. (2012) represented climate change issues through visual/performing arts (theatre, music, and dance) and it was concluded based on surveys that the arts should be harnessed to 'convey complex scientific information, promote new ways of looking at issues, touch people's emotions, and create a celebratory atmosphere'.

There are many mechanisms in which to influence public knowledge and attitudes with a wealth of research in which the arts have been used (referenced by Curtis et al. 2012), however, the most effective strategy in AMR PE is yet to be defined. For this reason, there is a call for the exploration of innovative AMR PE approaches, as long as they are published and evaluated to fast-track the development of a strong framework (Redfern et al. 2020).

9.2.1.3 Gaps in Current Antimicrobial Resistance, Public Engagement, and Outreach Activities

> Few of these events for a general public audience are documented or able to be sourced by those who might wish to repeat, adapt, or modify, particularly those events that are successful. 'Insider knowledge' appears to be the best search tool (Redfern et al. 2020).

The first systematic review summarising and analysing AMR-specific PE and PO activities was published in March 2020 by a Manchester Metropolitan University-based research group (Redfern et al. 2020), and it outlined the range of AMR PE activities delivered in the UK (2020). The study recognised that a wide range of creative AMR public engagement and outreach activities exist worldwide, however, there remains a lack of published papers on these activities, and those that are published rarely include success and impact evaluations. Their review concludes that 'implicit in all of these activities should be the identification of the key hypothesis/message and the particular audience, the design and delivery of the activity, and an evaluation of its success'. Only with critical, standardised, and published evaluation can activities be meaningfully compared and an evidence-based framework to adjust and strengthen future activity impact be built (Redfern et al. 2020).

Redfern et al. (2020) further note the broad range of AMR art-science interactive public activities that exist on a variety of platforms, including film, journalism, online games, festivals, and museum exhibitions. Although the goals of these AMR art-science activities are similar, their impacts are affected by different factors, including the audience type and key aim of the activity. The NCCPE suggests that tailoring an activity based on these factors is critical to effective PE and PO (NCCPE n.d.-b). Effective communication changes based on your target audience's preliminary understanding of the topic and established belief system (NCCPE n.d.-c).

9.3 Defining the Key Aim of AMR Art-Science Public Engagement and Outreach Activities

Redfern et al.'s conclusion, as mentioned above, identified that 'implicit in all of these (AMR public engagement and outreach) activities should be the identification of the key hypothesis/message and the particular audience, the design and delivery of the activity, and an evaluation of its success' (2020). Supported by the findings of Redfern et al. (2020), the WHO's global action plan on AMR (2015), and IACG (2019), we propose two key aims of AMR PE and outreach activities which will contribute to slowing down AMR rates, a key component of tackling the urgent global public health concern of AMR:

1. Awareness of AMR, specifically regarding common misconceptions in the public.
2. Long-term behavioural adjustment regarding misuse and overuse of antimicrobials.

The following section evaluates the effectiveness of existing AMR public engagement and outreach activities based on these key aims by interrogating existing 'behavioural interventions' attempted through AMR activities to result in effective AMR awareness. If a strong theme in method/methodology is identified, this could be used to formulate an AMR art-science activity framework.

9.3.1 How Effective Are Current AMR Activities to Long-Term AMR Awareness and AMR Stewardship?

Since the WHO global action plan was established, many AMR-related activities have been funded and developed with the overarching goal of effective 'behavioural intervention' of the public in two main ways: increasing the public's awareness of AMR and determining long-lasting AMR stewardship behaviour (Redfern et al. 2020; Price et al. 2018). AMR stewardship

(AMS) behaviour in the context of PE activities refers to responsible, AMR-informed actions that limit antibiotic overuse and misuse. Increasing the public's awareness of AMR can be tested through pre- and post-activity knowledge gain. The latter is more difficult to test because it involves a follow-up survey assessing long-term behavioural changes as a result of an intervention (Price et al. 2018). When discussing 'effectiveness of behavioural interventions' or mentioning 'effectiveness' in regards to AMR PE in the subsequent sections, the two criteria above established by Price et al. (2018) in their landmark analysis of evaluating the effectiveness of AMR interventions will be referred to.

In 2015, researchers from the Department of Health in Turin, Italy published findings suggesting that the general global population's knowledge and usage surrounding antibiotics is lacking by performing the first systematic review and meta-analysis of analysed surveys from January 2000 to November 2013 (Gualano et al. 2015). They found 33.7% of the sample subjects were unaware that antibiotics can treat bacterial infections, 53.9% of the sample did not know antibiotics are not useful against viruses, 47.1% of the sample subjects declaring that they stop taking antibiotics when they start to feel better, and though 59.4% of the sample 'was aware of antibiotics resistance, 26.9%' did not realise antibiotic misuse causes this resistance. The researchers recognise their limitations including the lack of standardised questionnaire administration and controlled sampling selection that may lead to biases.

McCullough et al. (2015) from the Centre for Research in Evidence-Based Practice in Australia, performed a similar systematic review of literature as Gualano et al. (2015) regarding the global public's knowledge about antibiotic resistance using a different methodology, and came to the same conclusion. They also address the limitations of their study being the lack of standardised evaluation methods. Nevertheless, these studies have since been referenced as evidence of the need for increased AMR public engagement as they suggest that there may be a lack of antibiotic resistance awareness in society, supporting the need for effective public engagement interventions to be explored and developed (Redfern et al. 2020).

A suggested lack of AMR awareness in the general public led to the landmark analysis by Price et al. in 2018, assessing if AMR behavioural interventions to date have resulted in significant improvements regarding the public's AMR knowledge and AMR stewardship behaviour (Price et al. 2018). This research project from the Institute for Applied Health Research at Glasgow Caledonian University compiled all existing 'interventions' and found that a majority targeting the general public were effective in improving AMR knowledge and antimicrobial stewardship behaviour. Some, however, did not show behaviour/knowledge improvement with two studies showing a decrease in AMR knowledge/antimicrobial stewardship behaviour. Price et al. (2018) recognised that these results suggest certain PE/PO methods may facilitate effective activities, and this is an area which should be further explored. Additionally, they recognised limitations of their analysed data including the findings being heterogeneous and complex with no distinction regarding PE and PO platform/method. They 'highlight the need to examine differences in content between these interventions targeting the general public' (Price et al. 2018).

Price et al. (2018) established that the results of behavioural interventions regarding AMR are mixed with the public continuing to show misconceptions and lack of knowledge, and a few months later, McParland et al. (2018), a research group containing several of the same researchers as in Price et al., responded by publishing the first formal analysis of AMR PE activities to determine 'the 'active ingredients' of interventions targeting the public's engagement with antimicrobial resistance' and how they might work. They found that national, European, and global AMR awareness campaigns used a wide range of methods including factsheets, posters, videos, social media, and interactive games, however, no studies 'identify exactly what works, why it works, for whom, when, and in which circumstances' making it 'impossible to determine which intervention components are typically used or indeed which are associated with effectiveness' which in turn makes it 'impossible for others to understand, evaluate, and replicate the active content of behaviour change interventions' (McParland et al. 2018).

Both research groups stated that a major limiting factor in their analyses was that most studies did not base their intervention designs on established behavioural change theories (BCT) in the field of Psychology and the studies that have were found only to have employed 15% of possible BCT's (McParland et al. 2018). BCTs are evidence-based interventions that are considered by the Medical Research Council as the key intervention components and the 'mechanisms of change' when understanding, developing, and evaluating complex interventions such as AMR (Craig et al. 2008). McParland et al.'s work concluded that BCTs should be referenced and employed when designing AMR PE activities with the goal of long-term behavioural change in participants as evidence-based methodology versus 'designed with implicit theoretical assumptions in mind'. McParland et al. (2018) suggest that BCTs in the most successful interventions include 'promoting beliefs about capability and reinforcing behaviour, as well as encouraging a commitment to change behaviour and imagining future outcomes if behaviour is not changed, monitoring behaviour (with and without feedback) and providing information on the antecedents of behaviour', but argue that due to the heterogeneity of the interventions found and a majority of BCT's yet to be harnessed, these factors are not definitive, and a 'new generation of different interventions' using different combinations of BCT's are needed.

9.3.2 Learning from Ineffective AMR National Engagement Campaigns

> The study highlights that there has been *no change in the status quo with respect to awareness of antibiotic usage and AR even after the implementation of several awareness campaigns* in England. Overall, public respondents residing in *affluent areas possessed much better knowledge and understanding on AR problems and antibiotic usage* and this was statistically significant.... In addition, public respondents *in deprived areas possessed much poorer knowledge on prudent antibiotic usage* compared to those in affluent areas (Mason et al. 2018).

In attempts to raise AMR awareness and reduce inappropriate antibiotic use in the general public, Public Health England launched several nationwide campaigns including one in 2008 and one in 2014 called 'Antibiotic Guardians', part of European Antibiotic Awareness Day (EAAD). However, there was no significant evidence of long-term behaviour change or increased AMR knowledge in the general public between these time frames (McNulty et al. 2010; Bhattacharya et al. 2017). In 2018, Mason et al. surveyed 768 residents of London and suggested that there was no significant change in the publics' AMR awareness since a study completed by McNulty et al. (2010) that was undertaken 10 years prior, despite implementation of Public Health England's campaigns. Analysis of the survey results suggested that 'any efforts to date to educate the general public have not made a remarkable impact' and 'exposure to the antibiotic awareness campaigns did not enrich knowledge of those respondents on how to take an antibiotic correctly, as the responses did not differ statistically to those who had not been exposed to campaigns' (Mason et al. 2018). This study also found a significant difference in AMR understanding between those from affluent areas versus deprived areas. Residents of affluent areas possessed greater knowledge of AMR compared to those of deprived areas. Despite these findings, both residents of affluent and deprived areas had misunderstandings of AMR (Mason et al. 2018). These findings highlight the need for additional toolkits concentrated on engaging the public, particularly in disadvantaged communities with lower socio-economic status and education, with the topic of AMR.

Utilising established engagement methods, how can AMR awareness campaigns be improved? The Public Health England campaign,

'Antibiotic Guardians' (AG), utilised informational pamphlets, posters, a website, and social media content in which people are prompted to sign up as an 'AG', to pledge their support to AMR and to promote discussion with their network (see https://antibioticguardian.com/). Despite pledges totalling '185,325' in 2023, a qualitative evaluation of this campaign by Kesten et al. found that a majority of AG's were healthcare professionals and those with pre-existing AMR knowledge with established supportive behaviours (2018). The study suggests that pledging as an AG succeeded in reinforcement of pre-existing behaviours and AMR understanding, however, it did not reach its intended 'general public' audience, those with limited AMR knowledge, and it did not significantly change behaviours. Furthermore, the perceived visibility of the AG campaign was referred to as 'lacking' among the public and among interviewed healthcare workers, with some failing to remember the 'one simple pledge' AG's signed up with. Several AG's interviewed commented that 'they were not the most appropriate target for the campaign and that the campaign is "preaching to the converted' with some GP's giving negative feedback, viewing the campaign as 'patronising' (Kesten et al. 2018). These findings suggest that engagement of non-scientist public is lacking and finding ways to widen reach using evidence-based methods must be addressed going forward.

Analysing the 'Antibiotic Guardian' website and resources available, the colourway is restricted to mainly black, red and white, and the educational materials are primarily blocks of text, defining terms and statistics with little to no visualisations or intuitive interactivity (Antibiotic Guardian 2023). The UK's National Health Service (NHS) states that '6 out of 10 adults struggle with health content that includes numbers and statistics' with there being a 'close link between socio-economic deprivation and low health literacy' (NHS 2021). To engage a wider audience, platforms that are inclusive and accessible are important to explore, especially those mediums that are research-backed as effective learning and integrate behaviour change techniques (BCT) in the non-science literate. The Office of Disease Prevention and Health Promotion (OASH) suggests in their research-based health literacy guide that to increase health literacy, interactivity, graphics, recall questions, and multimedia that meaningfully appeals to the targeted are essential when communicating public health issues and motivating those to take action (OASH 2016). The UK's National Institute for Health and Care Research (NIHR) emphasises OASH guidelines, for health information to be as widely accessible as possible, 'a crucial element of tackling health inequality and misinformation. This was demonstrated by the COVID-19 pandemic' (NIHR 2022).

There are many mechanisms in which to influence public knowledge and attitudes, however, the most effective strategy in AMR PE is yet to be defined. For this reason, there is a call for the exploration of innovative AMR PE approaches (Redfern et al. 2020).

In response to McParland et al.'s (2018) call for innovative combinations of BCTs, the need for AMR PE practitioners to cite evidence-based methodologies, and shifting methods utilised in art-science PE/PO activities in response to the SARS-CoV-2 pandemic, two case studies utilising art-science practice to increase AMR awareness are compared and contrasted below to provide insight into the benefits and limitations of such projects. One was developed and enacted pre-pandemic while the other was developed in response to pandemic restrictions, using a previously unexplored online platform in microbiology communication, to support microbiology PO. They both aim to challenge differing levels of understanding about AMR by developing narratives that delve into understanding norms, beliefs and practices about bacterial infections, antibiotic use and hygiene, through interactive art platforms.

9.4 Case Studies

The following case studies: *Antimicrobial Avengers!* and *Bacteria, the Future: A time travelling Twine game to save Hugh-manity!* are the results of artist-scientist collaborations. They are PE/PO activities developed to increase AMR awareness and promote microbiology literacy through evidence-based learning research, utilising playfulness, storytelling, and interaction to promote wonder and natural curiosity, aspects that facilitate learning (Davies 2019; Timmis et al. 2019). Each case study includes detail about the production of the activities; identifying successes, limitations, evaluation methods, and recorded impacts. Both cases are part of a wider intention to develop public engagement/outreach toolkits that may be utilised as a framework for effective future PE and PO art-science projects. The first case study took place pre-SARS-COV-2 pandemic restrictions, and the second case study was created with pandemic lockdown limitations.

9.4.1 Case Study 1: *Antimicrobial Avengers!*—Antimicrobial Resistance, Public Engagement with Comics, and Co-creative Activities

> Did you know that shark skin can prevent bacteria from sticking to it or that Komodo dragon blood is naturally antimicrobial? Liverpool University's D'Sa Lab and Liverpool School of Art and Design's MA Art in Science programme present the superheroes 'Shark Girl' and 'Komodo Dragon' versus the evil 'Superbugs' in the 'Antimicrobial Avengers!'

The aim of the *Antimicrobial Avengers!* project was to facilitate learning about microbes and antimicrobial resistance through co-creative participatory activities, which would encourage public dissemination of health information and research to friends and family members, and more widely within local communities.

9.4.1.1 A Co-creative Approach

Antonini (2021) defines co-creation as a cooperative process that brings people and professionals together to find innovative solutions for everyday problems. It is a common method used in arts-in-health practice, whereby arts-based activities can activate dialogue between stakeholders that are not based on assumptions (Padfield 2011). Unlike therapeutic arts-in-health practices where art-making is primarily for 'healing' (Tan 2020) and accessed when the individual is unwell (Gordon-Nesbitt 2017), *Antimicrobial Avengers!* instead adopted a participatory arts-based approach where activities, such as dance, sculpture, drawing, or photography are used in the conceptualisation, data collection, and dissemination of scientific research (Lapum et al. 2014) and the artist/researcher is the facilitator of the process (Archibald and Blines 2021).

Arts-based participatory approaches have been used to inquire about current understanding of health, to intervene in drivers of health problems, and to stimulate empathy by allowing people to situate themselves in specific scenarios (Bunn et al. 2020). Archibald and Blines (2021) state that 'arts-based health research offers unique opportunities to integrate evidence of patients' lived experience with other forms of research evidence to improve understanding and knowledge translation' and participatory arts-based health approaches have been noted to influence behaviour, and cognitive and linguistic competencies, especially with children (Gordon-Nesbitt 2017). Such approaches have been proven to successfully facilitate improved health literacy in communities that have varying concerns and understanding about health. An example includes using the arts to increase public health knowledge about the danger posed by the Ebola epidemic in sub-Saharan Africa and how to respond to symptoms (Bunn et al. 2020).

In addition, Melonio and Gennari (2012) advocate for the involvement of children in co-creative approaches, noting that 'adults do not experience the world as children do and do not have the same insights into the world as a child'. They also state that children are natural partners because they have 'unconventional viewpoints and are always ready to share ideas' (Melonio and Gennari 2012). The affordances of a participatory, arts-based, co-creative method led to this approach being implemented in the *Antimicrobial Avengers!* project.

9.4.1.2 A Graphic Medicine Approach

'Graphic Medicine' is a method of visual storytelling utilising comics. It shows promise as a valuable tool for presenting information on difficult or often misunderstood topics, particularly valued by adults with low literacy levels (Green and Myers 2010; Waite 2019). According to the National Literacy Trust, '16.4% of adults in England, or 7.1 million people can be described as having very poor literacy skills' meaning that reading from 'unfamiliar sources, or on unfamiliar topics' is difficult (National Literacy Trust 2019). To support a collective, global understanding of AMR, PE, and PO must target all levels of literacy, therefore practitioners must consider mediums accessible to lower literacy levels.

Comics are also increasingly utilised, viewed as a natural medium for public health education 'as a vehicle to present science in graphic narratives, harnessing the power of visuals, text, and storytelling in an engaging format' and as a medium established as having great accessibility for both children and adults (Kearns and Kearns 2020). In 2019, a group of microbiologists published an article citing research demonstrating the potential of comics in microbiology communication. They called for the development of comics communicating microbiology public health threats such as AMR with the caveat that, 'we have to make them publicly available and, most importantly, we have to evaluate their effectiveness in microbiology communication' (Scavone et al. 2019). Based on this information, *Antimicrobial Avengers!* sought to promote AMR public engagement through comics and co-creative activities.

9.4.1.3 Developing the *Antimicrobial Avengers!* Comics

Three comics were created as part of a wider AMR public engagement toolkit, and printed and digital online versions were produced. An artist in residence was hired to work alongside antimicrobial researchers in the D'Sa Laboratory at the University of Liverpool's School of Engineering to create a number of narratives that would explore antimicrobial biomaterials and resistance. Three research strands were chosen to be explored in the comics:

1. Researchers have developed coatings for high-touch surfaces to fight the spread and growth of microbes. For example, the skin of sharks is made up of a highly-ordered series of bars arranged in a diamond pattern that can fight off bacteria without antibiotics (Dundar Arisoy et al. 2018), and 'Sharklet', a product produced by Sharklet Technologies Inc. (https://www.sharklet.com/), is a coating that mimics shark's skin, reducing the ability of bacteria to stick to these surfaces.
2. Antimicrobial peptides (AMPs) are small molecules that are part of the immune system of various organisms, including humans, animals, and plants. They are one of the first lines of defence against invading pathogens. If you look at the saliva of Komodo dragons, they have a lot of pathogenic bacteria, but the dragons are unaffected, and this is because their blood contains AMPs that effectively fight off even the deadliest of bacteria (Bishop et al. 2017). These AMPs can be developed into drugs that can be added to our arsenal in the war against deadly infections.
3. Antibiotics were discovered approximately 100 years ago and are used to treat bacterial infections, however, the overuse and misuse of antibiotics have meant that bacteria have evolved in ways to become resistant to these drugs. The more antibiotic resistance that develops, the harder it becomes to defeat these stronger bacteria.

Following a literature review and detailed briefings between the artist in residence and the D'Sa Lab researchers, three partial storyboards comprising four 'frames' of a comic were produced for each research strand. These introduced each strand succinctly but left the endings unfinished. The intention was for a 'layered elaboration' approach for storyboarding and prototyping (Walsh et al. 2010) to be used in the next stage of the co-creation process. Working with local primary school pupils and the general public, the narratives would be developed in conjunction with the researchers and artists in residence.

Fig. 9.1 Drawing the endings to the three stories using partial storyboards during an outreach event in a South Liverpool primary school. Photograph by Prof. Raechelle D'Sa

Here, a spectrum of possibilities for the stories and their endings could be developed quickly, and without one being favoured over another (Walsh et al. 2010).

To co-create the comics, a number of events were scheduled to include diverse age and socio-economic groups in the co-creation and production of the stories. Two outreach events took place first in a local South Liverpool primary school in March 2019, where pupils could meet the researchers and artists in residence. The pupils watched presentations about the three research strands from D'Sa Lab researchers and were then provided the partially complete storyboards. They were asked to draw and/or write the endings to the three stories (Figs. 9.1 and 9.2) with opportunities to speak and learn from the researchers throughout. Comments from the school pupils and researchers likened the titular characters of each story as superheroes, a common comic book trope, and the *Antimicrobial Avengers!* name was chosen to group the three stories under one umbrella. The artist in residence then used the narratives provided by the school children to produce three mock 'completed' comic stories featuring each superhero that were to be used in the next public-facing event.

In May 2019, the research team and artist in residence took part in the public engagement event 'LightNight Liverpool' organised by Open Culture. Light Night is Liverpool's free one-night arts festival, where over 100 organisations collaborate annually to create an inspiring trail of events with talks, workshops, performance, live music, and more (https://lightnightliverpool.co.uk/). The *Antimicrobial Avengers!* Stand was installed in the Atrium Gallery at Liverpool John Moore's University's John Lennon Art and Design building. This building sits at the heart of Liverpool's Knowledge Quarter and was a central event hub during the evening, with over 500 visitors passing through the building. Here the public was encouraged to speak to the researchers, read the three comics and draw and write alternative endings to the stories produced by the school children (Figs. 9.3 and 9.4). It was hoped that this process would encourage and enhance learning beyond simply reading an already completed comic. Participants were also able to create their own microbes using craft materials (Fig. 9.5), touch sculptures of animals that have naturally antibacterial-resistant skin surfaces (Fig. 9.6), and take photographs with life-sized cut-outs of the *Antimicrobial Avengers!* characters (Fig. 9.7).

After the LightNight event, the three comic stories were refined and finalised, and made available in printed format and online. The final comics can be viewed in their entirety here: https://germwars.co.uk/, and Figs. 9.8, 9.9, 9.10,

Fig. 9.2 Drawing the endings to the three stories using partial storyboards during an outreach event in a South Liverpool primary school. Photograph by Prof. Raechelle D'Sa

Fig. 9.3 Drawing the endings to the three stories using partial storyboards at LightNight, Liverpool. Photograph by Mark Roughley

Fig. 9.4 Drawing the endings to the three stories using partial storyboards at LightNight, Liverpool. Photograph by Mark Roughley

Fig. 9.5 Creation of microbes using craft materials while learning about 'Bad' versus 'Good' bacteria. Photograph by Mark Roughley

Fig. 9.6 Touchable sculptures of animals that have naturally antibacterial-resistant skin surfaces by sculptor Helen Birnbaum. Photograph by Mark Roughley

Fig. 9.7 Members of the project team posing for photographs with the *Antimicrobial Avengers!* cut-outs at the John Lennon Art and Design Building on LightNight (Left: Artist in Residence, Jessica Irwin; Middle: Professor Raechelle D'Sa, Director of the D'Sa Lab; Right: Programme Leader for the MA Art in Science programme, Mark Roughley). Photograph by Mark Roughley

and 9.11 show the front pages of each comic and subsequently three pages from inside each comic. The comics have since been used as part of an AMR public engagement toolkit by the D'Sa Laboratory at a number of public engagement events including one at Alder Hey children's hospital, Liverpool in May 2019.

Fig. 9.8 The front pages of each of the comics in the *Antimicrobial Avengers!* series. Illustrations by Jessica Irwin. Reproduced with permission

Fig. 9.9 Three pages from 'The Adventures of Sharkgirl'. Illustrations by Jessica Irwin. Reproduced with permission

9.4.1.4 Reflecting on the *Antimicrobial Avengers!* Project

Over 500 people visited the John Lennon Art and Design building during the LightNight event on 17th Friday May, and of these visitors, over 250 hundred people directly engaged with the hands-on activities associated with *Antimicrobial Avengers!* Attendees were asked to complete a six-question anonymous paper-based feedback questionnaire after partaking in one or more of the activities. Twenty-four people responded and submitted the completed questionnaires in a post-box in the Atrium Gallery, however, not every question was answered on each returned questionnaire.

Using a 1–5 Likert scale with 1 being 'Not at all' and 5 being 'Loved it', 16 people answered with a 4 or 5 for the question *Did you like the Antimicrobial Avengers! characters?* For the question *Did you learn anything new about microbes and antimicrobial resistance?* sixteen

Fig. 9.10 Three pages from 'Komodo Dragons Save the World Illustrations by Jessica Irwin. Reproduced with permission

Fig. 9.11 Three pages from 'The Drugs Don't Work'. Illustrations by Jessica Irwin. Reproduced with permission

respondents chose 'Have learned a lot', six respondents chose 'Have learned a little' and zero respondents chose 'Haven't learned anything new'. When asked *What activity was your favourite?* six respondents wrote comments about the touchable sculptures, two wrote about taking selfies with the *Antimicrobial Avengers!* cut-outs, four wrote about drawing the endings to the comics, two stated it was reading the comics, and 12 noted it was creating microbes using craft materials.

For the questions *Do you think that events like this are a good way to learn about antimicrobial resistance?* and *Would you attend a similar event in the future?*, two respondents chose 'Maybe' and 21 chose 'Yes' from the 'No, Maybe, Yes' options available. The final question asked the respondents to *'Tell us three words that you associate with this event*'. The words written include 'educational; informative; interactive; entertaining; important; unexpected; imaginative; scary; creative'.

No feedback was gathered from the school children during the co-creation process, and while the questionnaire responses about the LightNight event were mostly positive about the activities and knowledge acquisition, the response rate is too low to make definitive conclusions about the success of the co-creation methods used in the hands-on activities and the AMR transfer of knowledge. It is hoped that by using partially complete storyboards for the comics at each event, the participants were not intimidated by an empty canvas and that learning about AMR would take place through 'playful exploration' of the AMR topics (Archibald and Blines 2021), and that word-of-mouth communication (WoMC) and dissemination of the science would occur naturally via social diffusion; reaching beyond those who directly took part in the activities (Compton and Pfau 2009). It is recognised that there is potential for misrepresentation of the science during WoMC to social networks including family members and peers, especially via children to adults, although simply stimulating conversation about health issues has been shown to be beneficial on the whole rather than detrimental, with a stronger number of messages that promote the accurate spread of intended information (Compton and Pfau 2009).

9.4.2 Case Study 2: Bacteria, the Future: A Time Travelling Twine Game to Save Hugh-Manity!

Bacteria, the Future: A time travelling Twine game to save Hugh-manity! is an interactive online storytelling adventure game devised by Natasha Niethamer in 2020 as her master's project for the MA Art in Science Programme at Liverpool School of Art in Design (Roughley et al. 2022). The project was developed over 3 months in collaboration with artists, scientists, and educators associated with this programme. It explores interactive fiction-based public outreach in which the participants embody a bacterium, imagine what it is like to be a bacterium within a colon, and explore a potential future in which AMR is not sufficiently addressed. The outcome can be viewed here: https://nniethamer.itch.io/bacteria-the-future.

Published research literature exploring AMR PE/PO suggests that there is a general lack of understanding of how and why certain public engagement methods work in certain contexts in this field. This project is built on several findings established in the literature review for this chapter:

1. It is important to critically evaluate activity in both its successes and its failures because this is lacking in the field of AMR public engagement and is restricting identification of factors that will ensure AMR PE is fulfilling its aims (Redfern et al. 2020).
2. It is important to make design, evidence-based methods, and reasonings accessible to support those who have an idea for this field yet are unsure how to proceed.
3. The aim of AMR public engagement is long-term behavioural change, however, this is hard to evaluate. Many call for the creation of a standardised framework which AMR PE practitioners should follow to develop stronger techniques. 'Behavioural Change Theory' (BCT) studies in the field of social science may provide critical insight when developing evidence-based, methodological frameworks in PE. This should be referenced and further explored.

9.4.2.1 Anthropomorphising Microbes Facilitates Learning

According to Timmis et al. (2019), it is essential for microbes to 'transit from abstraction to pictorial perception and substance and take up their rightful position in the human psyche' and they suggest that 'anthropomorphisation' is effective as a microbiology teaching approach. Kirk Dorion (2011), a researcher of science communication at the University of Cambridge, found that although most animals do not have human intentions and emotions, anthropomorphisation may facilitate learning of abstract concepts due to generating an emotional connection and therefore interest/moral concern. This suggests that generating a human, emotional connection to microbes through anthropomorphising them may better facilitate learning and long-term memory of what Timmis et al. (2019) describe as a largely invisible, abstract world.

9.4.2.2 Gamification as a Learning Tool

Gamified learning results in improved student engagement, motivation, confidence, attitude, perceived learning, and performance and due to these significant findings, gamification has been integrated for higher-learning and health-related purposes allowing material to be accessible to broader audiences (Subhash and Cudney 2018). Additionally, gamifying complex material may appeal to broader audiences including those of all ages and lower literacy levels to facilitate deeper understanding (De Freitas 2018). Using gamification in AMR PE and PO may encourage participants to consider and discuss AMR in the relaxed, light-hearted environment of their own home, with the aim of contributing to the generation of widespread support and intentional global policy changes needed to delay and prevent AMR from resulting in significant global death and economic burden.

9.4.2.3 Utilising Pandemic-Driven Online Technology for AMR Education

The SARS-CoV-2 pandemic transformed communication approaches for widespread public education. The lockdown restrictions forced educators to shift delivery from traditional, direct communication to one heavily utilising technology and online platforms. Educators were challenged to create engaging online content that facilitated learning despite the lack of face-to-face interaction. This has driven innovative approaches in online education. A technique utilised to promote greater student engagement and reinforce knowledge includes integrating more interaction through gamification, incorporating game motivation techniques and mechanics to non-game subjects (Joshi 2021; Lee and Hammer 2011).

Despite limitations to online learning, it is a major source of health literacy for the global public and is considered a valuable tool to connect with a wider audience and disperse public health information (Jia et al. 2021). With the pandemic driving innovative, interactive ways to better engage online, it is important to learn from these ideas and build upon identified strengths. With face-to-face interaction no longer restricted, continuing to utilise supplementary online platforms may provide a valuable resource to widen PE/PO activities' impact and reach from an isolated event to a globally accessible online resource. This is especially critical in regard to the public health threat of AMR, which, to prevent unneeded deaths and economic burden, urgently requires global, collective efforts, and widespread policy changes (Tarín-Pelló et al. 2022).

9.4.2.4 Developing Bacteria, the Future: A Time Travelling Twine Game to Save Hugh-Manity!

With these ideas in mind, this project was created with the aim of this work to explore alternative ways to communicate the public health threat of AMR and increase microbiology literacy. This activity was originally created as a public engagement project involving interactive theatrical performance (https://collaborativeumf.wordpress.com/), accepted as a physical PE activity to engage the public in Liverpool during LightNight 2020 (https://lightnightliverpool.co.uk/). Due to the SARS-CoV-2 pandemic, the festival was cancelled as well as any physical PE activities. This required the original performance script to be adapted into an online version. Research into novel interactive online platforms that are low-cost, accessible to non-coders, and achievable within the time constraint of 3 months, led to discovery and utilisation of 'Twine', an online interactive text platform allowing one to create a 'choose-your-own-adventure' game.

For this activity, participants were invited to take part in a text-based, choose-your-own-adventure, 'interactive fiction' game using the online platform 'Twine' (https://twinery.org/).

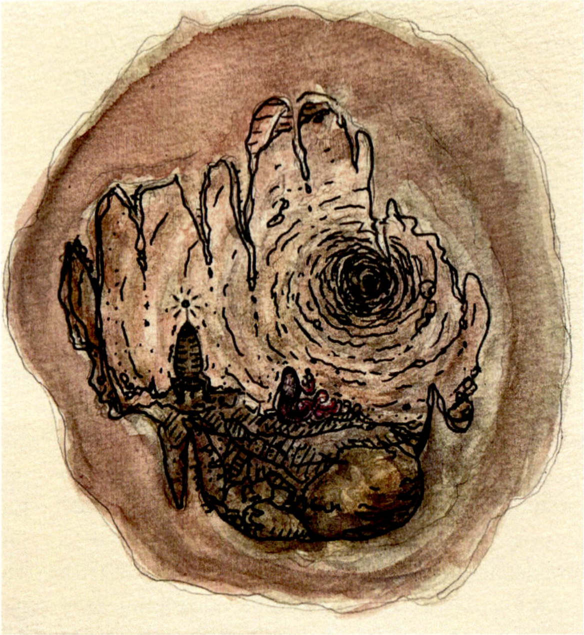

Fig. 9.12 *Bacteria, the Future* scene, showing the player (bacterial form) with their ship looking into the abyss of a human intestine, about to embark on a dangerous mission to save humanity. Illustration by Natasha Niethamer

The game involves the player entering a time machine in the year 2050. In 2050, AMR has wiped out much of the human population, and the player's consciousness has been chosen to inhabit the body of an *Escherichia coli* (*E. coli*) bacterium. In this bacterial body, the player travels back in time to 2020, and embarks on a mission to 'neutralise' AMR bacteria in the gut of human, Hugh Mungus (Fig. 9.12). The player's aim is to 'save the millions of future humans that will die from an evolved, stronger strain of the neutralised AMR bacteria'. However, there is a plot twist: the millions of humans that die include you. By succeeding with this mission, you are saving yourself from premature, AMR-caused death. However, does this come with consequences?

During this mission, the player explores the digestive system, questions the threat of AMR, and explores concepts surrounding AMR. A moral dilemma is raised when the player realises that the bacterial body they inhabit can trigger emotional reactions to the events occurring. The player comes face to face with current, top AMR bacterial health threats established by the Center for Disease Control and Prevention (CDC), antibiotic-resistant strains of *Clostridium difficile*, *Escherichia coli*, and *Salmonella* (CDC 2020). In 'neutralising' these AMR bacteria, the game raises questions about the misuse of antibiotics and imagines a world in the near future predicted to occur if antibiotic misuse is not addressed. It challenges the validity of demonising bacteria when in truth, bacteria are neither 'good' nor 'evil' (Fig. 9.17). Most gut bacteria are harmless or even helpful for your health. It leaves us humans in 2020 with the questions: Are we capable of preventing this apocalyptic future by adjusting our current behaviours? Can we fight AMR by working with bacteria versus wiping them out?

This project explored the effectiveness of non-linear interactive fiction using the 'Twine' platform on AMR public awareness. The 'Twine' platform (https://twinery.org/) is a free, open-source online game platform that allows those with limited coding knowledge to create non-linear (multiple endings are possible), interactive stories (hyperlinks, audios, pictures, and animations can be placed) in which the player can 'choose-their-own-adventure' while also playing in second person (Fig. 9.13). 'Twine' has been used as an educational tool in certain classrooms, and it is seen as a way to develop

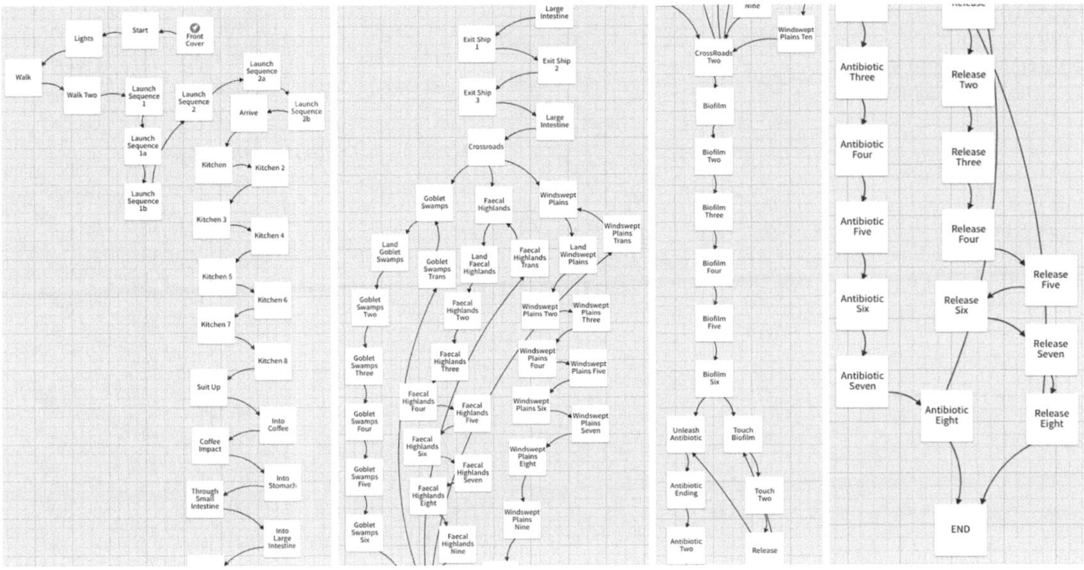

Fig. 9.13 *Twine* programme storyline flowchart displaying the choices the player is given in *Bacteria, the Future* in the form of branching plot points with resulting storyline consequences in sequential order from left to right. Created by Natasha Niethamer

empathy in the player as the platform allows the player to take part in the story in second person (Salter 2016). This platform, however, has not been used to communicate AMR. For this reason, this project was created to explore if this platform was an effective way to increase microbial literacy and AMR awareness.

Bacteria, the Future is a reference to *Back to the Future*, a science fiction film in which the protagonist is sent 30 years back in time. This interactive fiction game was inspired by a few different works including *Red Dwarf, Star Trek DS9/Next Generation*, the H.P. Lovecraft short story, *The Colour Out of Space*, and a Lovecraft-inspired video game, *Bloodborne*. *Red Dwarf* and *Star Trek* integrated time travel to change events of the future, also integrating humour/technology-based main characters. Lovecraftian descriptive words and literary style were integrated. *Bloodborne* inspired the bacterial tentacle monsters and the idea of the 'monsters' having previously been innocent if it was not for harmful community decisions (similar to the AMR bacteria being mutated, previously helpful bacteria). The structure of this game was inspired by researching choose-your-own-adventure games, traditional Twine game examples that were found on itch.io (http://itch.io) like *You Are Jeff Bezos* and *The Temple of No*, and puzzle games such as *Myst* and *Sam and Max*.

Digestive system locations were researched to make the setting realistic, using research referencing bacterial morphology descriptions, cells, such as the goblet cells in the large intestine which excrete mucus found at the 'Goblet Swamps' in Fig. 9.15, and details regarding the colonic landscape. Defecation was researched to ensure leaving the rectum in the game was realistic with its activation by stretch receptors. A BBC Earth Lab video of a pill camera moving through the digestive system was watched to describe the story's environments (BBC Earth Lab 2016). Mechanisms of antibiotics on bacteria were researched as well as information on how bacteria communicate/work together. This involved investigating academic papers because microbiology-based mechanisms as well as simplified inner organ descriptions were unable to be found by Google search.

Fig. 9.14 *Bacteria the Future* character development. Illustrations by Natasha Niethamer

Visual storyboards were then created as well as illustrations of bacteria (Fig. 9.14). Narratives were brainstormed, involving peer feedback and writing of non-linear thoughts. A backstory was created for the main character. Eventually, ideas were brought together in a linear, understandable way. The game was designed by creating flowcharts of plot point ideas on paper. A few story lines were explored as prototypes before deciding on the final form.

The script was then entered onto the Twine platform application, adjusting the ideas frequently to better fit gameplay. Coding elements were adjusted to ensure the game player would not revisit a page. When choosing fonts and styling the game, it was desired to make it more accessible to those who are dyslexic, suggested by a dyslexic collaborator and beta tester, so the font, 'open dyslexic' was used (https://opendyslexic.org/).

Niethamer, who had no coding experience at the beginning of this project, learned Twine-specific coding syntax by referencing the book, 'Writing Interactive Fiction with Twine' by Melissa Ford. This book takes one with no coding experience through traditional Twine game techniques and recommends persuasion/engagement elements to ensure an effective, engaging narrative design tailored to interactive fiction with a flowchart of checkpoints to structure a traditional Twine game (2016).

The website, 'itch.io' (https://itch.io), was used to upload the HTML code associated with the game. 'Itch.io' used the HTML to host the game on its website, where it hosts other online indie games. In 2020, this was the only free platform available to host Twine games.

Artists and scientists associated with the LJMU MA Art in Science programme played the prototype game, with gameplay excerpts shown in Figs. 9.15, 9.16, and 9.17. The game was adjusted in response to their feedback. Beta testers suggested the first script seemed too violent with antibiotics described as 'killing' bacteria, especially when role playing as the bacterium. On further discussion with beta testers, wording was changed from 'kill', 'destroy', and 'shoot' to 'medicate' and 'treat', words more suggestive (Fig. 9.16). Collaborating with a game enthusiast, physics educator, and SciArtist, Tom Hyatt, at LJMU, the physics of time travel was discussed as well as what makes a game effective. This resulted in adding realistic elements suggested by the collaborator such as 'Tachyon' drive, an element in physics believed to allow time travel. During this conversation, the game storyline was finalised. The final prototype game was then sent out to artists and scientists in LJMU with pre- and

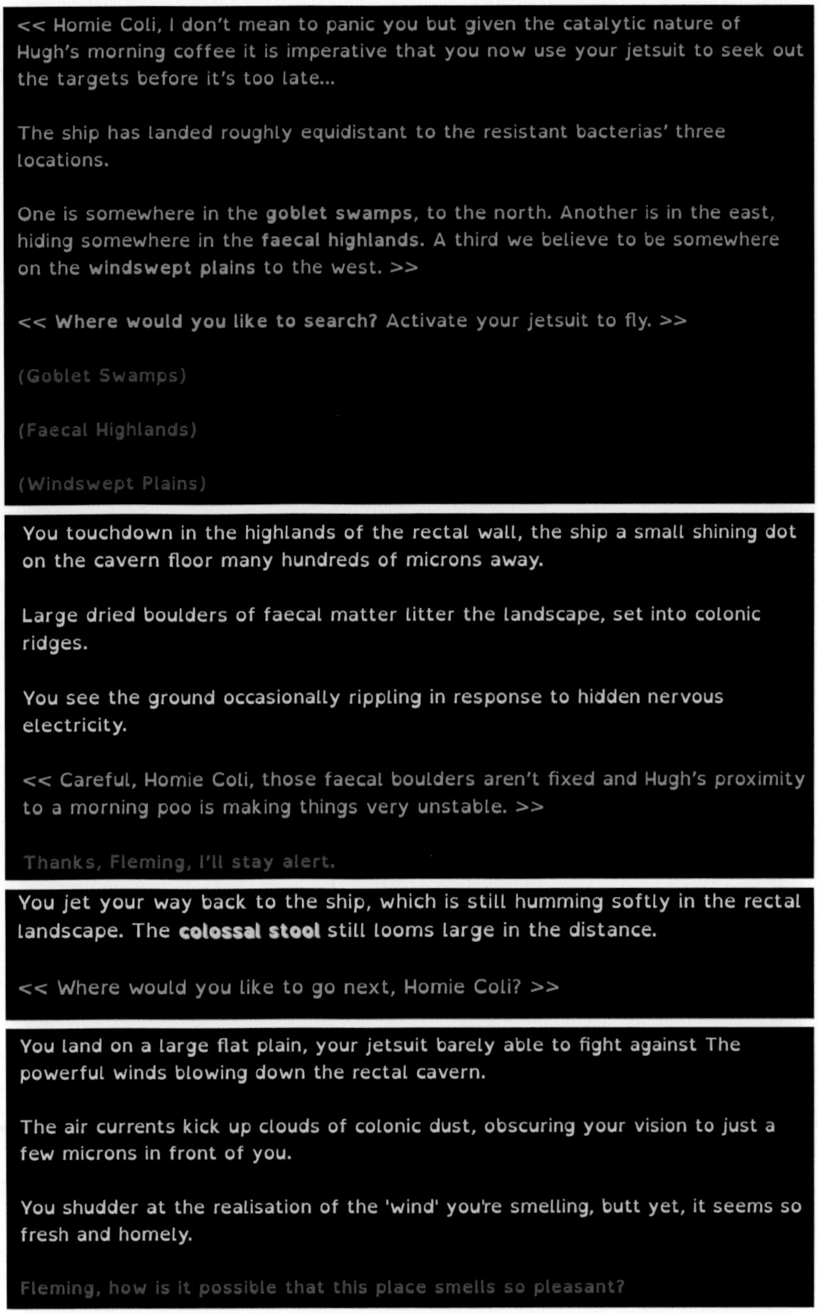

Fig. 9.15 Four gameplay screenshots from *Bacteria the Future*. Green text represents the ship's AI computer, Fleming, who assists during gameplay. Blue text is clickable to advance the game. Created on *Twine* by Natasha Niethamer

post-game surveys. The feedback allowed adjustment of coding errors, spelling/grammar errors, and timing errors. The game continues being improved based on feedback and updated scientific information.

9.4.2.5 Results from users of *Bacteria, the Future: A Time Travelling Twine Game to Save Hugh-manity!*

During final game prototype testing, participants were asked to complete online, pre- and post-

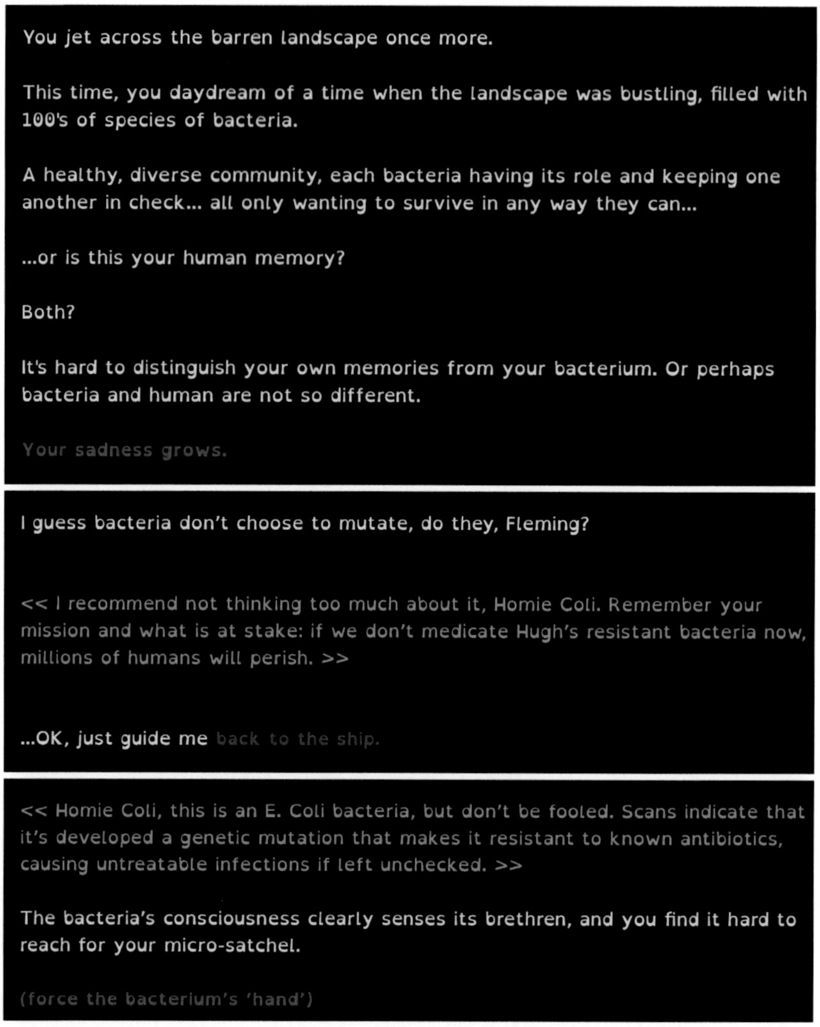

Fig. 9.16 Three gameplay screenshots from *Bacteria the Future*, in which the overuse and misuse of antibiotics is questioned. Sentences are revealed gradually. Green text represents the ship's AI computer, Fleming, who assists during gameplay. Blue text is clickable to advance the game. Created on *Twine* by Natasha Niethamer

activity feedback questionnaires. Pre- and post-game questionnaires were modelled after Ahmed et al. and their pre- and post-activity surveys to determine if there was an increase in AMR awareness (2020). A 1–5 Likert scale was used with 1 being 'Strongly disagree' and 5 being 'Strongly agree'. Google Forms were used to create the pre-game and post-game questionnaires. Questionnaires are available to view on the main gameplay page along with a participant information sheet for student research study permission.

Fifteen linked pre- and post-game survey responses were received. 11/15 participants claimed their knowledge about microbiology and antibiotics increased, with 4/15 participants reporting their knowledge about microbiology and antibiotics staying the same. 14/15 participants filling out the post-game form strongly agreed that AMR is a very serious problem with 1/15 agreeing AMR is a serious problem. In contrast, in the pre-game form, 7/15 respondents strongly agreed AMR is a very serious problem, 7/15 agreed AMR is a problem, and 1/15 answered neutrally on AMR being a serious problem. All respondents reported post-game that they disagree antibiotics will help improve their cold, cough, or sore throat more quickly, though 3/15 respondents were neutral about the issue beforehand with 1 respondent agreeing that antibiotics would help improve their cold,

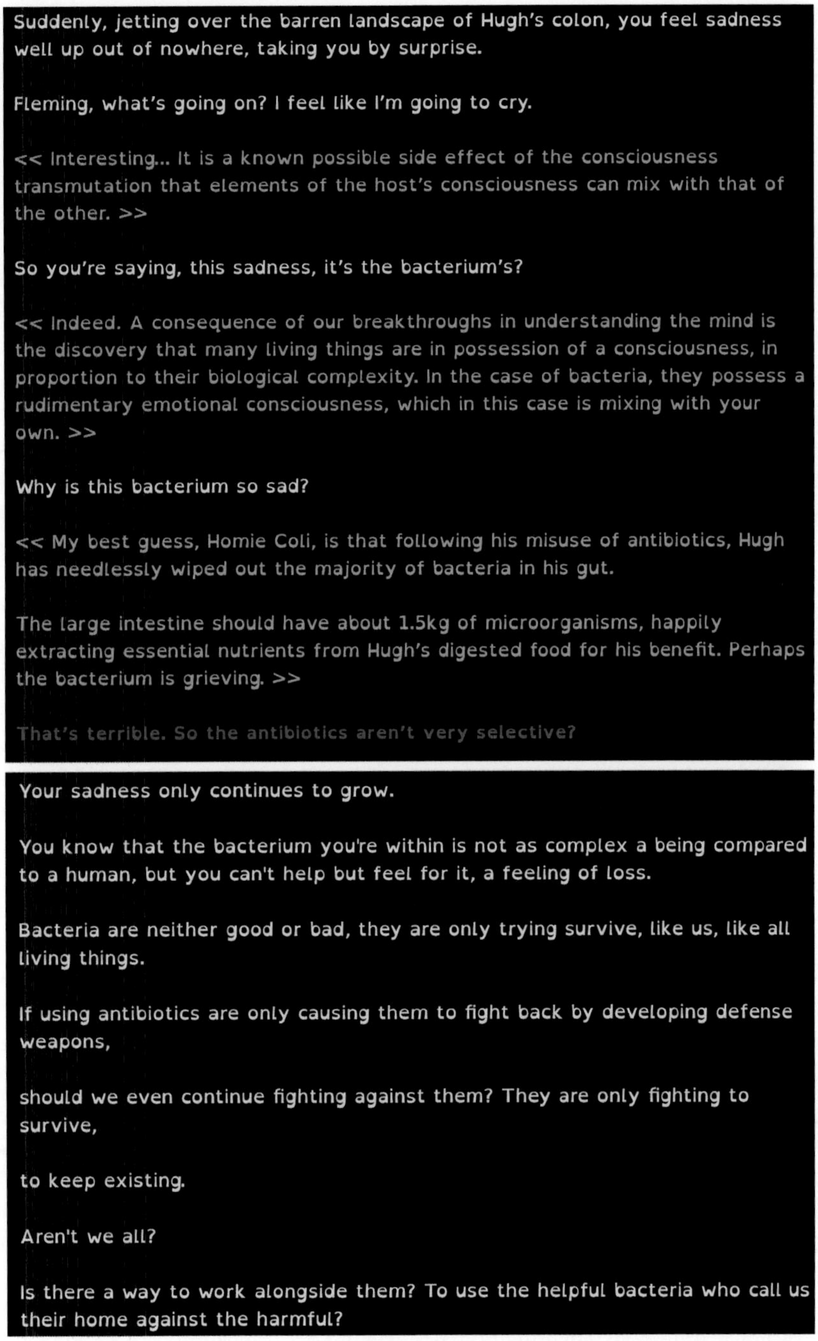

Fig. 9.17 Two gameplay screenshots from *Bacteria the Future*, in which the overuse and misuse of antibiotics are questioned. Sentences are revealed gradually. Green text represents the ship's AI computer, Fleming, who assists during gameplay. Blue text is clickable to advance the game. Created on *Twine* by Natasha Niethamer

cough, or sore throat more quickly. Most of the respondents post-game (13/15) strongly disagreed that you do not need to finish a prescribed course of antibiotics if you were feeling better. However, the remaining 2 respondents agreed with this statement. 14/15 respondents post-game strongly disagreed that you should use antibiotics left over from a prescription with one respondent responding neutrally.

In comparison, pre-game, one of the participants had strongly agreed with this statement, and 4/15 respondents changed their response from disagree pre-game to strongly disagree post-game. 13/15 respondents disagreed to the statement, *'If you have never taken antibiotics, you cannot get a drug-resistant infection'* both pre- and post-game. One respondent replied neutrally and 1 agreed to the statement both pre- and post-game. Most participants were neutral or agreed with the assertion that *'Your body becomes resistant to antibiotics'*. 14/15 respondents strongly agreed post-game that they could help slow down antimicrobial resistance by certain behaviours like handwashing with 1/15 responding neutrally. This is a shift from pre-game, where 8/15 agreed (4 of these strongly) with 5/15 giving a neutral response and 2/15 disagreeing.

9.4.2.6 Evaluating *Bacteria, the Future: A Time Travelling Twine Game to Save Hugh-manity!*

Feedback suggests that interactive fiction using Twine may be an effective platform to explore further facilitating awareness of antimicrobial resistance (AMR). In particular, comparing individual pre- and post-game survey results showed that a majority of game participants strengthened their agreement to the assertion of AMR being a serious problem, disagreed that antibiotics can improve a cold, cough, or sore throat, strongly disagreed that you do not need to finish a prescribed course of antibiotics if feeling better, and strongly agreed post-game that handwashing can slow down AMR.

However, other results were mixed with multiple reinforced misunderstandings. These topics should be addressed more clearly in the game for future adjustments. For example, all participants post-game thought that the body became resistant to antibiotics, the opposite of the desired response post-game. Participants were unsure if a human can get a drug-resistant infection if never taken antibiotics. Some participants expressed confusion with wording of the survey questions. These specifications are something to explore further with a larger sample size of participants as well as with a collection of demographic information to ensure conclusions are significantly meaningful.

Some responses were critical, which brings up ways to adjust the game in future versions. These criticisms included mention of better pacing with 'slow patches and fast to match the action'. The feedback also includes multiple suggestions for adding more audio: 'Sound/sound effects would have been good', 'Maybe some more animation/transitions in text would of been fun'. Two different participants mentioned creating a soundtrack with one saying, 'Would have loved a soundtrack. I want to hear what it's like to be Hugh's Anal Canal. The sounds of the bacteria and the soundscapes of Fecal Highlands and Windswept Plains'.

One limitation of this text heavy-based project is that it is not accessible to those of lower literacy levels. 73.3% of surveyed participants specified illustrations would improve the game with one respondent suggesting, 'Illustrations would really help to not get lost through the story' and 'some visuals of the bacteria would be cool so you could see what they look like'. This project may be made more appealing, understandable, and potentially reach a wider audience through heavier use of visuals, perhaps in the form of a graphic novel or adjusted, child-friendly comic book.

However, there were a few that expressed a positive preference for a text-based game, with comments including, 'I think not using any form of illustration or visualisation helps the viewer to create their own visual depiction of the game, which is really fun!' and one person commented that adding illustration may take away from the story—though stick figures may allow more interest: 'I thought it might be interesting to add pictures and graphics, but it might take away from our own imagination...although simple and fun stick figures might add to the humor'. The respondents of this study were not asked to specify their literacy level or preferred mode of learning, so a range of responses are expected. Due to the range of responses, specifying a target group and adjusting to their preferences would allow better future impact.

The post-game surveys highlighted some benefits regarding 'Twine' text-based medium. First, the second-person experience in which the reader plays as a main character of their choosing, and gender does not have to play a role. This may allow greater empathy from the player with one participant saying, 'Felt very immersive, truly believed I was Eliza Coli for a time'. This platform also allows a player to make choices or interact with the story by clicking through, allowing greater agency to the player. Post-game surveys mentioning, 'I enjoyed it, I wouldn't [sic] mind learning about other subjects in thus [sic] format', 'I enjoyed it, it was fun, better than reading another report or article', 'Like it—reminds me of reading goosebumps books as a kid! But also v good for learning experiences as keeps you engaged throughout. Liked the wiggly words when they moved around. Felt very immersive, truly believed I was Eliza Coli for a time and couldn't have done anything but not go for the satchel at the end!'.

Regarding the choices being made, one expressed that they liked that you could explore two different endings. Another recommended more choices saying, 'i liked the ability to choose different options, more of those choices would be nice'. Many respondents were positive about the enjoyability of the game: 13/15 participants expressed they enjoyed the game and would recommend it to a friend while two participants responded neutrally.

In future versions, based on this feedback, it seems adding more animation/transitions, simplistic, stick figure illustrations as well as some audio may be beneficial to interaction. To increase accessibility and widen reach, this story could be adapted into a visual graphic novel to appeal to low literacy adults, into a simplified, visual comic to appeal to kids or the current 'Twine' adventure can be transformed into a mostly visual version for those who prefer this. One respondent mentioned, 'I think it's a beautiful way of learning. I can really see this as a scientific app to help people learn more about their bodies on a micro level. You could also branch out into food and diet, how eating different foods affect your body and how important diet is to not just functionality, but also in aid of mental health'.

Based on observations during game creation, prototyping, and prototype feedback, there are a few things that would be included in future versions to increase interaction: more audio, illustrations, and more text-animation effects would be included. A guidebook to define important terms for the player in which the player is able to reference during gameplay would also be created. More options/agency would also be given to the player. Through this process, writing many story lines as an individual was a tough, time-intensive process, so this prevented the plan to explore as many options as originally proposed. Though low literacy levels may be a barrier to accessing this text-based platform, there is the option to adjust the level of complexity of the script in the future or to create illustrations/comics based on this story.

Unexpectedly, it was a learning process in itself to design a narrative in a scientifically realistic way. It was tough to find visual descriptions/mechanisms of certain chemicals, bacteria, and inner organ walls. The internet directs one to dense scientific papers, and as a result, definitions were frequently looked up and related scientists were consulted to simplify ideas.

Because two-way interaction is not possible with the public, this project is not considered public engagement but instead, public outreach, a one-way communication of research and higher education-based topics (Illingworth and Allen 2016). This is not seen as effective as public engagement, but it is a novel activity to consider when communicating AMR in an enjoyable way when comparing traditional methods of science communication, especially in a time in which face-to-face communication is not possible.

This form of interactive learning may have potential beyond AMR engagement. The outcomes of this project suggest that interactive fiction using Twine may be an effective platform to explore further in facilitating awareness of antimicrobial resistance (AMR), however, there are more respondents needed to create more meaningful results. In the following section, common themes from both case studies are compared and contrasted for use in strengthening future activities, as recommended by Redfern et al. (2020).

9.5 Discussion

The aim of this chapter was to determine whether art-science public engagement (PE) and public outreach (PO) activities are an effective method for increasing microbiology literacy in the public health threat of antimicrobial resistance (AMR) by exploring literature relating to interactive art-science activities. To inform this aim, two case studies involving AMR awareness art-science PE and PO activities created in response to the existing literature were detailed and analysed. The following questions were explored:

1. What is the key aim of an AMR-based public engagement (PE) and public outreach (PO) activities?
2. How do practitioners evaluate their activity to determine if the key aim was achieved?
3. Can a framework be developed for practitioners to utilise when developing effective art-science activities promoting microbiology literacy regarding AMR public engagement?
4. How has the COVID-19 pandemic and resulting public understanding/behavioural changes shifted art-science practice in AMR awareness?

When determining the key aim of AMR-based PE activities concluded by Redfern et al. (2020) as needing to be identified in AMR-based PE activities and supported by the findings of Redfern et al. (2020), the WHO's global action plan on AMR (2015), and IACG (2019), we proposed two key aims of AMR PE activities which will contribute to slowing down AMR rates, a key component of tackling the urgent global public health concern of AMR, which are awareness of AMR, specifically regarding common misconceptions in the public, and long-term behavioural adjustment regarding misuse and overuse of antimicrobials.

NCCPE, Illingworth, and Allen state in their determination of quality public engagement that evaluation is an essential way to critically self-reflect and contribute to strengthening future activities (NCCPE n.d.-c; Illingworth and Allen 2016). According to Redfern et al., evaluation is lacking in the AMR PE field (2020). For this reason, two AMR awareness art-science activities were explored. Practitioners administered pre- and post-activity surveys to participants with questions focused upon a pre-activity understanding of the field, misconceptions, and believed behaviour changes post-activity (Ahmed et al. 2020; Grimberg et al. 2019; Altay and Lakhlifi 2020). However, these activities do not assess long-term behavioural changes, and according to research by Storksdieck, Ellenbogen, and Heimlich in 2006, there must be subsequent reminders to ensure long-term behavioural change.

When assessing whether a framework may be developed for practitioners to utilise when developing effective art-science activities, specifically those regarding AMR public engagement, we identified research pulled from different disciplines such as psychology, science communication, public health and environmental sciences that may hold insight into promising, evidence-based methodology to be utilised. One example is the behavioural change theory (BCT) rooted in psychology and harnessed in public health public engagement activities as suggested by McParland et al. in 2018.

The SARS-CoV-2 pandemic transformed communication approaches for widespread public education, driving innovative approaches in online engagement. Online sources remain a major source of health information access for the global public (Jia et al. 2021). With face-to-face interaction no longer restricted, continuing to utilise supplementary online platforms may provide a valuable resource to widen PE/PO activities' impact and reach from an isolated event to a globally accessible online resource. This is especially critical in regard to the public health threat of AMR, which, to prevent unneeded deaths and economic burden, urgently requires global, collective efforts and widespread policy changes (Tarín-Pelló et al. 2022).

9.5.1 Comparing and Contrasting Case Studies

Both case studies had a small sample size of survey respondents preventing meaningful conclusions to be made, despite positive feedback. Gamification was utilised in both case studies, integrated due to its established role in facilitating learning through greater engagement and reinforced knowledge (Joshi 2021; Lee and Hammer 2011). One case study included pre- and post-activity surveys, as advised by AMR-based PE guidelines (Redfern et al. 2020). However, long-term learning impact and behavioural change were not tested. Player agency was integrated into both case study activities, allowing participants of '*Antimicrobial Avengers!*' to create their own endings to comics and allowing '*Bacteria, the Future: A time travelling Twine game to save Hugh-manity!*' players to choose their character name, order of location exploration, and allowing exploration of alternative endings. Both case study activities were adapted to a digital version, allowing expanded online reach. Collaboration with non-scientists and scientists were cited in both projects to be essential to its development.

The '*Antimicrobial Avengers!*' project ran pre-SARS-COV-2 pandemic restrictions, so this allowed participatory, public-facing co-creation-based activities at primary schools to drive the project. '*Bacteria, the Future*', created during pandemic restrictions, required utilising an online platform as no face-to-face contact was possible. No funding was available for the development of '*Bacteria, the Future*', with lockdown limiting resources to those accessible at home, with a time constraint of 3 months due to being a Master's research project. In contrast, '*Antimicrobial Avengers!*' was built on funding with multiple established multidisciplinary practitioners working together. Additionally, the two projects targeted two different audiences with '*Antimicrobial Avengers!*' tailored to kids of primary school age and '*Bacteria, the Future*' targeting adults. Furthermore, '*Bacteria, the Future*' being text-based, is not accessible to those of low literacy levels. In contrast, '*Antimicrobial Avengers!*' utilised 'Graphic Medicine', an established medium of accessibility to lower literacy levels (Green and Myers 2010; Waite 2019).

9.5.2 Proposed AMR Public Engagement and Outreach Framework

From reflections on both case studies, there are clear strengths to help develop an AMR public engagement and outreach framework. These include:

1. Integration of artist and scientist collaboration, ensuring feedback from science experts to directly discuss misconceptions.
2. Feedback from non-scientist public of diverse backgrounds at every step of development.
3. Utilisation of both online and public-facing activities to maximise reach.
4. Utilising evidence-based interactive learning methods like gamification and graphic medicine to increase accessibility, experimenting with different combinations of methods as well as seeking out novel platforms that show promise in other disciplines.
5. Consider an activity that can later be adapted to multiple literacy levels and audience targets.
6. Integrate pre- and post-activity feedback, and if possible, follow-up surveys to assess long-term behavioural change.
7. Share and publish details of your activity including development process, strengths, weaknesses, obstacles, successes, feedback, adjustments, and lessons learnt.

9.6 Conclusion

This chapter has established public engagement and outreach as a broad and complex field involving many different factors, aims, and methods. It is considered a complex practice influenced by many environmental and social factors, many still to be identified. For this reason, in agreement with Redfern et al.'s recent findings when evaluating

AMR PE and PO activities, there is not enough sufficient evidence to directly support SciArt interactions as one conducive to the long-term behavioural changes regarding AMR specifically. It is agreed, however, that it is important to continue informing the limited understanding of art-science activity effectiveness in PE and PO, improving AMR awareness and microbiology literacy in the public(s), and in doing so, contribute to slowing down AMR rates, believed to be a key component of tackling the urgent global public health concern of AMR.

Currently, few PE and PO activities are recorded let alone evaluated (Redfern et al. 2020). We hope this research begins the needed discussion in regard to creating effective AMR PE activities informed by evidence-based theories. We especially hope to inform those who would like to contribute to slowing AMR rates through art-science activities, but do not have a public engagement-related background. We want to ensure that strong, evidence-informed PE and PO activities are designed, developed, and shared to inform and strengthen future activities.

Acknowledgements The authors would like to acknowledge the support of the D'Sa Laboratory, the STEAM Lab at the University of Liverpool and the 2019-20 students of the MA Art in Science programme at Liverpool John Moores University in the development and delivery of the school outreach and public engagement events that made the *Antimicrobial Avengers* project possible. A special acknowledgement goes to the illustrator of the *Antimicrobial Avengers* comics, Jessica Irwin, whose interpretations of the science and public/school children's drawings made the stories come to life.
The author of *Bacteria, the Future* would like to acknowledge the support of the supervisors of the 2019-20 MA Art in Science programme at Liverpool John Moores University, Mark Roughley and Kathryn Smith, and the programme's students in game development. The thoughtful feedback and discussion that they provided at every step were essential in developing the final storyline. A special thanks to Tom Hyatt, a physics educator, SciArtist, and avid gamer, who used this expertise to help write and edit the final game script.

References

Ahmed R, Bashir A, Brown JE, Cox JA, Hilton AC, Hilton CE, Lambert PA, Theodosiou E, Tritter JQ, Watkin SJ, Worthington T (2020) The drugs don't work: evaluation of educational theatre to gauge and influence public opinion on antimicrobial resistance. J Hosp Infect 104(2):193–197. https://doi.org/10.1016/j.jhin.2019.10.011. Accessed: 25th May 2020

Altay S, Lakhlifi C (2020) Are science festivals a good place to discuss heated topics? Journal of Science Communication [online], 19(1):A07. Available at: https://doi.org/10.22323/2.19010207. Accessed 26 May 2021

Anderson M, Cecchini M, Mossialos E, North J (2020) Challenges to tackling antimicrobial resistance: economic and policy responses. Cambridge University Press, Cambridge. Accessed: 24th March 2023

Antibiotic Guardian (2023) Public resources [online]. Available at: https://antibioticguardian.com/public/. Accessed: 8th March 2023

Antonini M (2021) An overview of co-design: advantages, challenges and perspectives of users' involvement in the design process. J Des Think 2(1):45–60. https://doi.org/10.22059/jdt.2020.272513.1018

Archibald M, Blines J (2021) Metaphors in the making: illuminating the process of arts-based health research through a case exemplar linking arts-based, qualitative and quantitative research data. Int J Qual Methods 20: 1609406920987954. https://journals.sagepub.com/doi/pdf/10.1177/1609406920987954

BBC Earth Lab (2016) Pill Camera Swallowed | Follow Through Gut | Guts | Earth Lab [online video]. Available at: https://www.youtube.com/watch?v=5ufESc1bK78&ab_channel=BBCEarthLab

Bhattacharya A, Hopkins S, Sallis A, Budd EL, Ashiru-Oredope D (2017) A process evaluation of the UK-wide antibiotic Guardian campaign: developing engagement on antimicrobial resistance. J Public Health 39(2):e40–e47. https://doi.org/10.1093/pubmed/fdw059. Accessed: 24th March 2023

Bishop BM, Juba ML, Russo PS, Devine M, Barksdale SM, Scott S, Settlage R, Michalak P, Gupta K, Vliet K, Schnur JM (2017) Discovery of novel antimicrobial peptides from varanus komodoensis (komodo dragon) by large-scale analyses and de-novo-assisted sequencing using electron-transfer dissociation mass spectrometry. J Proteome Res 16(4):1470–1482. https://doi.org/10.1021/acs.jproteome.6b00857. Accessed: 24th March 2023

Bunn C, Kalinga C, Mtema O, Abdulla S, Dillip A, Lwanda J, Mtenga SM, Sharp J, Strachan Z, Gray CM (2020) Arts-based approaches to promoting health in sub-Saharan Africa: a scoping review. BMJ Global Health 5(5):e001987. Available at: https://gh.bmj.com/content/5/5/e001987.abstract. Accessed 24 Mar 2023

Carvalho GS, Lima N (2022) Public perception of microorganisms and microbiology education: a need for enhancing society's microbiology literacy. In: Importance of microbiology teaching and microbial resource management for sustainable futures [online chapter]. Academic Press: pp. 31-45. https://doi.org/

10.1016/B978-0-12-818272-7.00006-7. Accessed 29 Mar 2020

Craig P, Dieppe P, Macintyre S, Michie S, Nazareth I, Petticrew M (2008) Developing and evaluating complex interventions: the new Medical Research Council guidance. BMJ 337:a1655. https://www.bmj.com/content/337/bmj.a1655. Accessed 27 May 2020

Curtis DJ, Reid N, Ballard G (2012) Communicating ecology through art what scientists think. Ecol Soc 17(2). Available at: www.jstor.org/stable/26269030. Accessed 4 Mar 2023

Compton J, Pfau M (2009) Spreading inoculation: inoculation, resistance to influence, and word-of-mouth communication. Commun Theory 19(1):9–28. https://doi.org/10.1111/j.1468-2885.2008.01330.x

Davies SR (2019) Science communication as emotion work: negotiating curiosity and wonder at a Science festival. Sci Cult 28(4):538–561. https://doi.org/10.1080/09505431.2019.1597035. Accessed 29 Mar 2020

De Freitas S (2018) Are games effective learning tools? A review of educational games. J Educ Techno Soc 21(2):74–84. Available at: www.jstor.org/stable/26388380. Accessed 13th May 2020

Dorion K (2011) A Learner's tactic: how secondary students' anthropomorphic language may support learning of abstract science concepts. J Sci Educ Technol 12(2):1–22. Available at: http://ejse.southwestern.edu/article/view/8552. Accessed: 25th May 2020

Dundar Arisoy F, Kolewe KW, Homyak B, Kurtz IS, Schiffman JD, Watkins JJ (2018) Bioinspired photocatalytic shark-skin surfaces with antibacterial and antifouling activity via nanoimprint lithography. ACS Appl Mater Interfaces 10(23):20055–20063. Available at:. https://doi.org/10.1021/acsami.8b05066

Gordon-Nesbitt R (2017) Creative health: the arts for health and wellbeing. All-party parliamentary group on arts, health and wellbeing. Available at: http://www.artshealthandwellbeing.org.UK/appginquiry/Publications/Creative_Health_Inquiry_Report_2017_-_Second_Edition.pdf. Accessed: 24 Mar 2023

Green M, Myers K (2010) Graphic medicine: use of comics in medical education and patient care. BMJ 340:c863–c863. Available at: https://www.bmj.com/content/340/bmj.c863. Accessed 24 Mar 2023

Grimberg BI, Williamson K, Key JS (2019) Facilitating scientific engagement through a science-art festival. Int J Sci Educ 9(2B):114–127. https://doi.org/10.1080/21548455.2019.1571648. Accessed 29 Mar 2020

Gualano MR, Gili R, Scaioli G, Bert F, Siliquini R (2015) General population's knowledge and attitudes about antibiotics: a systematic review and meta-analysis. Pharmacoepidemiol Drug Saf 24(1):2–10. https://doi.org/10.1002/pds.3716. Accessed 29 Mar 2020

Guerilla Science (2018) The art of engagement. Available at: https://guerillascience.org/wp-content/uploads/2019/06/GS2018-Report.pdf. Accessed 1 Dec 2019

Hoffman SJ, Outterson K (2015) What will it take to address the global threat of antibiotic resistance? J Law Med Ethics 43(2):363–368. https://doi.org/10.1111/jlme.12253. Accessed 25 May 2020

Illingworth S, Allen G (2016) Chapter 5: outreach and public engagement. In: Effective science communication. IOP Publishing. Available at: https://iopscience.iop.org/book/978-0-7503-1170-0. Accessed 29 Mar 2020

Interagency Coordination Group on Antimicrobial Resistance (2019) No time to wait: securing the future from drug-resistant infections report to the secretary-general of the United Nations [online]. Available at: https://www.who.int/antimicrobial-resistance/interagency-coordination-group/IACG_final_report_EN.pdf?ua=1. Accessed 1 Dec 2019

Jensen EA, Buckley N (2014) Why people attend science festivals: interests, motivations and self-reported benefits of public engagement with research. Public Underst. Sci 23(5):557–573. https://doi.org/10.1177/2F0963662512458624. Accessed: 29 Mar 2020

Jia X, Pang Y, Liu LS (2021) Online health information seeking behavior: a systematic review. Healthcare 9(12):1740. https://doi.org/10.3390/healthcare9121740

Joshi LT (2021) Using alternative teaching and learning approaches to deliver clinical microbiology during the COVID-19 pandemic. FEMS Microbiol Lett 368(16): fnab 103. https://doi.org/10.1093/femsle/fnab103

Kearns C, Kearns N (2020) The role of comics in public health communication during the COVID-19 pandemic. J Vis Commun Med 43(3):139–149. https://doi.org/10.1080/17453054.2020.1761248. Accessed 20 Feb 2023

Kesten JM, Bhattacharya A, Ashiru-Oredope D, Gobin M, Audrey S (2018) The antibiotic Guardian campaign: a qualitative evaluation of an online pledge-based system focused on making better use of antibiotics. BMC Public Health 18:1–13. https://doi.org/10.1186/s12889-017-4552-9

Lapum JL, Liu L, Church K, Yau TM, Ruttonsha P, Matthews David A, Retta B (2014) Arts-informed research dissemination in the health sciences: an evaluation of peoples' responses to "the 7,024 th patient" art installation. Sage Open 4(1):2158244014524211. Available at: https://journals.sagepub.com/doi/pdf/10.1177/2158244014524211. Accessed 24 Mar 2023

Lee JJ, Hammer J (2011) Gamification in education: what, how, why bother? Acad Exch Q 15(2):146. Available at: https://www.researchgate.net/publication/258697764_Gamification_in_Education_What_How_Why_Bother

Mason T, Trochez C, Thomas R, Babar M, Hesso I, Kayyali R (2018) Knowledge and awareness of the general public and perception of pharmacists about antibiotic resistance. BMC Public Health 18(1):711. https://link.springer.com/article/10.1186/s12889-018-5614-3

McCullough AR, Parekh S, Rathbone J, Del Mar CB, Hoffmann TC (2015) A systematic review of the public's knowledge and beliefs about antibiotic resistance. J Antimicrob Chemother 71(1):27–33. Available at: https://academic.oup.com/jac/article/71/1/27/2363966. Accessed 21 May 2020

McParland JL, Williams L, Gozdzielewska L, Young M, Smith F, MacDonald J, Langdridge D, Davis M, Price L, Flowers P (2018) What are the 'active ingredients' of interventions targeting the public's engagement with antimicrobial resistance and how might they work? Br J Health Psychol 23(4): 804–819. https://doi.org/10.1111/bjhp.12317. Accessed 20 May 2020

McNulty CA, Nichols T, Boyle PJ, Woodhead M, Davey P (2010) The English antibiotic awareness campaigns: did they change the public's knowledge of and attitudes to antibiotic use? J Antimicrob Chemother 65(7): 1526–1533. https://doi.org/10.1093/jac/dkq126

Melonio A, Gennari R (2012) Co-design with children: the state of the art. KRBD Research Centre for Knowledge and Data. University of Bozen, Italy. Available at: https://www.academia.edu/download/41395015/KRDB13-1.pdf

Murray CJ, Ikuta KS, Sharara F, Swetschinski L, Aguilar GR, Gray A, Han C, Bisignano C, Rao P, Wool E, Johnson SC (2022) Global burden of bacterial antimicrobial resistance in 2019: a systematic analysis. Lancet 399(10325):629–655. https://doi.org/10.1016/S0140-6736(21)02724-0

National Coordinating Center for Public Engagement (n.d.-a) Evaluating public engagement [online]. Available at: https://www.publicengagement.ac.uk/do-engagement/evaluating-public-engagement. Accessed 25 May 2020

National Coordinating Center for Public Engagement (n.d.-b) What is public engagement? [online]. Available at: https://www.publicengagement.ac.uk/about-engagement/what-public-engagement. Accessed 20 Nov 2019

National Coordinating Center for Public Engagement (n.d.-c) About quality engagement [online]. Available at: https://www.publicengagement.ac.uk/do-engagement/quality-engagement/about-quality-engagement. Accessed 25 May 2020

National Health Service (2021) Health literacy [online]. Available at: https://service-manual.nhs.uk/content/health-literacy. Accessed 1 March 2023

National Institute for Health and Care Research (2022) Health information: are you getting your message across? [online]. Available at: https://evidence.nihr.ac.uk/collection/health-information-are-you-getting-your-message-across/. Accessed 1st Mar 2023

National Literacy Trust. (2019). Adult literacy: information on adult literacy in the UK and our books unlocked programme. [online]. Available at: https://literacytrust.org.uk/parents-and-families/adult-literacy/. Accessed 20 Feb 2023

The Office of Disease Prevention and Health Promotion (2016) Health literacy online. Available at: https://health.gov/healthliteracyonline/. Accessed 1st Mar 2023

Padfield D (2011) 'Representing' the pain of others. Health 15(3):241–257. Available at: https://journals.sagepub.com/doi/pdf/10.1177/1363459310397974

Price L, Gozdzielewska L, Young M, Smith F, MacDonald J, McParland J, Williams L, Langdridge D, Davis M, Flowers P (2018) Effectiveness of interventions to improve the public's antimicrobial resistance awareness and behaviours associated with prudent use of antimicrobials: a systematic review. J Antimicrob Chemother 73(6):1464–1478. https://doi.org/10.1093/jac/dky076. Accessed 26 May 2020

Salter A (2016) Playing at empathy: representing and experiencing emotional growth through Twine games. 2016 IEEE International Conference on Serious Games and Applications for Health (SeGAH) [online], pp. 1–8. Available at: https://ieeexplore.ieee.org/abstract/document/7586272. Accessed 15 August 2020

Subhash S, Cudney EA (2018) Gamified learning in higher education: a systematic review of the literature. Comput Hum Behav 87:192–206. https://doi.org/10.1016/j.chb.2018.05.028

Tarín-Pelló A, Marco Crespo E, Suay García B, Galiana Roselló C, Bueso Bordils JI, Pérez Gracia MT (2022) Innovative gamification and outreach tools to raise awareness about antimicrobial resistance. Front Microbiol 13. Available at: http://hdl.handle.net/10637/13962

Read B, McNulty CA, Verlander NQ, Moss N, Lecky DM (2022) Comparing public knowledge around value of hand and respiratory hygiene, vaccination, and pre-and post-national COVID-19 lockdown in England. Public Health 212:76–83. https://doi.org/10.1016/j.puhe.2022.08.015

Redfern J, Bowater L, Coulthwaite L, Verran J (2020) Raising awareness of antimicrobial resistance among the general public in the UK: the role of public engagement activities. JAC Antimicrob Resist 2(1):dlaa012. https://doi.org/10.1093/jacamr/dlaa012. Accessed 7 Mar 2023

Review on Antimicrobial Resistance (2016) Tackling drug-resistant infections globally: final report and recommendations [online]. Available at: https://amr-review.org/sites/default/files/160518_Final%20paper_with%20cover.pdf. Accessed 29 Nov 2019

Robinson JM, Cameron R, Jorgensen A (2021) Germaphobia! Does our relationship with and knowledge of biodiversity affect our attitudes toward microbes? Front Psychol 12:678752. https://doi.org/10.3389/fpsyg.2021.678752. Accessed 7 Mar 2023

Roughley M, Manetti V, Pettigrew A, Irwin J, York-Salmon G, Charuhas S, Tsenina I, Birnbaum H, Holmes E, Niethamer N (2022) A showcase of MA art in Science research projects (2016–2020) that explore the boundaries of art, anatomy, biology, medicine, microbiology and public health. J Vis Commun Med 45(1):52–58. https://doi.org/10.1080/17453054.2021.1975495

Scavone P, Carrasco V, Umpiérrez A, Morel M, Arredondo D, Amarelle V (2019) Microbiology can be comic. FEMS Microbiol Lett 366(14):fnz171. https://doi.org/10.1093/femsle/fnz171

SfAM (2019) Microbiology literacy knowledge framework. Available at: https://sfam.org.uk/knowledge/policy/applied-microbiology-matters/microbiology-literacy-knowledge-framework.html. Accessed 14 May 2020

Tan MKB (2020) Towards a caring practice: reflections on the processes and components of arts-health practice. Arts & Health 12(1):80–97. https://doi.org/10.1080/17533015.2018.1494452

Timmis K, Cavicchioli R, Garcia JL, Nogales B, Chavarría M, Stein L, McGenity TJ, Webster N, Singh BK, Handelsman J, de Lorenzo V (2019) The urgent need for microbiology literacy in society. Environ Microbiol 21(5):1513–1528. https://doi.org/10.1111/1462-2920.1461

United Nations General Assembly (2016) Draft political declaration of the high-level meeting of the General Assembly on antimicrobial resistance [online]. Available at: https://www.un.org/pga/71/wp-content/uploads/sites/40/2016/09/DGACM_GAEAD_ESCAB-AMR-Draft-Political-Declaration-1616108E.pdf. Accessed 25 May 2020

Waite M (2019) Writing medical comics. J Vis Commun Med 42:144–150. https://doi.org/10.1080/17453054.2019.1575641

Walsh G, Druin A, Guha ML, Foss E, Golub E, Hatley L, Bonsignore E, Franckel S (2010) Layered elaboration: a new technique for co-design with children. In: Proceedings of the SIGCHI Conference on Human Factors in Computing Systems (pp. 1237-1240). Available at: Layered elaboration | Proceedings of the SIGCHI Conference on Human Factors in Computing Systems (acm.org)

World Health Organization (2015) Antibiotic resistance: multi-country public awareness survey. World Health Organization. Available at: https://apps.who.int/iris/handle/10665/194460. Accessed 20 Feb 2023

World Health Organization (2018) Antimicrobial resistance. Available at: https://www.who.int/en/news-room/fact-sheets/detail/antimicrobial-resistance. Accessed 1st Dec 2019

Collaged Creatures: Using Collage Art Practice and Posthuman Ethics to Discuss Bioengineering Hybrids

10

Alice Thickett

Abstract

Research into the bioengineering of hybrid human and non-human animals for human organ donation is progressing forward, but it is a contentious issue, debated in ethics panels and the media. This chapter discusses using posthuman methodologies and collage art practice to critically reflect on emerging human and non-human animal hybrids in contemporary bioengineering. Human perceptions of human and animal hierarchies have been culturally enacted across time in cave drawings, non-human-animal metaphors, domestication, and medical experimentation. To critically engage with dominant ideologies of the animal-human dualism, I review the human uneasiness felt around animal hybridity and discuss how we might use posthuman methodologies to decentre the human and reconsider the historically held perceptions that shape contemporary bioethics. In addition, I reflect on the mimicking of biotech practices such as CRISPR through cut and paste methods used for analogue collage, and review collaboration with other human and non-human animals, creating a space for the discussion of bioethics and the critiquing of othering of organisms that are difficult to define.

A. Thickett (✉)
Manchester Metropolitan University, Manchester, UK

Keywords

Collage · Edges · Hybrids · Ethics · Posthuman · Bioengineering

10.1 Background

The word 'hybrid' describes something that is made by combining two different elements. In biology, the term is used to describe the offspring of animals or plants from differing species. The word chimera can also be used to describe hybrid organisms. In science 'chimera' describes an organism that has genetic material from two or more sources. In Greek mythology and generally in visual culture, a chimera is the name for a specific mythological creature which is a mix of a lion, a goat, and a snake. Hybrid animals exist naturally when offspring occurs as the result of one species breeding with another. One of the most common and oldest known hybrid animals is a mule—which is the offspring of a male donkey and a female horse (offspring of a female donkey and a male horse is a hinny) (Babb 2021), and there are hundreds of naturally occurring, or 'naturally' bred animals on the planet, including ligers, wolfdogs, and grolars.

Since ancient times, humans have purposefully bred hybrids to assist them in daily activities such as carrying heavy loads, however, with the development of genetic engineering in the 1970s, researchers have been able to achieve

M. Roughley (ed.), *Approaches for Science Illustration and Communication*, Biomedical Visualization 4,
https://doi.org/10.1007/978-3-031-41652-1_10

hybridisation synthetically. In 1984, scientists created a sheep-goat chimera by combining the embryos of the two animals (Fehilly et al. 1984). With the invention of the Clustered Regularly Interspaced Short Palindromic Repeats (CRISPR) tool (Martin et al. 2012), often referred to as 'genetic scissors' (Cohen 2020) in 1987 (addgene n.d.) and its adaption in 2008, scientists could inject stem cells from one species more easily into the embryo of another, leading to the birth of mouse-rat chimeras in 2017 (Reardon 2017). Since then, monkey-human, pig-human, and mice-human embryos (amongst others) have been successfully grown in labs (Schneider 2019).

The success of these hybrid experiments has elicited much caution around experimentation with chimeric embryos as there is concern about the consequences of such hybridisation, for example the unknown result of human cells potentially entering the non-human's brain (Ball 2021). For this reason, until recently worldwide regulations (policies vary between countries) dictated that animal-human hybrid embryos must be destroyed after 14 days, however, in 2019 the Japanese education and science ministry published new guidelines which overturned the ban on the practice of approving stem cell researcher Dr. Hiromitsu Nakauchi's experiments to allow pig-human embryos to be brought to term, for the purpose of human organ donation (Cyranoski 2019).

Statistics from February 2023 suggested that in the UK alone, nearly 7000 people are waiting for an organ transplant (NHS 2023), and the global shortage of organs results in the unnecessary deaths of patients who could have been saved if organ donations had been available to them (Kupiec-Weglinkski 2022). Therefore, research into creating organs through bioengineering processes is of the upmost importance to help minimise deaths due to organ shortages. Engineering pigs with human hearts is a hypothesis to solve the problem.

Breeding hybrid organisms for human benefit requires ethical considerations. Not only do we not fully understand the biological consequences of creating hybrid creatures with human stem cells as mentioned above, but we also do not know the philosophical, psychological, and sociological consequences of bringing human and non-human hybrids into the world. Geneticist George Church, who has been working on breeding pigs with human organs with the help of CRISPR techniques says, '. . .the more powerful, or the more rapidly-moving the technology, the more cautious we need to be, the bigger the conversation involving lots of different disciplines, religion, ethics, government, art, and so forth' (Pelley 2019).

Trained in deciphering and making visual culture, I feel compelled to join the conversation on animal hybridity that affects medical research, to understand more fully the ethics involved with chimeric experimentation. My interest in this topic stems from the visualisation of human and non-human animal hybrids. Visualising animal hybrids is not new; it is believed that humans started visualising animal-human hybrids more than 40,000 years ago on the walls of caves (Aubert et al. 2019). Mythology from around the world includes stories and depictions of hybrids, such as the chimera mentioned above, the Minotaur from Greece (who had the body of a man and the head of a bull) and Horus the Egyptian deity (depicted as a man with a falcon's head). Contemporary art culture, therefore, includes a wide range of representations of animal hybrids, for example Patricia Piccinini's sculptures of imagined hybrids including *The Young Family* (2002) (which is one of the first images to appear in a Google search of 'pig human'), Max Ernst's reoccurring human bird character—Loplop (*Loplop Introduces Loplop,* 1930), and Paula Rego's *Dog Woman* (1994), which is a depiction of a Portuguese fairy tale of a woman who eats her pets one by one (representing the animality of the human).

Without the reality of human, non-human animal hybrids, humans have visualised them. On the surface, it could be reasoned that humans visualise themselves as having non-human appendages because we appreciate the features that other species possess, however, humans (particularly those in the West, which is my own context) have varied relationships with non-human animals, and this relationship paired with the approval of hybrid embryos being bought to term is the focus of my artistic research.

Theorist Donna Haraway (2016) argues that speculative fabulation and everyday storytelling (or 'worlding') are important for analysing current issues and allow us to move away from anthropocentric thinking. By using my collage art practice as a research method (Frayling 1993) and posthumanism as my methodology (which I will explain in a subsequent section) to engineer creative narratives, I aim to engage in the ethical discussions around human and non-human animal hybridity in a non-anthropocentric way. In engaging in research where I use the edges of collage art practice as a tool for investigating the edges of the human, I push the boundaries of what defines collage as well as contributing to the understanding of hybrids bred for medical purposes, as they are no longer just visualised, but become a genetically engineered reality.

10.1.1 Human Perceptions

Some humans are biological chimeras which occur if the foetus absorbs its twin in the womb, or if a person undergoes a bone marrow transplant, as the stem cells from the donor will enter the blood of the patient (Rettner 2016). However, if hybridity was to be created using human and non-human DNA it may be considered unnatural. The difference in this latter case is that homo sapiens is the species to which all humans belong and therefore all non-human organisms are from a different species, and so they could not naturally breed.

This cross-species hybridisation involving humans, therefore, creates unease. Around the time of the CRISPR adaptation, a 2007 survey was conducted by the Human Fertilisation and Embryology Authority (HFEA) in the UK, which found that a quarter of participants opposed animal-human hybrid embryo research and 61% of the public supported the research once they had received an explanation of the processes and intended goals. Taddeo and Robert's analysis of the HFEA's survey rounds up, that the public's concerns included '...what constitutes a human, the integrity of what it means to be human, allowing odd and confusing human-animal chimeras to mature to birth...' (Taddeo and Robert 2014).

Humans may have depicted themselves as hybrid with non-human animals since the palaeolithic period, but the reality of combining DNA cross-species brings into question what a human is, and therefore seemingly peaks the public's anxiety. This is understandable considering the perceptions Western society has of our standing alongside other animals and the varied relationship we have with other beings. For example, we keep some animals as pets, some we eat, some we turn into clothing, some are reintroduced into the wild after they are nearly wiped out by hunting, and some we use for science experimentation. Although humans are also taxonomically animals, our laws dictate that we must not hunt, kill, or eat, other humans, yet the ethical lines are drawn in different places for our treatment of other animal species, because of how we have historically perceived the hierarchies between humans and other species of animals.

It is still debated as to when humans first domesticated animals. Until recently it was believed to be around 15,000 years ago, but new fossil finds suggest it could have been more than 36,000 years ago, when the interactions between humans and wild wolves changed the wolves' DNA until they evolved to become what we know taxonomically as dogs (Pilcher 2020). Researchers have been unclear as to why the first wolves became domesticated, although it has been suggested by researchers Maria Lahtinen et al. (2021) that wolves would eat surplus meat from a human's hunt and in time both species began to live together. Once domesticated, dogs were used by humans to guard and carry supplies (Nuwer 2021), and this relationship with dogs continues today with them being trained for police work, as assistive animals, and for herding livestock. Not only are dogs used for work, but they have also been labelled as 'man's best friend', are bought as pets, and some are considered members of the family. Humans have bred dogs until they are no longer recognisable as decedents of wolves, and some have been bred so excessively for their desirable features (desirable from a human aesthetic perspective) that they are at risk of not even being considered dogs (Ray 2022)—a strange consequence of being a best friend to man.

In Western society changes are starting to occur, that destabilise our long-held perceptions of animals that are not human, and to consider their agency differently. For instance, as recently as May 2021 the UK introduced new legislation that formally recognises vertebrate animals as sentient which could make society consider animals beyond the services they could provide humans (Department for Environment, Food & Rural Affairs and Lord Goldsmith 2021). So, how do these perception changes shift how humans perceive engineering biology synthetically?

The engineering of organisms, food, and medicine is generally a contentious issue often debated beyond ethics panels in the media. Articles sceptical of genetic engineering have been known to label modified produce with the prefix 'franken' to highlight the idea of scientists 'meddling' with nature (Lewis 1992), as Dr. Frankenstein is believed to have done so when he created his monster in Mary Shelley's seminal gothic novel *Frankenstein* ([1818] 1831). The 'franken' prefix suggests a general wariness from the public about genetic engineering, and most recently this was evidenced in the spread of misleading information about the COVID-19 vaccine and its 'ability' to turn humans into 'genetically modified organisms' (Schwarcz 2020).

It seems that whilst society is in some way comfortable with 'natural' engineering of organisms, there is a general wariness (although not complete nonacceptance) of engineering in labs. Jon Turney in *Frankenstein's Footsteps: Science, Genetics and Popular Culture* says, 'The science *is* extraordinary but we are consistently ambivalent about its application. We recognise that there may be great benefits, but also see less well-defined threats in biological technology's ability to break down old boundaries, and dissolve taken-for-granted categories' (author's emphasis) (1998, p. 2).

More specifically, David A Jones' article, 'What does the British public think about human-animal hybrid embryos?*'* (2008), presents a sceptical analysis of surveys collecting the perceptions of hybrids. Jones suggests that the public may seem to agree with hybrid research only when they are given specific information about the reasons behind carrying out the research; but the surveys do not capture, potentially leave out, and influence the public's real feelings towards the prospect of animal-human hybridity. Jones argues that representative-sample polling can '. . .elicit conditional approval of the research but combined with continuing underlying unease'. Here, Jones states an interesting point, that these surveys do not often analyse what it is about hybrids that humans find difficult.

As recently as 2020, researchers published a paper claiming that 'The American Public Is Ready to Accept Human-Animal Chimera Research' and that 83% of the respondents found chimera research acceptable (Crane et al. 2020). What is notable is that again the researchers provided the participants with clear purposes for the embryo research and organ transplantation, which perhaps shows that the *informed* public is comfortable with hybrids, particularly if they are of *benefit* to humans, but what these surveys do not show, is how society perceives the consequences of merging human and non-human DNA in a wider context, and how previous notions of humanness could change if animal-human hybrids were bought to term.

Western scientific ethics panels are made up of humans who have been influenced and shaped by a human history that has regarded non-human animals as other or inferior. This othering creates hierarchies that skew our relationship with the world in which we are embedded, as Haraway says, 'It matters what matters we use to think other matters with... what thoughts think thoughts...' (2016, p. 12). The human perceptions of ourselves, and non-human animals, and the perceived hierarchies between humans and other species are important to question and analyse, as these perceptions affect the decisions that we make about bioengineering research and its applications, which in turn shapes our whole society. It *matters* with what thoughts, knowledge, and understanding we are making ethical decisions with. When we analyse our inherited perceptions, evermore questions are exposed, and we gain the opportunity to consider questions of humanness beyond our current understanding. When is an animal a *human*? If it speaks? If it has a human brain? If it has a soul (whatever that is)? At what point would an animal become human *enough* to be classified as human? And is this latter question problematic, when—as previously stated—the UK has recognised *all* vertebrates as sentient? To echo Turney's earlier point, when biotechnology breaks down old boundaries, what will we *become*?

10.2 Posthuman Methodology and Creative Methods

> Posthuman is just the question... the answer is what we are capable of becoming.—Rosi Braidotti (2019a, p. 98)

Current theorists in posthumanism challenge anthropocentric closed definitions of what a human is, attempt to deconstruct the notion of the human, and re-conceive the definition of human as fluid (Ferrando 2017). Posthumanism posits humans are not superior to non-human-animals and that we should move away from humanism and speciesism (Wolfe 2010). Unlike transhumanism (Bostrom 2003; Sandberg 2014) and discourses on the 'posthuman' (Rugo 2020) which focus on the driving forward and evolution of the human, posthumanism is concerned with critical analysis in the Anthropocene—a geological epoch so named, because of the significant impact humans have had on the planet (Lewis and Maslin 2015). At this point, in time we are also met with the multiple challenges of, the acceleration of climate change, rapid advancement of technology, and the potentiality of the Sixth Extinction (2019b).

Posthuman theorists posit that to work through the challenges of the Anthropocene, we must find new ways of knowing that decentre what we have historically perceived as human. Posthumanist theorists, therefore, define their own type of ethics that not only *account for* voices that have previously been unheard, but support the mobilisation of the sexualised (woman & LGBTQ+), racialised (black and indigenous), and naturalised (animals, plants, earth) others. In the posthuman glossary MacCormack explains, '...posthuman ethics seeks relations based on imperceptible and infinite difference, that is, not based on anything in advance and based on all possible expressions and affects' (MacCormack 2018, p. 346).

Posthuman ethics requires creative experimentation with alternative ways of relating to ourselves and those that we share the world with. The theory opposes power-driven interpretations that *presume* the nature of relations and the entities involved (ibid.). These presumptions are evident in the sciences when decisions are often made based on financial backing and historical perceptions and biases (Turner and Chubin 2019., Straw 2020). These biases are based on a culture in which *other* humans do not represent the Vitruvian man (male, white, able-bodied, heterosexual etc.)—a vision of the human created by the historical excluding of certain groups of people (women, black, disabled, homosexual etc.). Therefore, as Church says, when discussing bioengineering and particularly genetic research involving human stem cells, it is important to involve a range of voices, as the decisions we make now, as mentioned above, will shape what kind of humans we become, not just evolutionary but societally and culturally.

When critically analysing bioengineering from a posthuman perspective, Braidotti explains that, it is itself, an assemblage, 'The current economy is not only about technological mediation, or exclusively about gene-editing and stem-cell engineering, or only about the exploitation of labour, or just about climate change and the spectre of species extinction. It is about all these factors at the same time' (Braidotti 2021, p. 60). By involving artist-researchers and posthumanism theorists in discussions of bioethics, it allows for novel ways of approaching rapidly advancing technologies—and the potential ethical consequences—by considering a wider range of ideas of thinking and being with the world.

If decisions concerning hybrid organisms are evolved from a humanistic, anthropocentric ontology, then there is a risk that the hybrids would fit our Frankensteinian notions of the monstrous *other* that needs to be used, shunned, or killed to preserve the sanctity of the human. Stacey Alaimo (2010, p. 16) contemporises this notion by pointing out '. . .the multitude of horror movies that shock us with the prospect of monstrous human-animal hybrids. . .only to conclude with the triumphant transcendence of "man"'. Alaimo goes on to suggest that this notion can be critiqued if the human is decentralised and considered '. . .always already part of an active, often unpredictable, material world' (2010, p. 17).

Using posthuman methodologies in my research, I ask—can organisms that are a biological assemblage of human and non-human animal cells make us address the human, non-human animal relationship? Therefore, I exchange the focus from what the human *is*, which has preoccupied us culturally and scientifically, and instead investigate how we intra-act (Barad 2007). Intra-action is a term coined by Posthumanist Karen Barad, which posits that separate entities emerge through their entanglement, in contrast to 'interaction' which assumes the 'existence of separate individual agencies' (Murris 2021). When the research focus is on intra-action and becoming *with* multiple species, human exceptionalism can be reconsidered (Haraway 2016).

Intra-action is useful when critically analysing human, non-human, animal dualisms, but it can also be used for thinking through the making of artwork. In *Vibrant Matter: A Political Ecology of Things*, Jane Bennet argues that objects possess a 'strange ability' to exceed their status as objects, in her work she uses the term 'vital materialism' which considers the relationship between 'things' and humans (Bennett 2010). As I will go on to explain, my research with collage is not about my mastery over paper, but the intra-action I have with the material and its edges. If I consider that my art materials have agency, is my art practice itself a kind of becoming-with where I decentralise myself as, a knowledge-making-human or art-creating-artist and reconfigure *myself* as an assemblage?

10.2.1 Collage and Its Edges

Collage is multiplicitous and fluid, it is neither defined as a medium nor a genre and it is often classified as 'mixed media' (Kadour n.d.). Some artists use collage in sketchbooks before making a final piece, for other artists a collage is the end product. Collage's origins are disputed and change between publications, depending on the author's standpoint on giving credit to, perceived art 'masters' (Cubists—Picasso and Braque), female artists (Dadaist Hannah Hoch), scrapbookers (housewives in the fifteenth century), or the invention of paper (Ts'ai Lun in China around AD 105) (Elliot 2019, p. 10). The unclear starting point and lineage of collage only add to collage's rebellious history.

Collage made out of cut and pasted war propaganda was used by the Dadaists in the 1920s to shock the bourgeois. Surrealists soon after, adopted collage with the hopes of unlocking the subconscious, Pop artists in the 1950s such as Richard Hamilton, used collage to critique consumerism, and in the 1970s artists such as Linder Sterling were using collage on punk record covers. Until recently, collagists have used collage against the zeitgeist to create revolution and protest, however, more recently, and with the advent of social media and neoliberalism, collage has been used by large-scale marketing campaigns such as Adidas (DR ME 2021) and Hermes (The Daily Splice 2022).

With the saturation of visual media online especially on social media streams, collage is now more likely to be used by designers and corporations than to critique them (Rose and Kahn 2011). Curator and Art Historian, Yuval Etgar suggests that '. . .the task ahead of us [collage artists] is to understand what it is that collage actually does. . . what are the effects that it produces. . .' (Etgar 2017, p. 11). I agree it is pertinent for collage artists to expand on the ideas of what collage can be, but I would suggest that this is not to define collage, but to further expand its edges.

My collage art process involves the analogue cutting of photographic images from books and magazines using scissors, placing the cut images together and gluing them in place to make a new image. Taking one image from a particular source and merging it with a different type of image from a separate source, simultaneously blurs and highlights the edges of the disparate images, we know that the images 'do not belong' together but despite this, we accept the new whole image, in its new configuration. The collaging of images, therefore, affects the edges of the images physically, as well as the edges of meaning of the original sources. For example, an advertisement for a washing machine could become a critique of capitalism and domestic life when the washing machine is cut out and pasted onto the head of a woman.

Writer Ali Smith poetically defines an edge as '. . .the difference between one thing and another. It's the brink. It suggests keenness and it suggests sharpness. It can wound. It can cut. It's the blade – but it's the blunt part of the knife too' (2012, p. 126). An edge according to Smith is a provocation ('*it* suggests'), a weapon ('*it* can wound'), a tool ('*it* can cut'), the edge is sharp but also the opposite of sharpness (two sides of a knife), and it is the point at which something occurs (the 'brink'). Edges are therefore a useful tool because the edge is the point at which something *happens.*

As a critical posthuman artist, I recognise that I am materially '. . .embedded and embodied, with the power to affect and be affected by. . .' (Braidotti 2019a, p. 54). The collage itself affects and is affected through the process, as whilst I am arranging the images, new configurations are made visible to me that change the aesthetics of the collage and the reading of the images. I am also aware that I bring my own understandings and readings to the original sources, which will be changed by the cutting and pasting of source images into a new piece, and the work becomes a further assemblage when viewed by an audience who bring their own meanings to the viewing.

The process of making collage is particularly tactile and visceral because of the utilisation of tools such as a scalpel or scissors to create edges. It has been likened by journalists and prolific collage artists such as Linder Sterling and John Stezaker, to the cutting and pasting of body parts, as photographs of humans are a regular motif in photographic collage. Commentators often make a comparison to the process in which Dr. Frankenstein makes his creation (the 'monster') (Dillon 2011, Sterling 2017). As highlighted above, in a science context the comparison to *Frankenstein* is often unfavourable, however, the artist as pseudoscientist is a generally accepted notion (bioartists excepted as they lie at the disciplinary border of art and science).

In the context of the age of bioengineering and not vivisections, the art practice of hand-cut collage mimics the cutting and editing of genes. DNA is cut by CRISPR/Cas9 much like the scissors a collage artist uses to dissect images from their original source. The DNA then heals or 'glues' itself back together, creating an altered organism; similar to photographic images from two disparate sources 'merging' together in the new context that they share—the collaged artwork. In both instances a human hand is being used to slice something apart to engineer something new, and in both hybrid engineering research and my art research the 'something new' is a hybrid organism. However, the collaged images do not *represent* the genetically engineered hybrids (the pigs with human hearts, for instance, should—in reality—look like regular pigs), but instead the collage *enacts* hybridity. The collage is in constant flux between its disparate parts and its whole, it is in flux with myself as artist (the heterogenous assemblage I am), and also the audience (the heterogenous assemblage they are)—we are collectively on the edge.

10.2.2 Collaging Hybrids

I have made hybrid creatures before, anthropomorphised non-human animals like a fashion magazine model with a cow's head or a yogi with pigeon wings. However, these hybrids only imitated characters that are already familiar in mythology or popular culture. I wanted instead to make collages that suggested a merging of non-human-animals and humans that did not favour human physiology. I realised that this anthropomorphosis was in part due to my collaging process as I had already grouped cut-out images into boxes of 'body parts' (i.e. human body parts) and 'animals'. The taxonomies that I had applied to my images—just like taxonomies in biology—had created dualisms and ways of working that privileged the human and reduced the non-human animal parts of the collage to metaphors. However, regular reflection on my relationship with non-human-animals revealed my nonchalant physical and metaphysical consumption of them.

Whilst being with non-human animals, I started to think about fleshiness, and bought images of flesh and meat into my collages. These meaty images (Fig. 10.1), alongside and merged with images of non-human-animal and human limbs create a jarring effect. Is the creature (if read as a whole creature at all), deformed? Is it sewn together or grown together? Could I eat it? Would it eat me? The questions that are formed from the difficulty we have of placing the hybrid in a category, because of its blurred boundaries between human and non-human animal, flesh and meat, start to break down perceived dualisms.

Kelly Oliver, in discussing Merleau-Ponty's thoughts on shared embodiment, describes a fundamental kinship between humans and non-humans because of their consciousness that they are not separated from the body (2009, p. 217). Oliver goes on to posit that Merleau-Ponty's ideas of fleshy kinship may change our ethics, 'How can we imagine a kinship that is based on embodiment and not generation?... How can animals be or become our kin?' (2009, p. 224). If these questions are raised from a critique of embodiment and flesh, can my collages provoke the same kind of questioning?

To push the fleshiness further, I made my 2D paper collage into a 3D material model. The 3D model was made by printing the individual collage elements onto fabric, which I then cut out and reassembled as a 3D collage. Not wanting to turn my creatures into objects however, the models were 3D scanned and a 3D digital model was made (Fig. 10.2), with the future aim to animate it. This digital 3D model is more than a representation of a hybrid creature—through the aid of technology (just like bioengineering), the hybrids are beginning to take form.

To challenge perceptions of hybrid creatures it is important to me that the creatures are not only not objectified, but they have agency. To enable this, I will speculate and reflect with others by inviting a range of humans who have experience with ideas of hybridity, animals, or philosophy, to join me in storytelling. They will be asked to imagine that they have encountered a hybrid creature and to describe their interaction with the animal. Collaborating with others in this imaginative way means we can reflect and speculate in an embodied and embedded way, without the need-to-know definitive answers. These stories will inform the collages I make going forward and the way I expand my practice.

Haraway implores in *Staying with the Trouble* that to deal with disturbing times (such as the Anthropocene) we should act out 'sympoiesis' or 'make-with', 'who and whatever we are, we need to make-with –become-with, compose-with –the earth-bound...' (2016, p. 102). By speculating with others, I am working in a posthuman way, acknowledging that other voices should be heard within the project. I cannot bring the 'voice' of non-human animals into the research through human methods, and I am aware that by guessing animal responses without being with them, that I will inevitably uphold human biases. Therefore, my aim is to observe non-human animals and reflect on my own relationship with them. Being in sympoiesis with other humans, non-human animals, and collage practices to create the collaged monsters, encourages discussions about our long-held perceptions of humans and their boundaries that are heterogenous and less anthropocentric. Being engaged in research in this way will allow me to push collage art practice and critique the boundaries between non-human animals and humans in an affirmatively ethical way.

Fig. 10.1 *Humig* [2D analogue collage, 29.7×42 cm]. Image by Alice Thickett

Fig. 10.2 Humig [3D digital model, detail 1]. Image by Alice Thickett

10.3 Final Thoughts

There is no denying that the images that I am creating are jarring and perhaps monstrous. Posthumanism acknowledges that we have a long history of monstering *others* that do not conform to our perceived ideas of human. We are therefore at risk of monstering future biologically engineered organisms that do not conform to our established categories. In 'Promises of Monsters: A Regenerative Politics for Inappropriate/d Others' (1992), Haraway dissects the monster—as it were—and what we consider to be 'nature'. She argues that the physical and the imagined are interconnected, and therefore nature is a topic of public discourse where many of our ideas around the 'other' have been formed by the way we have structured meaning through binary oppositions, for example male and female are the binary positions, therefore anything in-between is 'other'.

Haraway disrupts these dualistic ideas by using imaginative methods to map out and find points of interruption (examples) between the extremes of the binary meta-narratives, that are collective and connected, and therefore oppose notions of dominance. Haraway concludes by redefining the monster, suggesting that we are *all* monsters—incongruous beings, that are not just comprised of ourselves but a collection of human and non-human others.

By using fleshy collages, I attempt to interrupt dualistic perceptions by bringing together images of non-human and human animal parts and reconfiguring them as a whole organism—they are not quite human, and they are not quite non-human animal, therefore they lie somewhere between the binary narrative of human and animal. I ask the viewers of the collaged hybrids to recognise parts of themselves within them, visually and metaphorically, so that they can consider the potential humans have for hybridity. If we perceived humans as multidirectional beings that are affected and affecting, and if we saw ourselves as already interconnected with non-human and human others, how would we shape our bioethics? What are the possibilities if we recognise *ourselves* as hybrids?

References

addgene. (n.d.) CRISPR history and development for genome engineering. addgene. [Online] https://www.addgene.org/crispr/history/. Accessed 27 Jun 2021

Alaimo S (2010) Bodily natures: science, environment, and the material self. Indiana University Press, Bloomington

Aubert M, Lebe R, Oktaviana AA, Tang M, Burhan B, Hamrullah J, Abdullh A, Hakim B, Zhao J, Geria IM, Sulistyarto PH, Sardi R, Brumm A (2019) Earliest hunting scene in prehistoric art. Nature 576:442–445

Babb D (2021) History of the mule. American Mule Museum [Online] https://www.mulemuseum.org/history-of-the-mule.html. Accessed 27 Jun 2021

Ball P (2021) Mixed messages: is research into human-monkey embryos ethical? The guardian. [Online] https://www.theguardian.com/science/2021/may/15/mixed-messages-is-research-into-human-animal-hybrids-ethical-chimera. Accessed 27 Jun 2021

Barad K (2007) Meeting the universe Halfway: quantum physics and the entanglement of matter and meaning. Duke University Press, Durham

Bennett J (2010) Vibrant matter: a political ecology of things. Duke University Press, Durham and London

Bostrom N (2003) Human genetic enhancements: a transhumanist perspective. Journal of Value Inquiry 37(4):493–506

Braidotti R (2019a) Posthuman knowledge. Polity Cambridge, UK, Cambridge

Braidotti R (2019b) The transversal Posthumanities. Philos Today 63(4):1181–1195

Braidotti R (2021) Posthuman feminism. Polity Cambridge, UK, Cambridge

Cohen J (2020) CRISPR, the revolutionary genetic 'scissors,' honored by Chemistry Nobel. Science Mag. [Online] https://www.sciencemag.org/news/2020/10/crispr-revolutionary-genetic-scissors-honored-chemistry-nobel. Accessed 20 Nov 2020

Crane AT, Shen FX, Brown JL, Cormack W, Ruiz-Estevez M, Voth JP, Sawai T, Hatta T, Fujita M, Low WC (2020) The American public is ready to accept human-animal chimera research. Stem Cell Reports 15(4):804–810. https://doi.org/10.1016/j.stemcr.2020.08.018

Cyranoski D (2019) Japan approves first human-animal embryo experiments. Nature [Online] https://www.nature.com/articles/d41586-019-02275-3. Accessed 13 Jan 2021

Department for Environment, Food & Rural Affairs and Lord Goldsmith (2021) Animals to be formally recognised as sentient beings in domestic law. GOV.UK. [Online] https://www.gov.uk/government/news/

animals-to-be-formally-recognised-as-sentientbeings-in-domestic-law. Accessed 27 Jun 2021

Dillon, B. (2011) John Stezaker: what a carve up. The guardian. [Online] https://www.theguardian.com/artanddesign/2011/jan/29/john-stezaker-whitechapel-gallery. Accessed 15 Oct 2019

DR ME (2021) [Artwork] [Online] https://www.instagram.com/p/CR9SgTIMqe2/?hl=en. Accessed 3 Feb 2023

Elliot P (2019) Collage over the centuries. In: Elliott P, Gowrley R, Etgar Y (eds) Cut and paste: 400 years of collage. Trustees of the National Galleries of Scotland, pp 9–23

Ernst M (1930) Loplop introduces Loplop. Oil and collage on wood, 100×180cm, Private collection

Etgar Y (2017) The ends of collage. Luxembourg & Dayan, London

Fehilly CB, Willadsen SM, Tucker EM (1984) Interspecific chimaerism between sheep and goat. Nature 307(5952):634–636

Ferrando, F. (2017) 3. What is POSTHUMANISM? Dr. Ferrando (NYU)–course "the Posthuman" lesson n. 3. Posthumans. [online video] https://www.youtube.com/watch?v=hm9zkvHVkL4. Accessed 27 Jun 2021

Frayling C (1993) Research in art and design. Royal College of Art Research Papers 1(1):1–5

Haraway D (1992) Promises of monsters: a regenerative politics for inappropriate/d others. In: The Haraway reader. Routledge, New York

Haraway D (2016) Staying with the trouble. Duke University Press, USA

Jones DA (2008) What does the British public think about human–animal hybrid embryos? J Med Ethics 35(3): 168–170. https://doi.org/10.1136/jme.2008.026336

Kadour RK (n.d.) Editorial: how do we talk about collage? Kolaj. [Online] http://kolajmagazine.com/content/content/articles/editorial-how-do-we-talk-about-collage/. Accessed 27 Jun 2021

Kupiec-Weglinkski JW (2022) Grand challenges in organ transplantation. Frontiers [Online] published 6 May 2022 https://doi.org/10.3389/frtra.2022.897679. Accessed 22 Mar 2021

Lahtinen M, Clinnick D, Mannermaa K et al (2021) Excess protein enabled dog domestication during severe ice age winters. Sci Rep 11(7):7. https://doi.org/10.1038/s41598-020-78214-4

Lewis P (1992) Mutant foods create risks we Can't yet guess; since Mary Shelley. New York times. p. 24

Lewis S, Maslin M (2015) Defining the anthropocene. Nature 519:171–180

MacCormack P (2018) Posthuman ethics. In: Braidotti R, Hlavajova M (eds) Posthuman glossary. Bloomsbury Academic, London, pp 345–346

Martin J, Chylinski K, Fonfara I, Hauer M, Doudna JA, Charpentier E (2012) A programmable dual-RNA–guided DNA endonuclease in adaptive bacterial immunity. Science 337(6096):816–821

Pelley S (2019) A Harvard geneticist's goal: to protect humans from viruses, genetic diseases, and aging. CBS News. [Online] https://www.cbsnews.com/news/harvard-geneticist-george-church-goal-to-protect-humans-from-viruses-genetic-diseases-and-aging-60-minutes-2019-12-08/. Accessed 13 Jan 2021

Murris K (2021) A glossary for doing postqualitative, new materialist and critical posthumanist research across disciplines. Routledge, Oxford

NHS. (2023) Organ donation and transplantation activity data: United Kingdom. Blood and transplant. [Online] https://nhsbtdbe.blob.core.windows.net/umbraco-assets-corp/28506/nhsbt-united-kingdom-summary-report-dec-22.pdf. Accessed 22 Mar 2023

Nuwer R (2021) Dog domestication may have begun because paleo humans couldn't stomach the original paleo diet. Scientific American [Online] https://www.scientificamerican.com/article/dog-domestication-may-have-begun-because-paleo-humans-couldnt-stomach-the-original-paleo-diet/. Accessed 22 Mar 2023

Oliver K (2009) Animal lessons: how they teach us to be human. Columbia University Press, New York

Piccinini P (2002) The young family. In: Silicone, fiberglass, leather, human hair, and plywood, 85.1×149.9×120cm, National Museum of women in the arts, Washington

Pilcher H (2020) Life changing: how humans are altering life on earth. Bloomsbury Publishing Plc, London, UK

Ray H (2022) Pugs cannot be considered "a typical dog" due to dire health issues, study finds. CBS NEWS [Online] https://www.cbsnews.com/news/pugs-dog-breed-health-issues-study/. Accessed 22 Mar 2023

Reardon S (2017) Hybrid zoo: introducing pig–human embryos and a rat–mouse. Nature [Online] published 26 January 2017. https://doi.org/10.1038/nature.2017.21378. Accessed 27 Jun 2021

Rego P (1994) Dog woman. Pastel on canvas, 120×160cm

Rettner R (2016) 3 human chimeras that already exist. Scientific American. [Online] https://www.scientificamerican.com/article/3-human-chimeras-that-already-exist/. Accessed 27 Jun 2023

Rose A, Kahn M (2011) Collage culture: examining the 21st Century's identity crisis. JRP|Ringier, Zurich

Rugo D (2020) Posthuman, In Oxford research Encyclopaedia of literature. Oxford University Press, pp. 1–19

Sandberg A (2014) Ethics of brain emulations. J Exp Theor Artif Intell 26:439–457

Schneider D (2019) The five: chimeras created by science. The Guardian. [online] https://www.theguardian.com/technology/2019/aug/11/the-five-chimeras-human-monkey-hybrid-genetic. Accessed 27 Jun 2023

Schwarcz J (2020) The vaccine is here but so is the fear. McGill [Online] https://www.mcgill.ca/oss/article/covid-19-health/have-no-fear-vaccine-here. Accessed 27 Jun 2021

Shelley M ([1818] 1831) *Frankenstein*. Modern Publishing Group

Smith A (2012) Artful. Hamish Hamilton, Penguin Press, London

Sterling L (2017) Performance artist and photomontage queen linder sterling tells us about keeping curious. *In* (ed.) Hanra, H., I-D magazine. [Online] https://thefifthsense.i-d.co/en_us/article/performance-artist-and-photomontage-queen-linder-sterling-tells-us-about-keeping-curious/. Accessed 13 Jan 2021

Straw I (2020) The automation of bias in medical artifical intelligence (AI): decoding the past to create a better future. Artif Intell Med 110:101965

Taddeo, S. Robert, J.S. (2014) '"Hybrids and chimeras: a report on the findings of the consultation" by the Human Fertilisation and Embryology Authority in October, 2007.' Embryo Project Encyclopedia. [Online] https://embryo.asu.edu/handle/10776/8240. Accessed 28 Jun 2021

The Daily Splice (2022) [Artwork] [Online] https://www.instagram.com/p/Cb4vfreAn_6/?hl=en. Accessed 3 Feb 2023

Turner S, Chubin D (2019) What's lost when research is driven primarily by funding. Scientific American. [Online] https://blogs.scientificamerican.com/observations/whats-lost-when-research-is-driven-primarily-by-funding/. Accessed 23 Mar 2023

Turney J (1998) Frankenstein's footsteps: science, genetics and popular culture. Yale University Press, New Haven and London

Wolfe C (2010) What is Posthumanism? University of Minnesota Press, Minneapolis

GPSR Compliance

The European Union's (EU) General Product Safety Regulation (GPSR) is a set of rules that requires consumer products to be safe and our obligations to ensure this.

If you have any concerns about our products, you can contact us on ProductSafety@springernature.com

In case Publisher is established outside the EU, the EU authorized representative is:

Springer Nature Customer Service Center GmbH
Europaplatz 3
69115 Heidelberg, Germany

Batch number: 10756355

Printed by Printforce, the Netherlands